MW00785513

Orchids of Australia

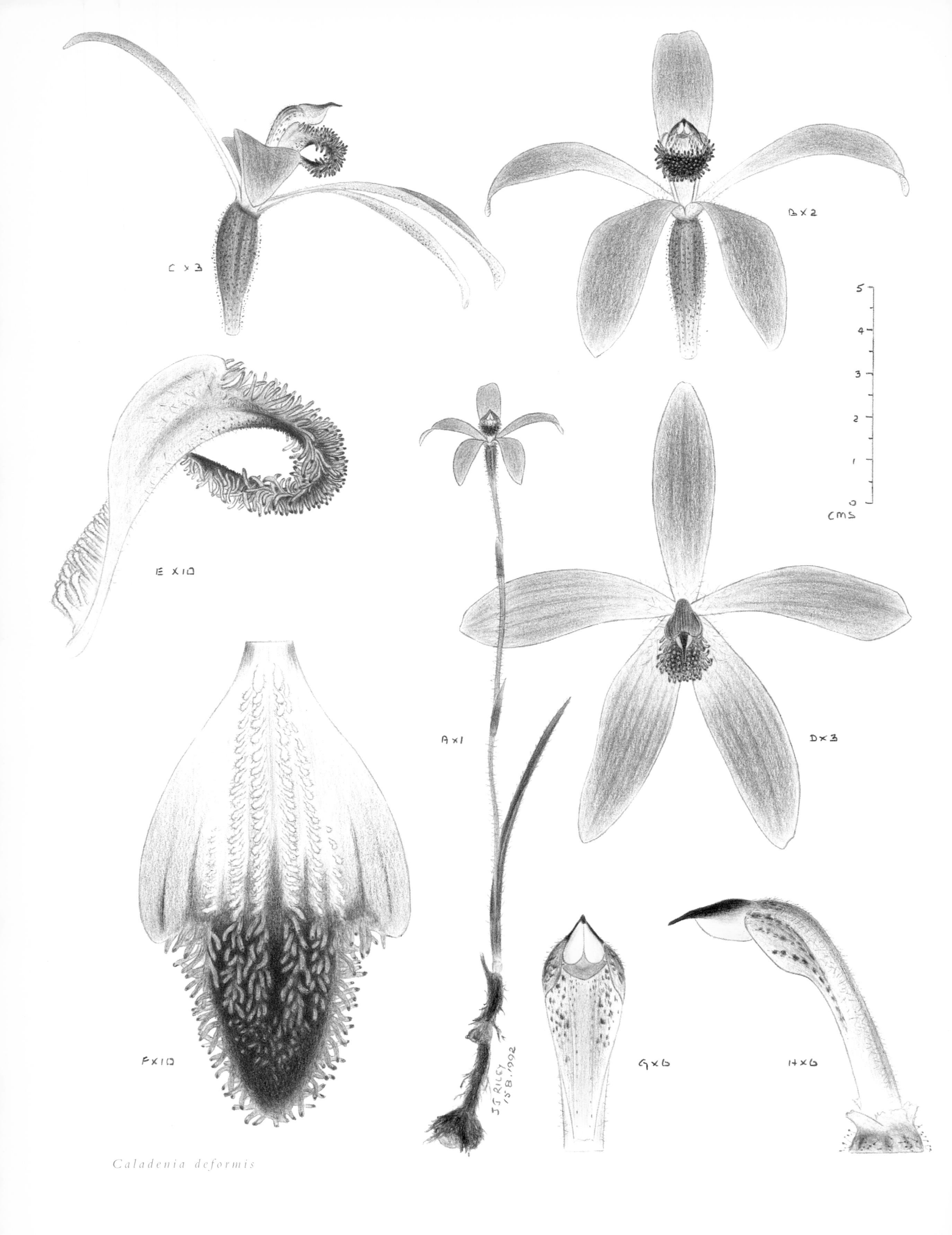

Caladenia deformis

Orchids of Australia

John J. Riley and David P. Banks

Illustrations by John J. Riley

Princeton University Press
Princeton and Oxford

contents

Published in North America, South America, and the European Union by
Princeton University Press
41 William Street
Princeton, New Jersey 08540

First published in 2002 by
University of New South Wales Press Ltd
University of New South Wales
UNSW Sydney NSW 2052
AUSTRALIA

ISBN 0-691-11490-0

Library of Congress Control Number 2002111071

This book has been composed in Bembo

www.pupress.princeton.edu

Printed by Everbest Printing Co Ltd, China

10 9 8 7 6 5 4 3 2 1

foreword

More than 80 per cent of Australia's 1200 or so species of native orchid are not found anywhere else in the world. With their delicate beauty, often brilliant colours, bizarre shapes and wondrous biological relationships, they have attracted a dedicated following of passionate devotees. Within artistic circles, however, Australian orchids have received very limited exposure. Of the Australian orchid publications featuring botanical artistry only two great works come to mind. These are Robert David Fitzgerald's internationally famous and multi-award-winning *Australian Orchids* (1875–94), and William Henry Nicholls' *Orchids of Australia* (1951 and 1969), both long since out of print and very expensive collectors' items.

Now in 2002, the artistic delights of John James Riley can be revealed. Every so often an individual emerges with such talents as to rise like a beacon above the rest of the flock. Among botanical artists, John Riley is such a person and we are indeed fortunate that he has devoted his talents to the faithful portrayal of Australian orchids. John draws what he sees with no artistic embellishments and his coloured drawings are masterpieces of meticulous accuracy and beauty.

This book will not only add to the assemblage of fine books currently available on botanical illustration but will also have significant appeal for orchid enthusiasts. They, in particular, will appreciate the magnificent drawings — which will be a great aid in identification — together with the accompanying pertinent and accurate text which complements, but in no way detracts from, the drawings. The text was largely compiled by David Banks, an accomplished author, orchid journal editor and dedicated orchidologist, with significant contributions from John, who spends much time studying the plants he portrays in their natural state.

It is with great pleasure that I contribute the Foreword to this publication, as I consider myself fortunate to count both authors as colleagues and friends. This publication is the first volume in a series that I hope the publishers will have the courage to see through to completion. It will be wonderful to have John Riley's magnificent drawings freely available to the public and I applaud both John Riley for his artistic ability and David Banks for his enthusiasm and editorial skill. Their combined talents have produced a beautifully illustrated and authoritative book which I am sure will have tremendous appeal for all those who appreciate fine work.

David L. Jones
Canberra

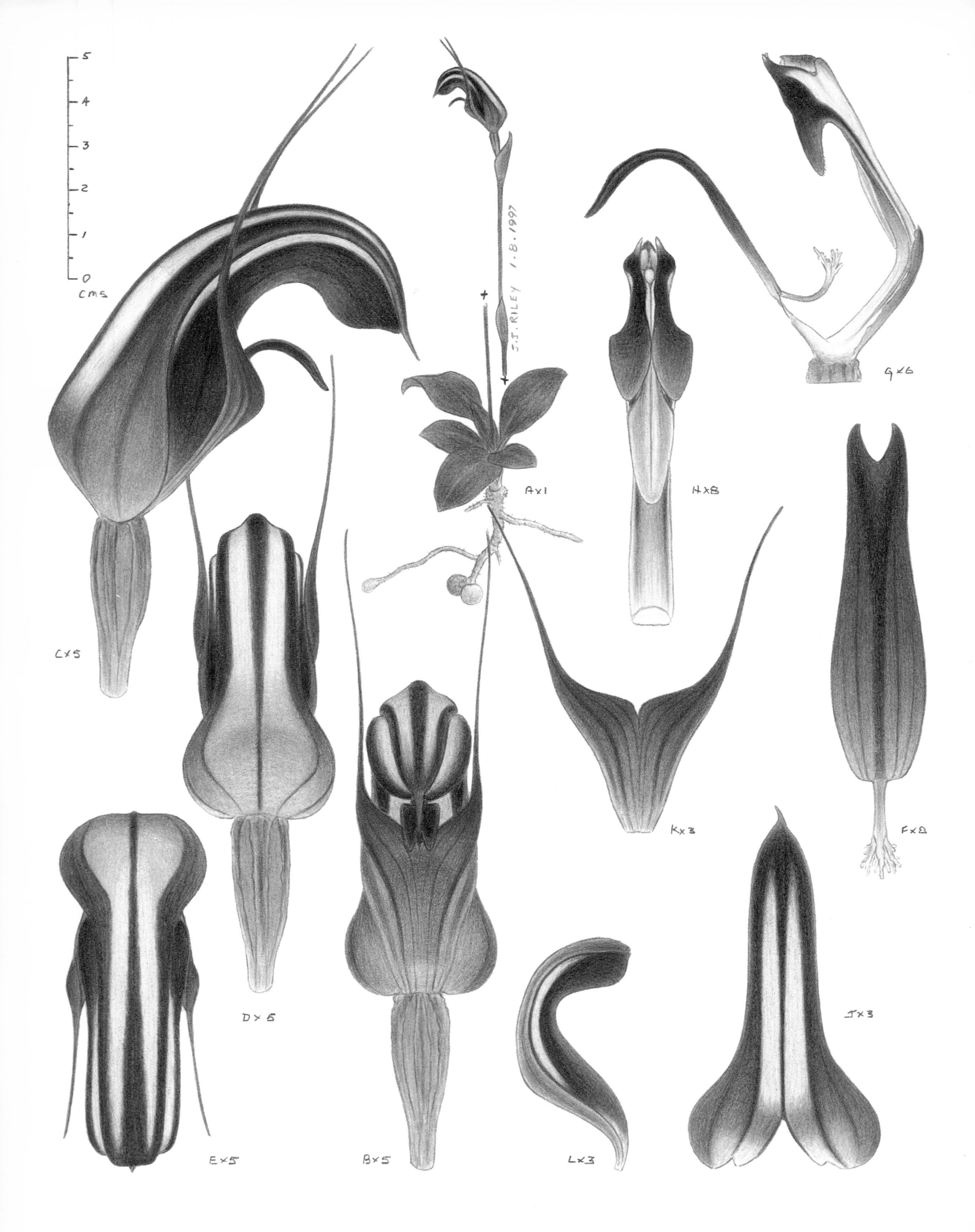

5
4
3
2
1
0
cms
J.J. RILEY 1.8.1997
A×1
H×8
G×6
C×5
K×3
F×8
D×5
J×3
E×5
B×5
L×3

preface

There have only been two serious attempts to botanically illustrate Australian orchids with colour drawings or paintings. Both of these are long out of print, even in their facsimile editions. I am referring to R.D. Fitzgerald's opus *Australian Orchids* originally published in the 1870s, and W.H. Nicholls' landmark *Orchids of Australia* published in the 1950s. Predictably, much of the nomenclature used back then is obsolete today. By today's standards, some of these illustrations are somewhat stylised, but they have been most useful in illustrating plants from specific localities, and in Fitzgerald's case, a number of type specimens. Unfortunately localities are only given in a minority of instances and some plates depict more than one taxon.

This series of volumes will present a cross section of different orchid genera and species from various parts of Australia. It will include many of the well-known and widespread taxa as well as many that are extremely rare, some of which have only been recently discovered and described. David Jones and Mark Clements, respected botanists from the National Herbarium in Canberra, have described many of these newer species in the *Orchadian*, in the period coinciding with my editorship. These included a full botanical description, accompanied by quality line drawings and, often, colour photographs. New species continue to be described in this journal, and it is a must for those who want to keep up-to-date with the latest in the taxonomy and nomenclature of Australian orchids. Details of this periodical and the cultivation of Australian orchids can be found at the Australasian Native Orchid Society's website at www.anos.org.au.

I vividly remember my first encounters with native orchids in the wild. As a child, my father took me for bushwalks, just up the road from my grandparents' holiday house near Greenpoint on the Central Coast of New South Wales. He pointed out numerous epiphytic species to me, including *Cymbidium suave*, *Dendrobium speciosum* and what is now known as *Dockrillia linguiformis*, all growing naturally. However, it was finding a patch of the Flying Duck Orchid, *Caleana major* that really got me in. I could not believe this was a flower. I could not believe it was real, particularly after Dad showed me how the duck's head snapped shut once touched! I was hooked. We regularly revisited our special piece of bushland, and always turned up different species throughout the year. Later that day we also found a large colony of the King Greenhood, *Pterostylis baptistii*, which decades later is still growing happily on that hillside.

Many of our indigenous epiphytic and lithophytic orchids have become popular subjects in horticulture, both here and abroad. There has been much written on their cultivation requirements, with the genera *Bulbophyllum*, *Cymbidium*, *Dendrobium*, *Dockrillia* and *Sarcochilus* being the most popular and amenable to cultivation. Sadly, many of the showy-flowered, solitary, deciduous terrestrial species from the genera *Caladenia*, *Calochilus*, *Pterostylis* and *Thelymitra* have proved difficult or impossible to maintain in cultivation. Colony-forming species have proved to be the best performers in cultivation, with many multiplying in number annually. Some native orchid groups have tuber banks to disperse excess material, and concentrate on promoting specific colony-forming and vigorous members of the genera *Chiloglottis*, *Corybas* and *Pterostylis*

It should be remembered that all orchid species are protected and should not be collected from the wild. Hopefully most of our species are secure in national parks and designated nature reserves, yet there are still unscrupulous individuals who raid protected sites for their personal gain. I have been fortunate to have seen and photographed many of our indigenous orchids in the wild, including some very rare species, and I hope that future generations may also get this opportunity.

I first met John Riley well over a decade ago, and was immediately impressed with his style, wisdom and enthusiasm. He is the sort of likeable person that you wish you had met years beforehand, as your knowledge would have been further enriched. This was even before I had seen his drawings, which had become the talking point within the native orchid community.

I was simply in awe when I first saw examples of his work. For an untrained artist, John would put many commercial and professionally trained botanical artists out of business; such is the realism he puts in his faithful and perfectly colour-matched illustrations. He has a remarkable eye for detail and is extremely observant when in the field. It still amuses me the way John has to remove his thick glasses before looking at the finer details of a bloom. If you consider his drawings beautiful, then that is a bonus, as he refers to them as botanical drawings, not stylised works to impress art critics. He draws what he sees. This was emphasised some years back, when I photographed a plant that John was drawing, an undescribed species from Kurnell related to *Pterostylis plumosa*. My photo is a replica of John's illustration, right down to the markings on the leaf and the chunks taken out of it by a hungry insect. John will only draw the tubers on the deciduous terrestrial species if he has seen them, and will certainly not guess what they may look like. Readers will also be able to see how John's drawing style has evolved over the past decade, as he has refined his skills even further, particularly with the smaller species.

In the field, John is simply unbelievable. I am sure he knows every tree, blade of grass and native orchid in the bush on a first name basis. He has a photographic memory for orchid sites in the bush, yet struggles in suburbia with a street directory! It has been a pleasure to have spent many hours on numerous occasions in the bush with John, and he has shared localities of some extremely rare and restricted taxa, many of which are undescribed. I have certainly learnt so much more about our terrestrial orchids and other aspects of Australiana, thanks to John.

The idea of putting John's drawings in a book was conceived about five years ago. In the interim, John has continued to draw and add to his collection of over 500 illustrations of native orchids, drawing them with watercolour pencils. Each separate work takes between 18 and 24 hours to complete.

I feel truly honoured to have been asked by John to co-write this series of books with him, as I believe that John James Riley, apart from being a true gentleman, is also one of the finest botanical artists in Australia's history.

David P. Banks
Seven Hills, NSW
dpbanks@ozemail.com.au

acknowledgments

The following people have played such a significant role with their advice, expertise and support, that this book would not have been possible without their involvement. Bruce Dalyell, George Hillman, David Jones, Dennis Sinclair, Ron Tunstall and Alan Williams.

David Jones has provided tremendous support for this project and we thank him for writing the foreword. As one of Australia's most knowledgeable and prolific botanists, we are grateful for the time he spent checking the manuscript and providing additional information, and for some controversial discussion on the distributions of the species presented in this volume. We have also been privy to the latest taxonomic research and sighted many unpublished papers.

We also wish to thank the following people who assisted in a variety of ways, including sharing information and locations, accompaniment in the field, discussions on specific species, providing specimens for drawing, and encouragement. Reg Angus†, Gary Backhouse, Graeme & Lynette Banks, Sid Batchelor†, Doug Binns, Tony Bishop†, Gary Borthwick, Col Bower, Graeme Bradburn, Margaret Bradhurst, Boris Branwhite, Peter Branwhite, Bill Brinsley, Wayne Burns, Keith Bursill, Sid Burton, David Butler, Leo Cady, Geoff Carr, Ray Clement, Mark Clements, Steve Clemesha, George Colthup†, Terry Cooke, Jim Cootes, Murray Corrigan, Ralph Crane†, Steve Deards, Jim Dennison, Paul Dennison, Alick Dockrill, Kev Dodt†, Bill Dowling, Tania Duratovic, John English, Peter Eygelshoven, Len Field, Ron Formby, Everett Foster, Vern Frampton†, Chris French, Ted Gregory, Richard Hanman, Mike Harrison, Andrew Harvie, Dick Hepper, Doug & Sue Herd, Peter Hind, Ron Howlett, Colin Hunt, John Hynds, Bill Jackson, Shane Kammera, Mabs & Bill Keppie, John Larkey, Eric Lielkajis, Jim Lykos, Les McHugh, Lorraine Marshall, Brian Milligan, Bob & Helen Morton, John Moye, Bruce Mules, Chris Munson, Bob & Janet Napier, Thelma O'Neill, Steve & Alison Pearson, Andrew Perkins, Mick Price, Helen Richards, John Roberts, Geoff Robinson, Dean Rouse, Reg Sheen†, Norm Shipway, Darryl Smedley, Greg Steenbeeke, Alan Stephenson, Norm Stockton, Steve Summerville, Brian & Ann Tindall, Matthew Tiong, David Titmuss, Wal Upton, David Wallace, Gerry Walsh, Lyn Walsh, Hans & Annie Wapstra, Bill Whiteford, Brian Whitehead, Ron Williamson, Les Winch and Ron Worrell†.

We would also like to express our appreciation to John Elliot, the publishing manager at UNSW Press, for his enthusiasm and for providing the opportunity to present this book.

† Deceased

about this book

This work is the first volume in a planned series of books that will showcase the diversity and beauty of Australia's unique orchid flora, and also the rare talent of illustrator J.J. Riley. Each volume will contain 150 illustrated species, covering both terrestrial and epiphytic orchids from all parts of Australia. It will include many well-known and widespread species, and many extremely rare species, some only recently discovered and described.

For every species, a page of text faces a full-page colour illustration. Page numbers appear only on the pages with text, so not to distract from the impact of the drawings. The orchid genera have been arranged alphabetically. These genera, where appropriate, have been separated into informal groupings, with the species then appearing in chronological order, starting with the first-named. Therefore, closely related taxa will be grouped together, for easier comparison.

We do not provide formal botanical descriptions for the taxa in this volume, as there are many fine reference works that already provide this information. The detailed botanical illustrations will provide, in visual form, the diagnostic characteristics to aid positive identification. The text is arranged as follows.

NAME

This is simply the name of the species, in italics, followed by the abbreviated author's name. Interestingly, most scientific names used in biology are derived from a combination of Greek and Latin terms. This is followed by the year the species was described under that name and where the original description was published. The year cited frequently does not reflect the period of the plant's discovery, as species that have been reclassified as either separate taxa or raised to a different generic rank will have a later date.

TYPE LOCALITY

This is the location from which the specimen/s used for the original descriptions were furnished. Species named in recent times usually have quite detailed and specific location data. In colonial times, however, place names referred to significantly larger regions than we interpret today, and this needs to be taken into account when studying distributions. Where plant specimens were collected in Australia and subsequently shipped back to England, either as live plants or herbarium material and subsequently named, the location was often just given as a State or as Australia. Robert Brown named many Australian species, and frequently used Port Jackson as a type locality. Today we refer to this basically as Sydney Harbour, but it had a broader interpretation back in 1810. Port Jackson referred to the area from Sydney north to Newcastle and the Hunter River, across to the Hunter Range, down the eastern slopes of the Great Dividing Range back to Sydney. Similarly, Ferdinand von Mueller named many species from Rockingham Bay in northern Queensland. Back in 1870, this referred to a large area from the coast of Ingham to Cairns, including the Atherton Tableland. Today Rockingham Bay relates to the body of water north of Hinchinbrook Island and Cardwell. Similarly, the Swan River in Western Australia represented a much larger area than is defined today. Where possible, the illustrated specimens come from the type localities or representative samples from nearby.

ETYMOLOGY

This gives an insight into why the specific name was chosen. Frequently the Latinised names are derived from distinguishing features of the bloom, but can also be named after the habitat, location or individuals.

FLOWERING TIME

The main blooming season is given in months, for flowering plants in the wild. Specimens in flower may occasionally be observed outside these times, as this factor is highly influenced by temperature, rainfall and fires. Many of the deciduous terrestrial species become dormant almost straight after seed dispersal or a month or so after blooming, as flowering is the pinnacle of their annual growth cycle. After this they lie dormant as underground tubers until recommencing their growth sequence some weeks to months later.

DISTRIBUTION

This encompasses the known natural range for the species under discussion. This has been gleaned from our first hand field observations, respected texts and floras, reliable informants and herbarium records. The information provided here may contradict other published works as we have used a narrower concept of the species. What were once considered variable species that previously enjoyed a wide distribution have since been dissected into separate taxa. The distribution maps will also reflect this. These maps are provided to give a snapshot of the natural distribution, as it is impossible to give finer details on such a small scale, particularly when a species is known from one small site or occurs in a very narrow coastal band. Obviously, further research and new recordings may increase or decrease these limits.

ALTITUDINAL RANGE

The altitudinal range is given in metres above sea level. Occasionally plants may be seen outside these limits.

DISTINGUISHING FEATURES

Here we have provided some key recognition tools to assist identification when used with the illustration. The use of a hand lens may be required when studying live material for observing some of the important botanical features, especially those with smaller blooms.

HABITAT

The preferred habitat/s are provided for each species. Sporadic plants may be observed in other vegetation communities.

CONSERVATION STATUS

The details given here are very brief, with the main emphasis to determine if the taxon is conserved within a national park or a secure nature reserve and its abundance, rarity or threats in the wild.

DISCUSSION

Here we provide miscellaneous observations about the species, including historical notes and other points of interest where appropriate. The reproductive biology of the taxon and known pollinating agents are also provided.

ILLUSTRATION DETAILS

Includes the name of the species, the location the depicted specimen was found, and the date drawn. All illustrations are drawn to the scale that appears on every drawing, with a key to all the floral parts. Note that the letter I is not used. All of the colour illustrations are reproductions from original works by John J. Riley.

GLOSSARY

This appears towards the back of the book and includes terms used throughout the text. Obviously the meanings of most of these words have been used in a botanical context.

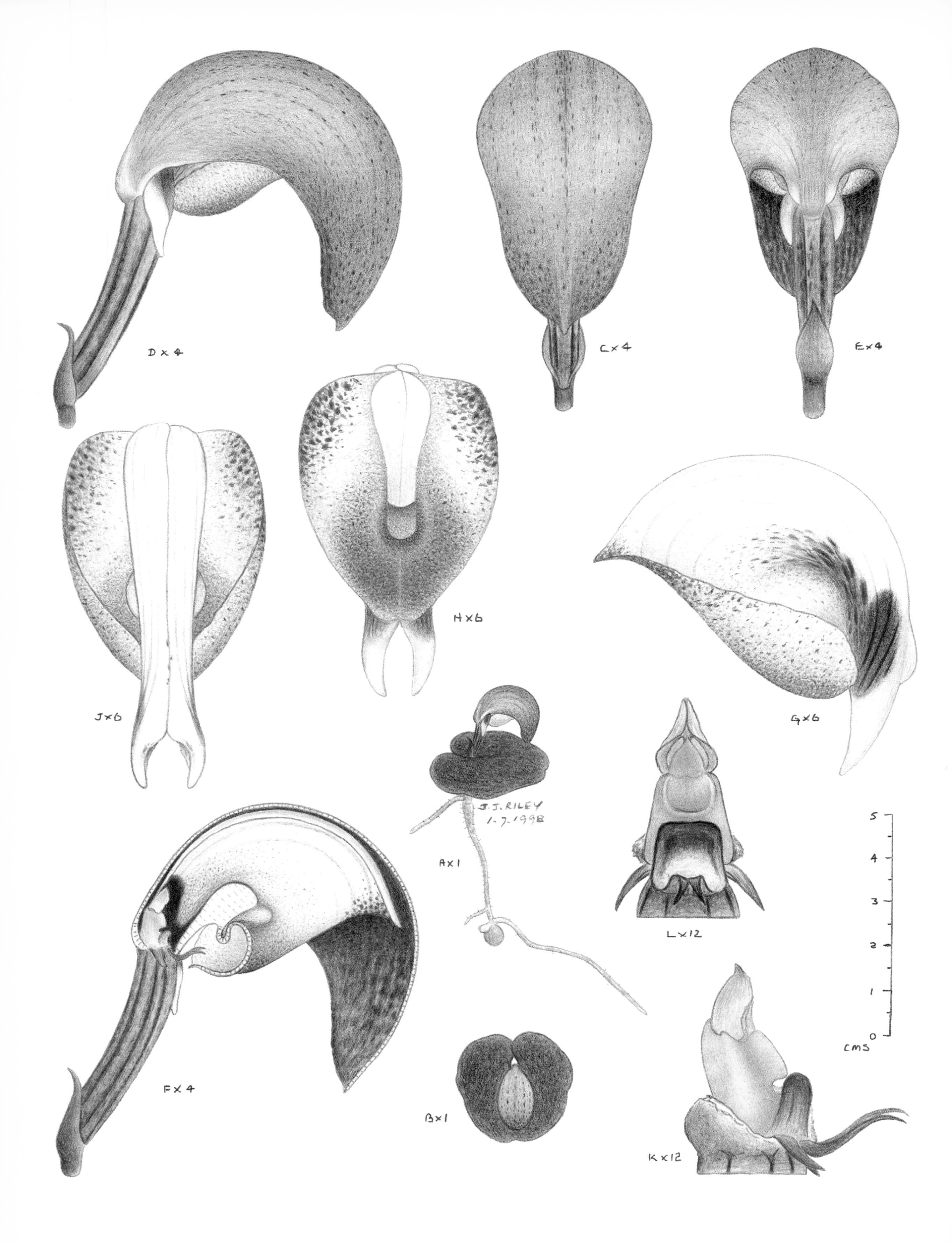
D x 4
C x 4
E x 4
H x 6
J x 6
G x 6
J. J. RILEY
1. 7. 1998
A x 1
L x 12
5
4
3
2
1
0
CMS
F x 4
B x 1
K x 12

Orchids of Australia

Species descriptions

Acianthus fornicatus R. Br.

1810 | *Prodromus Florae Novae Hollandiae*: 321

TYPE LOCALITY Port Jackson, New South Wales

ETYMOLOGY *fornicatus* — bent, peaked

FLOWERING TIME April to August

DISTRIBUTION From the South Coast of New South Wales, north to Fraser Island off the Queensland coast. Mostly coastal but sometimes extending to nearby ranges.

ALTITUDINAL RANGE Up to 1000 m

DISTINGUISHING FEATURES *A. fornicatus* has a broad dorsal sepal. The labellum is deeply concave, greenish-maroon at the base, greener towards the apex, with reddish-maroon, serrated margins. The labellum callus is very papillose. The flowers are a translucent pale green with maroon markings.

HABITAT This widespread species grows in a range of habitats, from coastal, stabilised and vegetated sand dunes to moist gullies in heavy shade. It does not seem to have a preferred soil type or plant community association.

CONSERVATION STATUS Common. Well represented in national parks and reserves. Secure.

DISCUSSION *A. fornicatus* is one of our most commonly encountered terrestrial orchids. It can form large colonies, reproducing vegetatively from daughter tubers formed on the end of stolonoid roots. There is some colour variation in this taxon; all, however, are translucent. The plant illustrated is representative of the most common colour form. Some clones can be a lot darker while, less commonly, may have few or no maroon markings, and are a translucent pale green. It is possible for all colours to be present in the same location. A complex of taxa exists around *A. fornicatus*, with some of these (e.g. *A. apprimus*, *A. borealis*, *A. collinus* and *A. exiguus*) recently given specific rank. Further research may show that more taxa are involved in this complex.

Acianthus fornicatus
Fingal Bay, New South Wales

30 June 1999

A plant
B flower from side
C flower from front
D flower from above
E labellum from above — flattened
F labellum and column from side
G labellum and column from front
H dorsal sepal
J lateral sepal
K petal

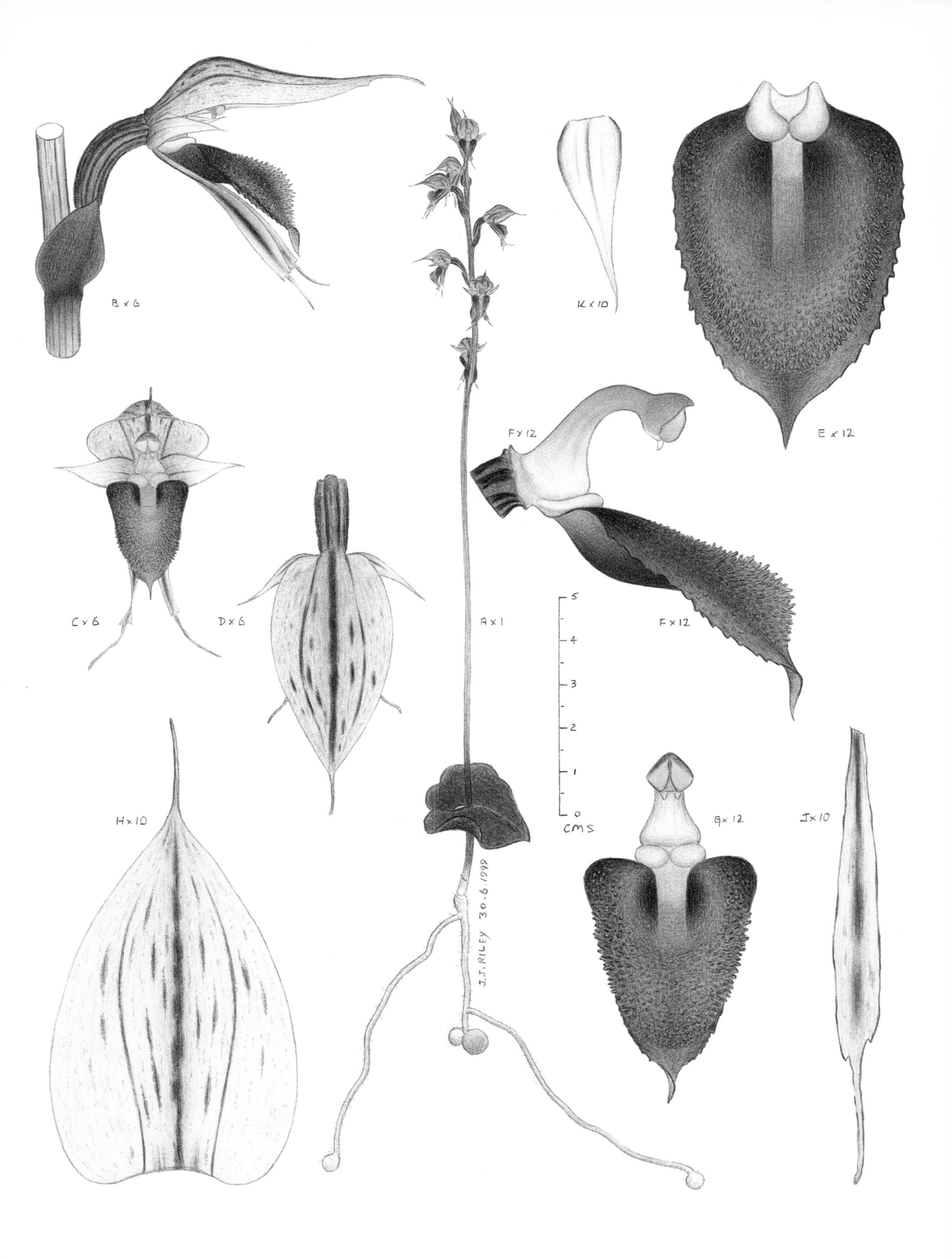
B x 6
K x 10
E x 12
F x 12
C x 6
D x 6
A x 1
F x 12
5
4
3
2
1
0
CMS
H x 10
G x 12
J x 10
J.J. RILEY 30.6.1999

Acianthus collinus D.L. Jones

1991 | *Australian Orchid Research* 2:6

TYPE LOCALITY Conimbla National Park, New South Wales

ETYMOLOGY *collinus* — living in the hills

FLOWERING TIME June to August

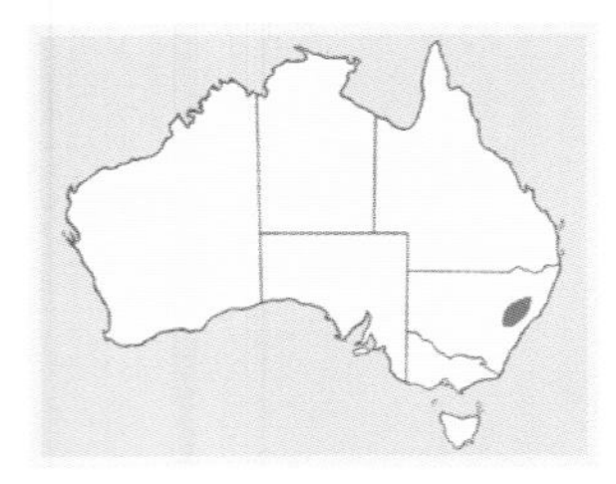

DISTRIBUTION This is an inland species growing on the western slopes of the Great Dividing Range, northwards from Young to the Manilla district, New South Wales. An isolated disjunct population occurs in the Beechworth area of Victoria.

ALTITUDINAL RANGE 450 m to 800 m

DISTINGUISHING FEATURES *A. collinus* has a narrower dorsal sepal than *A. fornicatus*. The labellum is not as papillose or concave and is reddish-maroon with the serrations on the margins not as prominent. The flowers are usually darker with more maroon markings, and are less translucent.

HABITAT It grows on the ranges and ridges of the central slopes of New South Wales. In sheltered moist areas, colonies can be large and the leaves quite robust. On the drier, more exposed ridges, plants are more scattered and the leaves are smaller. In this habitat, it often grows under the white cypress pine (*Callitris glaucophylla*), co-existing with other terrestrial orchids from the genera *Glossodia*, *Caladenia*, *Diuris* and members of the *rufa* group of *Pterostylis*.

CONSERVATION STATUS Common. Well represented in national parks and reserves. Secure.

DISCUSSION *A. collinus* forms colonies from daughter tubers grown on the end of stolonoid roots. This species, being recognised at specific level in 1991, is part of the *A. fornicatus* complex.

Acianthus collinus
Conimbla Range, New South Wales

15 May 2001

A plant
B flower from side
C flower from front
D flower from above
E labellum from above — flattened
F labellum and column from side
G labellum and column from front
H dorsal sepal
J lateral sepal
K petal

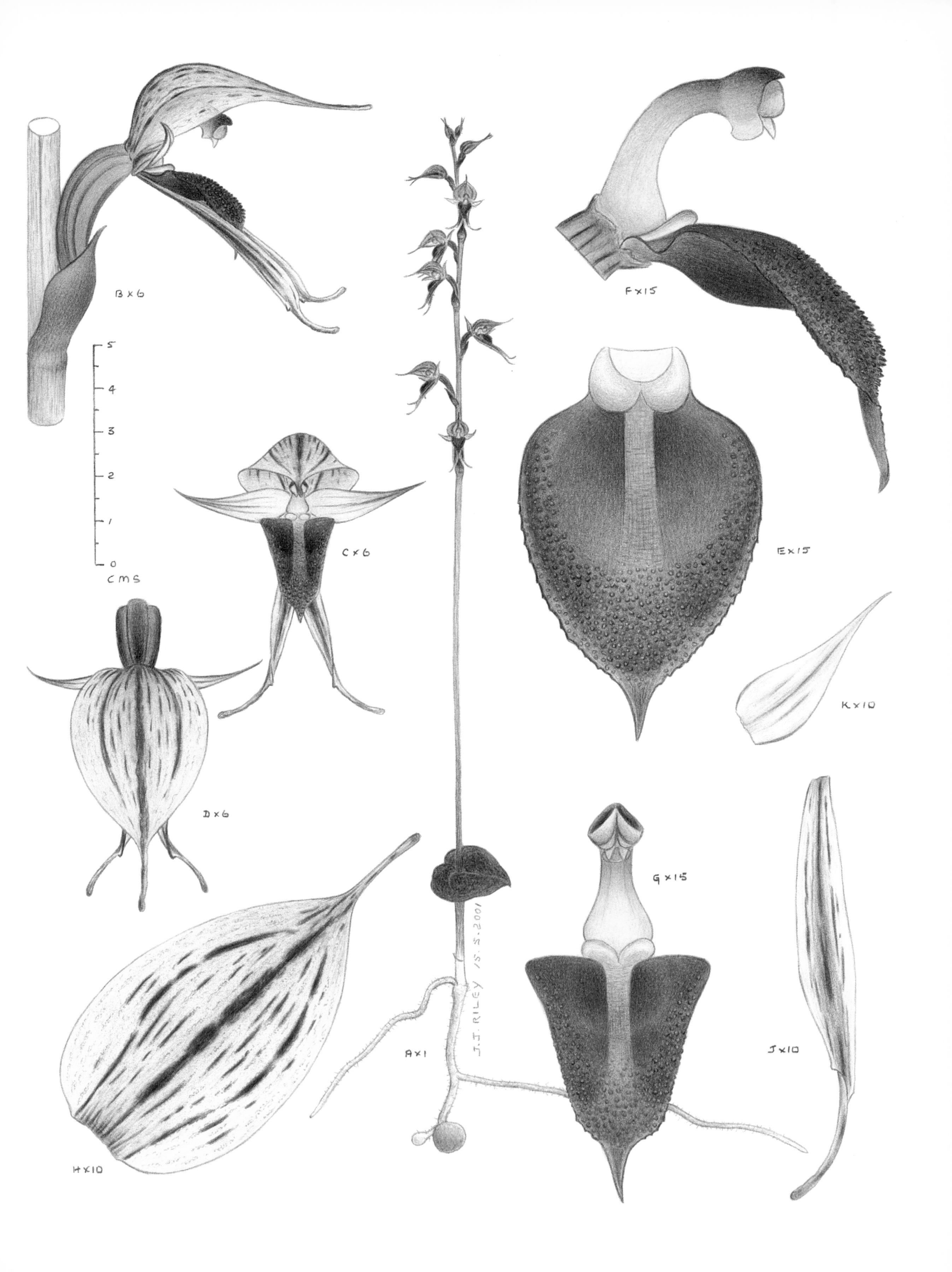
B x 6
F x 15
5
4
3
2
1
0
CMS
C x 6
E x 15
D x 6
K x 10
G x 15
J.T. RILEY 15.5.2001
A x 1
J x 10
H x 10

Acianthus apprimus D.L. Jones

1991 | *Australian Orchid Research* 2:5

TYPE LOCALITY Mount Wilson, New South Wales

ETYMOLOGY *apprimus* — first, early blooming

FLOWERING TIME January to March

DISTRIBUTION Occurs sporadically on the higher (sub-alpine) areas of the Great Dividing Range, from the Boyd Plateau, northwards to the New England National Park, New South Wales.

ALTITUDINAL RANGE 900 m to 1550 m

DISTINGUISHING FEATURES The flowers of *A. apprimus* are narrower and more nodding than those of the sister species, *A. fornicatus* or *A. collinus*. The labellum is narrow and reddish-maroon with prominently folded edges that are entire. The blooms are a pinkish-green colour. The other obvious difference is the summer flowering period.

HABITAT This is a species from cool montane areas, growing in rich, well structured, dark loams, generally of volcanic origin. It is mostly found in moist open woodland among grasses and ferns. It is particularly at home growing at the base of large trees, and around fallen logs and rotting timber.

CONSERVATION STATUS Uncommon. Well represented in national parks and reserves. Secure.

DISCUSSION *A. apprimus* is an early-flowering species, that frequently grows with other, unrelated montane orchid species, such as *Pterostylis coccina*, *P. decurva*, *P. abrupta* and *Corybas* and *Chiloglottis* species. It forms large colonies from daughter tubers grown on the end of stolonoid roots.

Acianthus apprimus
Boyd Plateau, New South Wales

4 March 1996

A plant
B flower from side
C flower from front
D flower from above
E labellum from above — flattened
F labellum and column from side
G labellum and column from front
H dorsal sepal
J lateral sepal
K petal

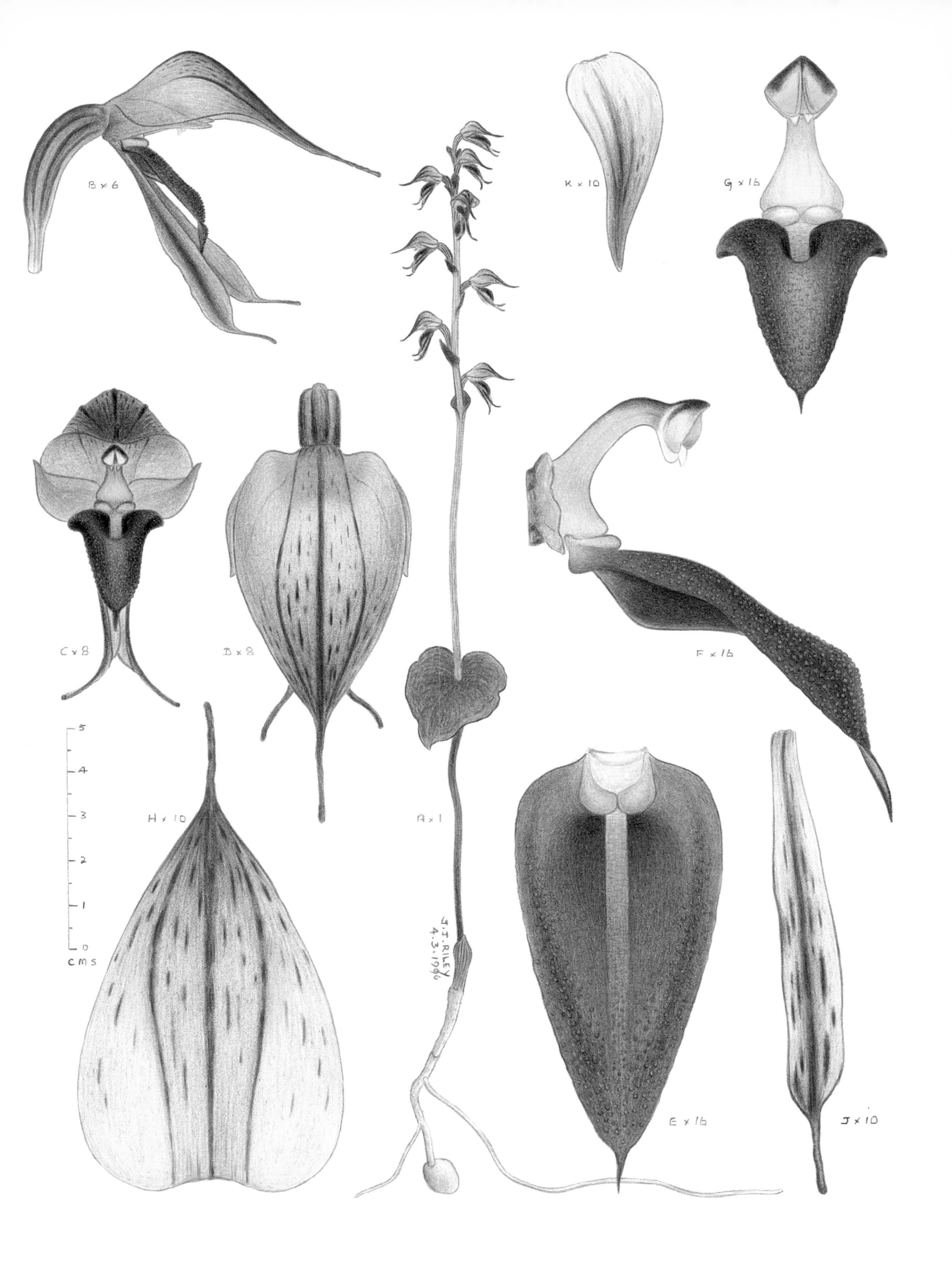
B x 6
K x 10
G x 16
C x 8
D x 8
F x 16
H x 10
A x 1
5
4
3
2
1
0
CMS
J.T. RILEY
4.3.1996
E x 16
J x 10

Acianthus borealis D.L. Jones

1991 | *Australian Orchid Research* 2:5

TYPE LOCALITY Atherton Tableland, Queensland

ETYMOLOGY *borealis* — northern

FLOWERING TIME March to June

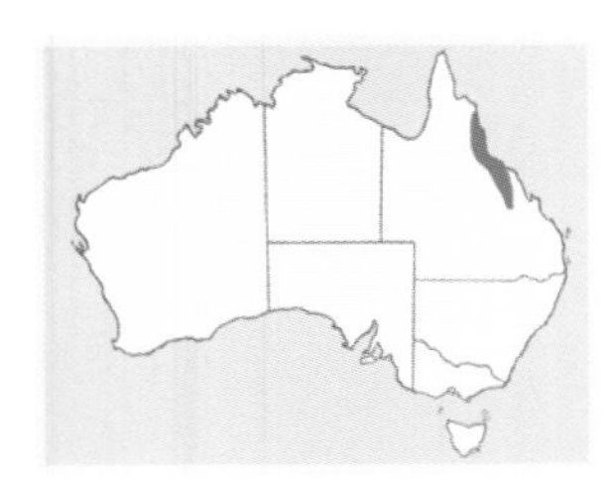

DISTRIBUTION Found on most of the ranges north from the Blackdown Tableland to just south of Cooktown, Queensland.

ALTITUDINAL RANGE 100 m to 1200 m

DISTINGUISHING FEATURES *A. borealis* has smaller flowers than *A. fornicatus*, a narrower ovate labellum (which is less papillose) and dorsal sepals. The blooms have fewer maroon markings and the labellum is greener.

HABITAT With such a wide distribution range, this species is found in a number of different habitats. In dry sclerophyll forest, it is often found on the moister slopes, sheltering under grass trees (*Xanthorrhoea* spp.) or close to large rock outcrops. In more favourable areas, large colonies grow on sheltered banks and gullies. It is also quite common in moist sclerophyll areas on the fringes of rainforest, often growing in association with *Pterostylis*, *Corybas* and *Chiloglottis*.

CONSERVATION STATUS Common. Conserved in national parks and reserves. Secure.

DISCUSSION *A. borealis* forms large colonies, reproducing from daughter tubers grown on the end of stolonoid roots. The distribution of *A. borealis* does not overlap any of the other named members of the *A. fornicatus* complex.

Acianthus borealis
Eungella, Queensland

3 July 1996

A plant
B flower from side
C flower from front
D flower from above
E labellum from above — flattened
F labellum and column from side
G labellum and column from front
H dorsal sepal
J lateral sepal
K petal

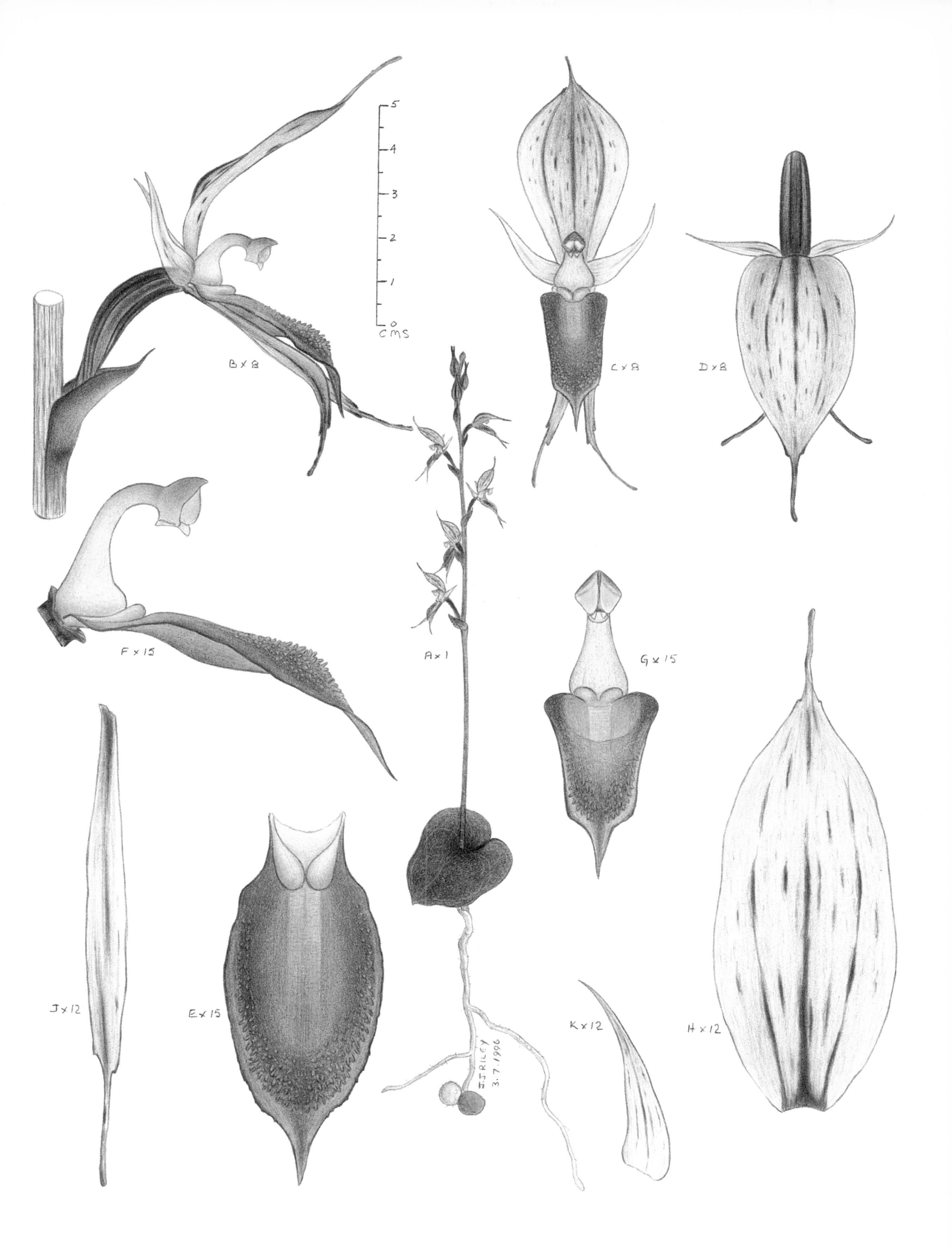

5
4
3
2
1
0
CMS
B × 8
C × 8
D × 8
A × 1
F × 15
G × 15
J × 12
E × 15
K × 12
H × 12
J.J. RILEY
3.7.1996

Acianthus exiguus D.L. Jones

1991 | *Australian Orchid Research* 2:7

TYPE LOCALITY Wardell, New South Wales

ETYMOLOGY *exiguus* — small, weak, little

FLOWERING TIME May to June

DISTRIBUTION Found north from Urunga to Ballina, New South Wales. This species always grows near the coast.

ALTITUDINAL RANGE Up to 100 m

DISTINGUISHING FEATURES *A. exiguus* is the smallest member of the *A. fornicatus* complex. The individual flowers are 5–7 mm tall and the flowering inflorescence is less than 10 cm high. The labellum has entire, unfolded margins and is only slightly papillose.

HABITAT *A. exiguus* grows in white-grey, peaty sand and sandy clay loams. It may be found on the fringes of, and often in, littoral rainforest, or the dry heath areas around coastal swamps and adjacent moist sclerophyll forest.

CONSERVATION STATUS Uncommon. Represented in national parks and reserves. Secure.

DISCUSSION This species is not very well known. It can form colonies, reproducing vegetatively from daughter tubers formed on the end of stolonoid roots. Some clones of this species appear to be self-pollinating. Isolated populations, which further research may prove to be this species, occur rarely in the Port Stephens district.

Acianthus exiguus
Pimlico, New South Wales

13 July 2000

A plant
B flower from side
C flower from front
D flower from above
E labellum from above — flattened
F labellum and column from side
G labellum and column from front
H dorsal sepal
J lateral sepal
K petal

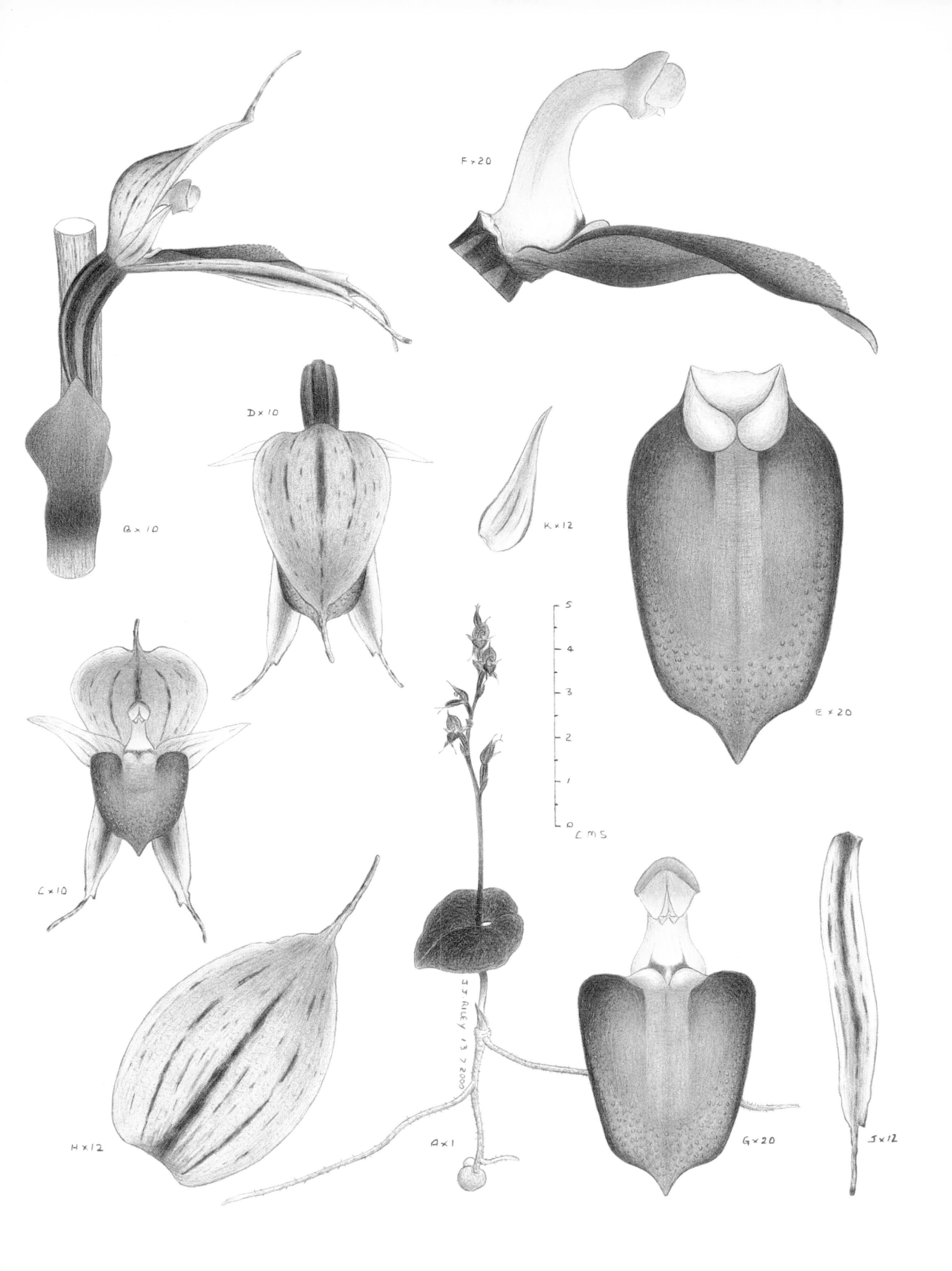
F×20
D×10
B×10
K×12
E×20
C×10
5
4
3
2
1
0
CMS
J.J. RILEY 13 7 2000
H×12
A×1
G×20
J×12

Adenochilus nortonii Fitzg.

1876 | *Australian Orchids* 1(2)

TYPE LOCALITY Mount Victoria, New South Wales

ETYMOLOGY After original collector, James Norton

FLOWERING TIME November to January

DISTRIBUTION Endemic to New South Wales, being recorded from Fitzroy Falls, the upper Blue Mountains, Barrington Tops and New England National Park.

ALTITUDINAL RANGE 850 m to 1550 m

DISTINGUISHING FEATURES *A. nortonii* is a very distinctive species, the blooms are superficially similar to some white-flowered *Caladenia*. The tuber is greatly elongated and resembles a root-like stolon.

HABITAT The Blue Mountains populations grow along the cliff lines near waterfalls. They can be seen in black soils with mosses and ferns under wet overhangs that are constantly dripping, often growing with the Forked Sundew (*Drosera binata*) and Native Violet (*Viola hederacea*). On the Barrington Tops and New England National Park, the species grows mostly in *Sphagnum* hummocks along and in small creeks, often under Antarctic Beech (*Nothofagus moorei*).

CONSERVATION STATUS Uncommon. All known populations occur within national parks. Secure.

DISCUSSION In the Blue Mountains, *A. nortonii* is commonly found growing below *Rimacola elliptica*, which is found in the rock crevices above. This species reproduces from seed and slowly, by vegetative means, from very brittle root-like tubers. Interestingly, the flowers are attracted to light and will track the sun's path. A healthy colony in full bloom is an impressive sight.

Adenochilus nortonii
Barrington Tops, New South Wales

17 December 1989

A plant
B flower from front
C flower from side
D flower from rear
E labellum from side
F labellum from front
G labellum top view — flattened
H column from side
J column from front

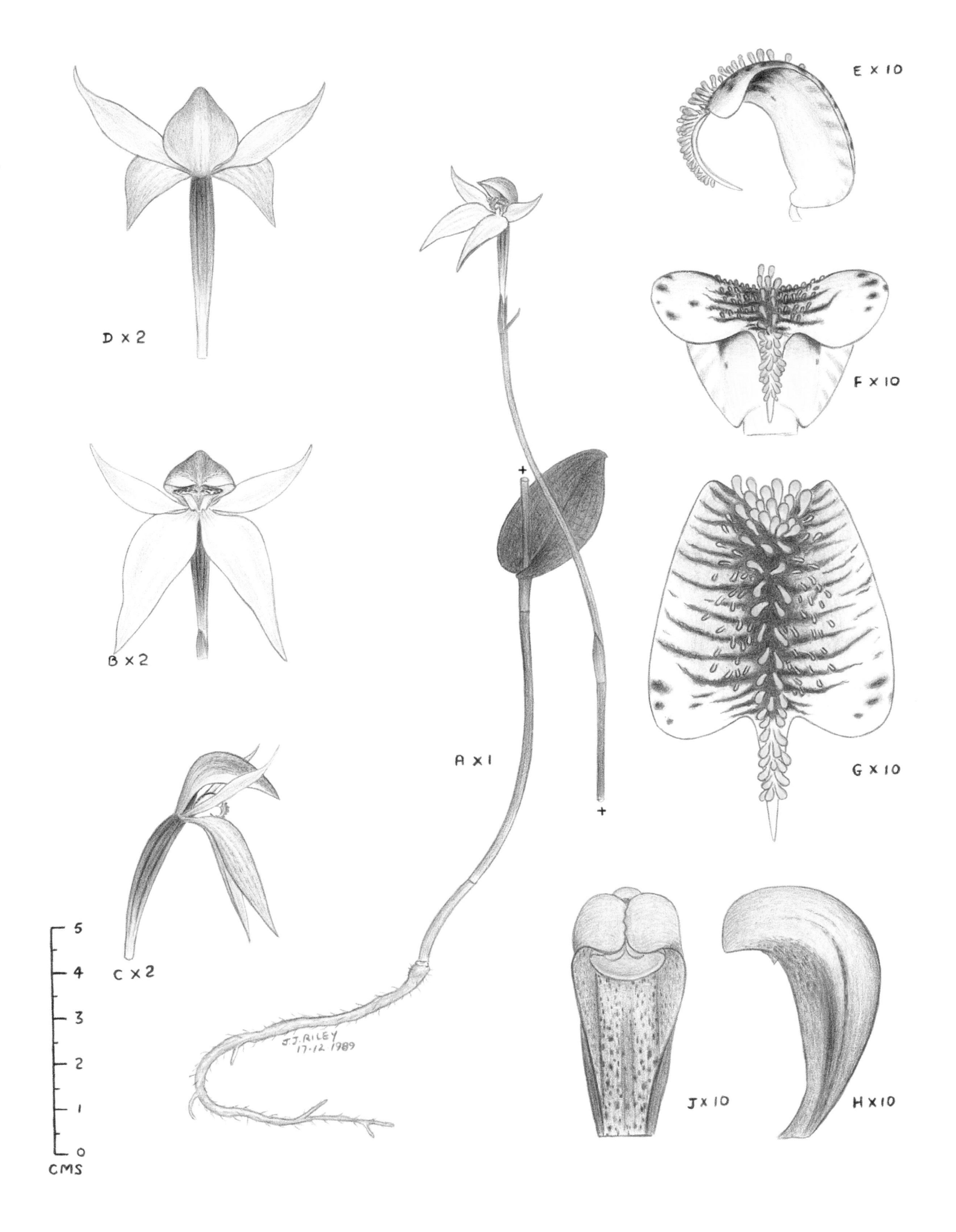
E x 10
D x 2
F x 10
B x 2
G x 10
A x 1
C x 2
5
4
3
2
1
0
CMS
J.J. RILEY
17.12.1989
J x 10
H x 10

Arthrochilus huntianus (F. Muell.) Blaxell

1972 | *Contributions from the New South Wales National Herbarium* 4:277

TYPE LOCALITY Tingerinji Mountain, New South Wales

RECENT SYNONYMS *Spiculaea huntiana* (F. Muell.) Schltr.
Drakaea huntiana F. Muell.

ETYMOLOGY After Robert Hunt

FLOWERING TIME November to March

DISTRIBUTION From the Central Tablelands and Blue Mountains of New South Wales, south to East Gippsland and along the coast and ranges to the Grampians in Victoria. Also recorded once from Flinders Island in Bass Strait.

ALTITUDINAL RANGE Up to 1200 m

DISTINGUISHING FEATURES *A. huntianus* is a summer-flowering, leafless species, which is saprophytic. *A. huntianus* subsp. *nothofagicola* (which has recently been reclassified as *Thynninorchis nothofagicola*) only occurs in central, southern Tasmania.

HABITAT In New South Wales, this distinctive species occurs in montane to sub-alpine open forest with a grassy understorey. In Victoria, it grows from the coast to the ranges in open woodland and areas of heath. Plants grow in bare ground or leaf litter among grasses or small shrubs.

CONSERVATION STATUS Occurs, and is conserved, in national parks and flora reserves in New South Wales and Victoria. Secure.

DISCUSSION *A. huntianus* is a leafless saprophyte with a thin and wiry habit, which makes it difficult to locate among the understorey as it blends into the background. It has up to ten long-lasting blooms on an erect inflorescence. Robust specimens may be in bloom for many weeks, as the flowers are produced sequentially. *A. huntianus* is pollinated by male thynnid wasps during pseudocopulation. It often occurs as isolated individuals and reproduces from seed. Was recently reclassified as *Thynninorchis huntianus* (F. Muell.) D.L. Jones & M.A. Clem.

Arthrochilus huntianus
Boyd Plateau, New South Wales

24 December 1989

A plant
B flower from front
C flower from side
D labellum from side
E labellum from above
F labellum from below
G column from side
H column from front
J column from rear

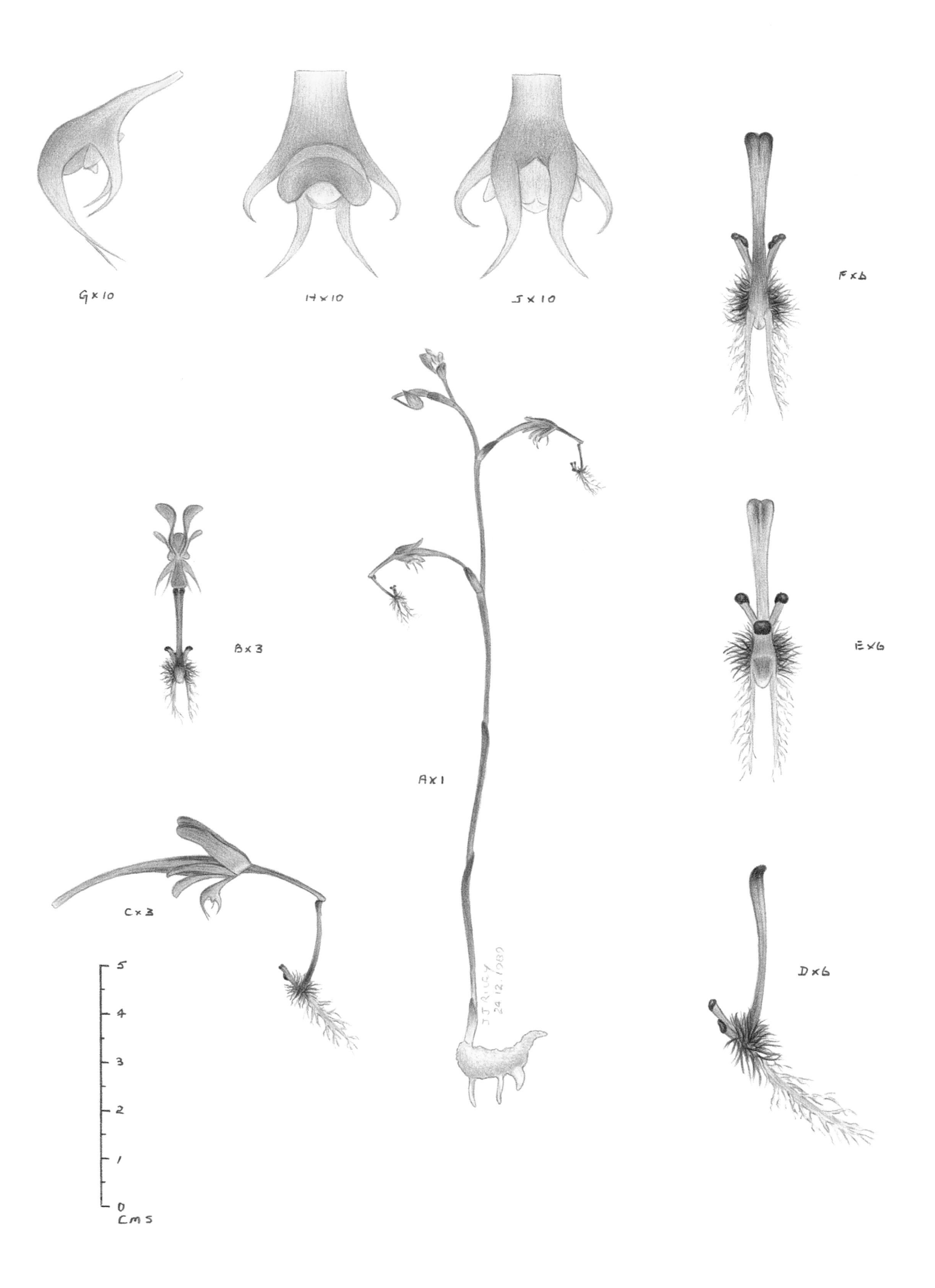
G x 10
H x 10
J x 10
F x 6
B x 3
A x 1
E x 6
C x 3
D x 6
5
4
3
2
1
0
cms
J J RILEY

Arthrochilus prolixus D.L. Jones

1991 | *Australian Orchid Research* 2:10

TYPE LOCALITY Bellangry, New South Wales

ETYMOLOGY *prolixus* — stretched out, relating to the long labellum calli

FLOWERING TIME November to March

DISTRIBUTION This is a coastal species, found from the northern suburbs of Sydney, New South Wales, to Gladstone in Queensland.

ALTITUDINAL RANGE Up to 200 m

DISTINGUISHING FEATURES *A. prolixus* has fewer, longer, linear lamina calli, while *A. irritabilis* has massed, strongly clavate lamina calli. These species overlap in distribution in south-eastern Queensland.

HABITAT Occurs in a wide range of habitats, from open *Eucalyptus* forest to areas dominated by *Casuarina* and *Melaleuca* spp. Usually growing along moist drainage patterns and in the vicinity of swamps and other seasonally damp areas.

CONSERVATION STATUS Uncommon but widespread. It is conserved in a number of national parks and reserves. Secure.

DISCUSSION *A. prolixus* has up to 20 flowers per plant and the basal rosette leaves are absent until late in the flowering. The leaves of non-flowering plants appear much earlier. The individual blooms are long-lasting and open progressively. Individual plants may be in flower for many weeks. *A. prolixus* is pollinated by males of a species of thynnid wasp during pseudocopulation. It reproduces from seed and vegetatively from numerous daughter tubers growing on the ends of stolonoid roots. In favourable conditions, this species can form large colonies.

Arthrochilus prolixus
Bulahdelah, New South Wales

15 January 1995

A flowering plant
B non-flowering plant
C flower from side
D flower from front
E labellum
F labellum and column from side
G column
H dorsal sepal
J lateral sepal
K petal

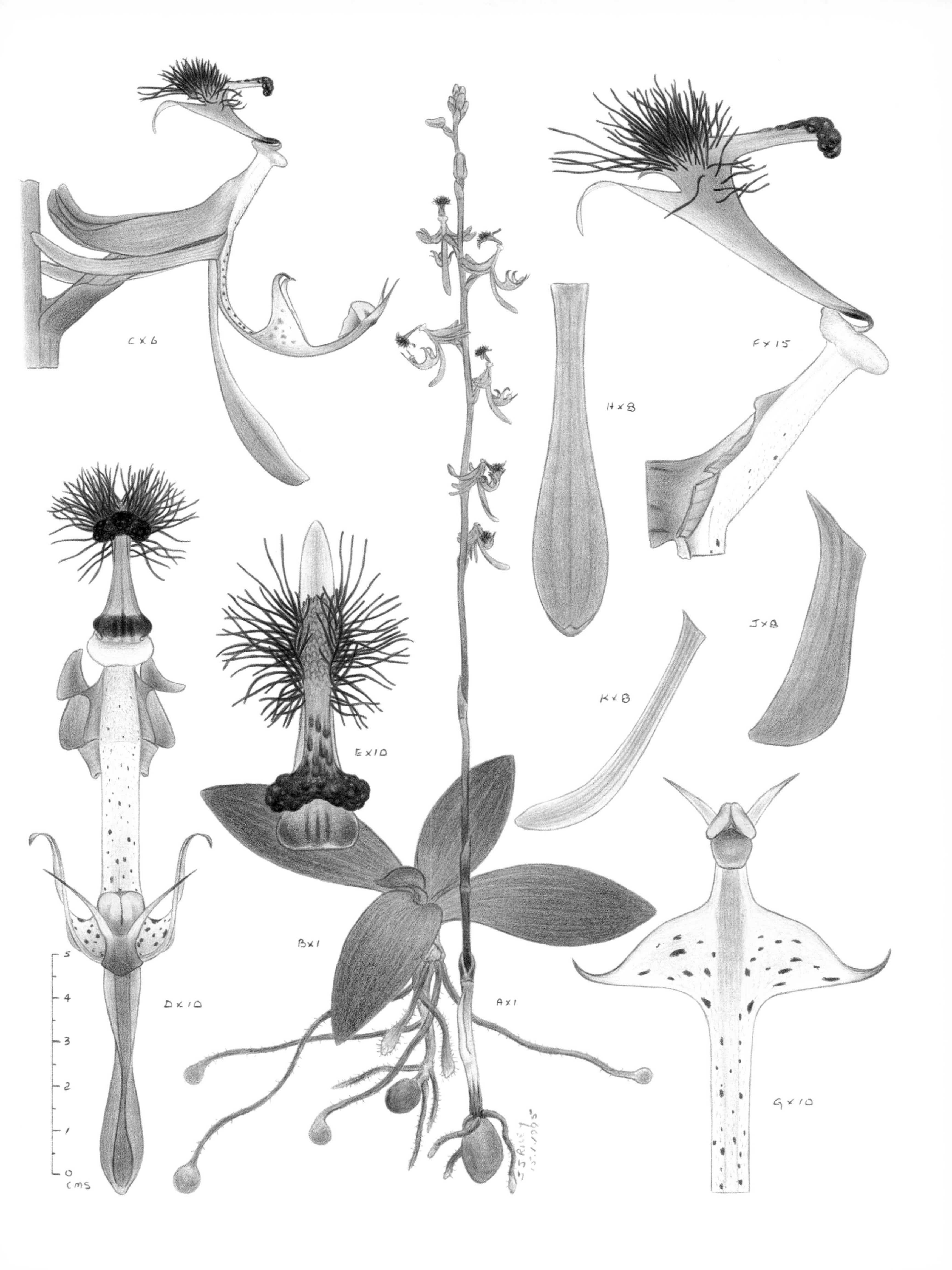

C × 6
F × 15
H × 8
J × 8
K × 8
E × 10
D × 10
B × 1
A × 1
G × 10
5
4
3
2
1
0
cms

Bulbophyllum exiguum F. Muell.

1860 | *Fragmenta Phytographiae Australiae* 2:72

TYPE LOCALITY Illawarra, New South Wales

ETYMOLOGY *exiguum* — small, weak, little

FLOWERING TIME March to May

DISTRIBUTION North from Ulladulla, New South Wales, to near Gympie, Queensland. It is found from the coast to the tablelands and ranges.

ALTITUDINAL RANGE Up to 1000 m

DISTINGUISHING FEATURES *B. exiguum* has small pseudobulbs widely spaced on a wiry rhizome with a single leaf. The inflorescence is multiflowered with pale, translucent, greenish-yellow blooms.

HABITAT Grows on trees in rainforest and in moist, sheltered areas in both wet and dry sclerophyll forest. In these habitats it may be found as both an epiphyte and a lithophyte. It is frequently seen growing on rocks in sheltered gullies in the Sydney Sandstone region, often with *Liparis reflexa* and *Dockrillia linguiformis*.

CONSERVATION STATUS Widespread. Common and well represented in national parks and reserves. Secure.

DISCUSSION This species can form quite large mats when growing on rock faces. Although the individual flowers are small, a large clump in bloom is an impressive sight. A particularly robust form occurs in the Watagan State Forest, New South Wales.

Bulbophyllum exiguum
Nattai, New South Wales

9 April 1993

A plant
B flower from front
C flower from side
D flower from above
E labellum from front
F labellum and column from side
G labellum from above
H column from front
J dorsal sepal
K lateral sepal
L petal

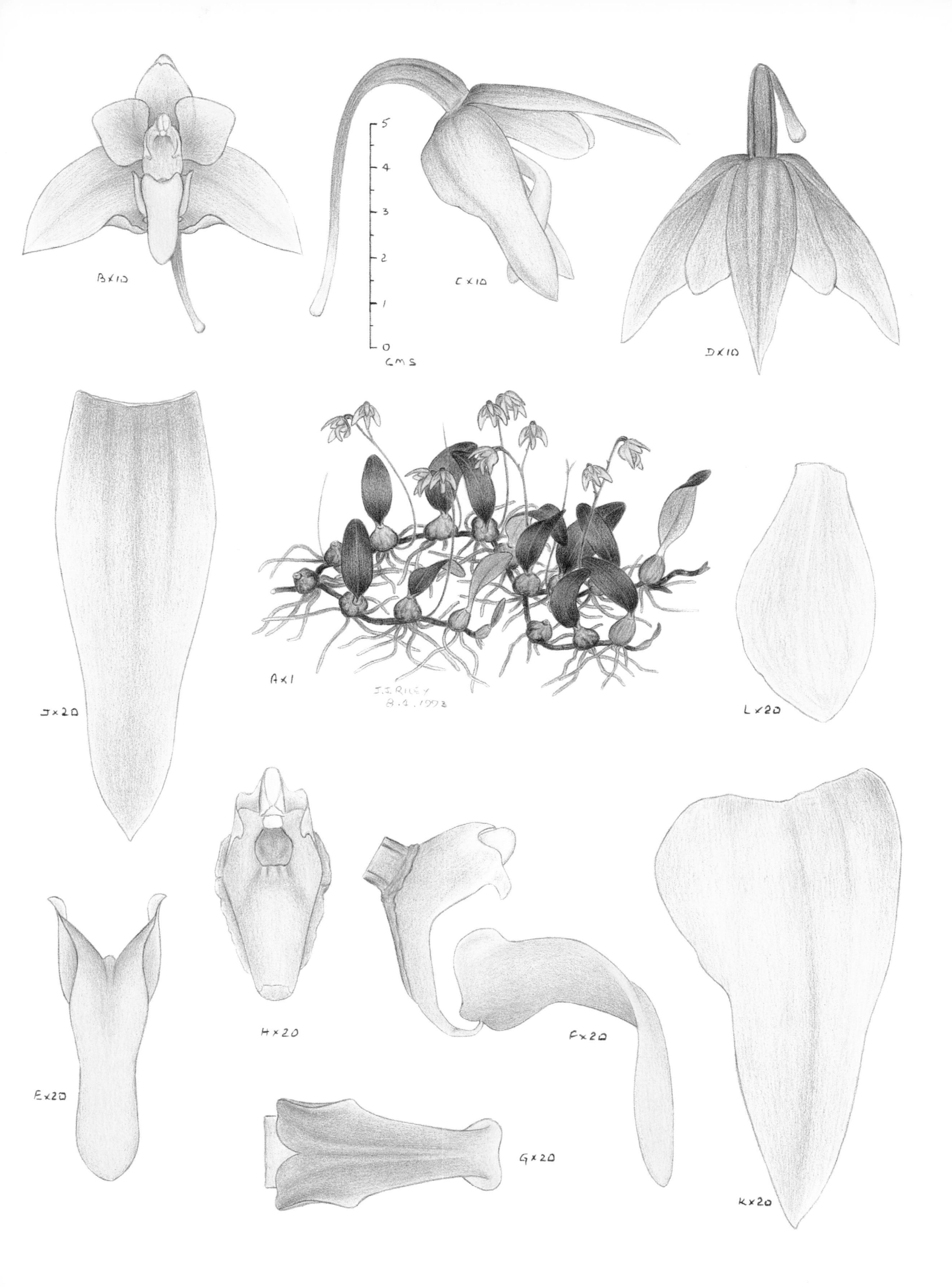
B×10
C×10
D×10
5
4
3
2
1
0
CMS
J×20
A×1
J.J. RILEY
8.4.1993
L×20
H×20
F×20
E×20
G×20
K×20

Bulbophyllum elisae (F. Muell.) Benth.

1873 | *Flora Australiensis* 6:289

TYPE LOCALITY Blue Mountains, New South Wales

ETYMOLOGY After Elisa Kearn

FLOWERING TIME June to December

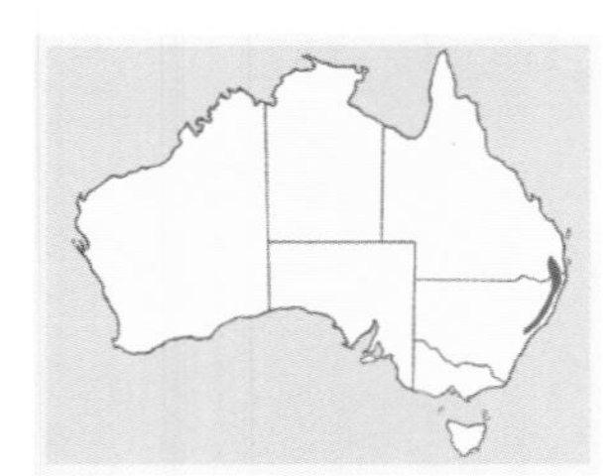

DISTRIBUTION *B. elisae* can be found from the upper Blue Mountains of New South Wales, northwards along the ranges to the Bunya Mountains in southern Queensland.

ALTITUDINAL RANGE 650 m to 1300 m

DISTINGUISHING FEATURES *B. elisae* is a very distinctive species that can be readily identified even when not in bloom. It has wrinkled pseudobulbs and an inflorescence of predominantly green blooms that are nodding with long lateral sepals.

HABITAT This species is found on rocks and trees in, or on the fringes of, rainforest often in areas of high rainfall. Can also be found in wet sclerophyll forest growing on large boulders or cliff faces. In some locations at higher elevations, plants may experience light dustings of snow in winter.

CONSERVATION STATUS Widespread. Conserved in national parks and nature reserves. Secure.

DISCUSSION The flowers of *B. elisae* are carried on erect inflorescences with up to a dozen flowers on the inflorescence. The flowers, that do not open widely, are usually bright apple green, with a dark reddish-brown labellum, facing the same direction. Occasional plants have blooms with purplish-red suffusions. It can sometimes form large clumps when growing as a lithophyte, however, epiphytic plants, which favour smaller limbs and branches, rarely grow into large specimens. Plants occur as scattered individuals rather than in colonies.

Bulbophyllum elisae
Barrington Tops, New South Wales

9 December 2001

A plant
B flower from front
C flower from side
D labellum from side
E labellum from above
F column from side
G column from front
H dorsal sepal
J lateral sepal
K petal

B×4
F×20
K×10
C×4
G×20
H×10
J×4
A×1
D×20
E×20
CMS
J.J. RILEY 9.12.2001

Caladenia deformis R. Br.

1810 | *Prodromus Florae Novae Hollandiae*: 324

TYPE LOCALITY Port Dalrymple, Tasmania

RECENT SYNONYMS *Cyanicula deformis* (R. Br.) Hopper & A.P. Brown

ETYMOLOGY *deformis* — deformed, referring to the labellum

FLOWERING TIME May to October

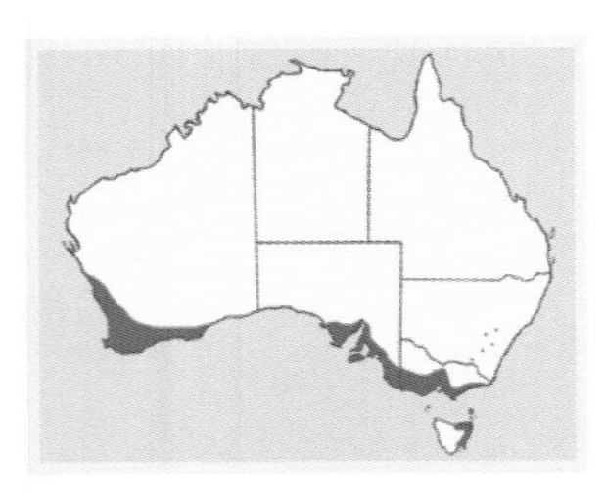

DISTRIBUTION Widespread in the south-west of Western Australia from north of Kalbarri to east of Israelite Bay. Widespread in South Australia from the south-east corner to the lower Flinders Ranges and Eyre and York Peninsulas. Widespread in the western half of Victoria (excluding the mallee) and rare in eastern Victoria. Uncommon and widespread in eastern Tasmania and on Flinders Island. Very rare in New South Wales, occurring around the Cowra–Orange–Mudgee–Rylstone area, and West Wyalong to near Mulwala in southern New South Wales.

ALTITUDINAL RANGE Up to 1000 m

DISTINGUISHING FEATURES *C. deformis* has blue flowers and a distinctive labellum, with numerous calli, mainly towards the apex.

HABITAT This widespread species grows in a range of habitats. Some of these include between granite outcrops, sandy heathland, stony ridges and open forest in a range of soil types.

CONSERVATION STATUS Widespread, common and conserved in national parks and nature reserves in Western Australia, South Australia and Victoria. Uncommon in Tasmania, and conserved in national parks. Rare and conserved in New South Wales.

DISCUSSION *C. deformis* is an attractive orchid that in some areas forms clumps of several individuals. While not requiring fire to stimulate blooming, areas burnt the previous summer will have abundant flowering plants. This species mainly reproduces from seed, however, robust specimens will slowly multiply vegetatively. This species has recently been reclassified as *Pheladenia deformis* (R. Br.) D.L. Jones & M.A. Clem.

Caladenia deformis
Willetton, Western Australia

15 August 1992

A plant
B flower from front
C flower from side
D flower from above
E labellum from side,
F labellum from front — flattened
G column from front
H column from side

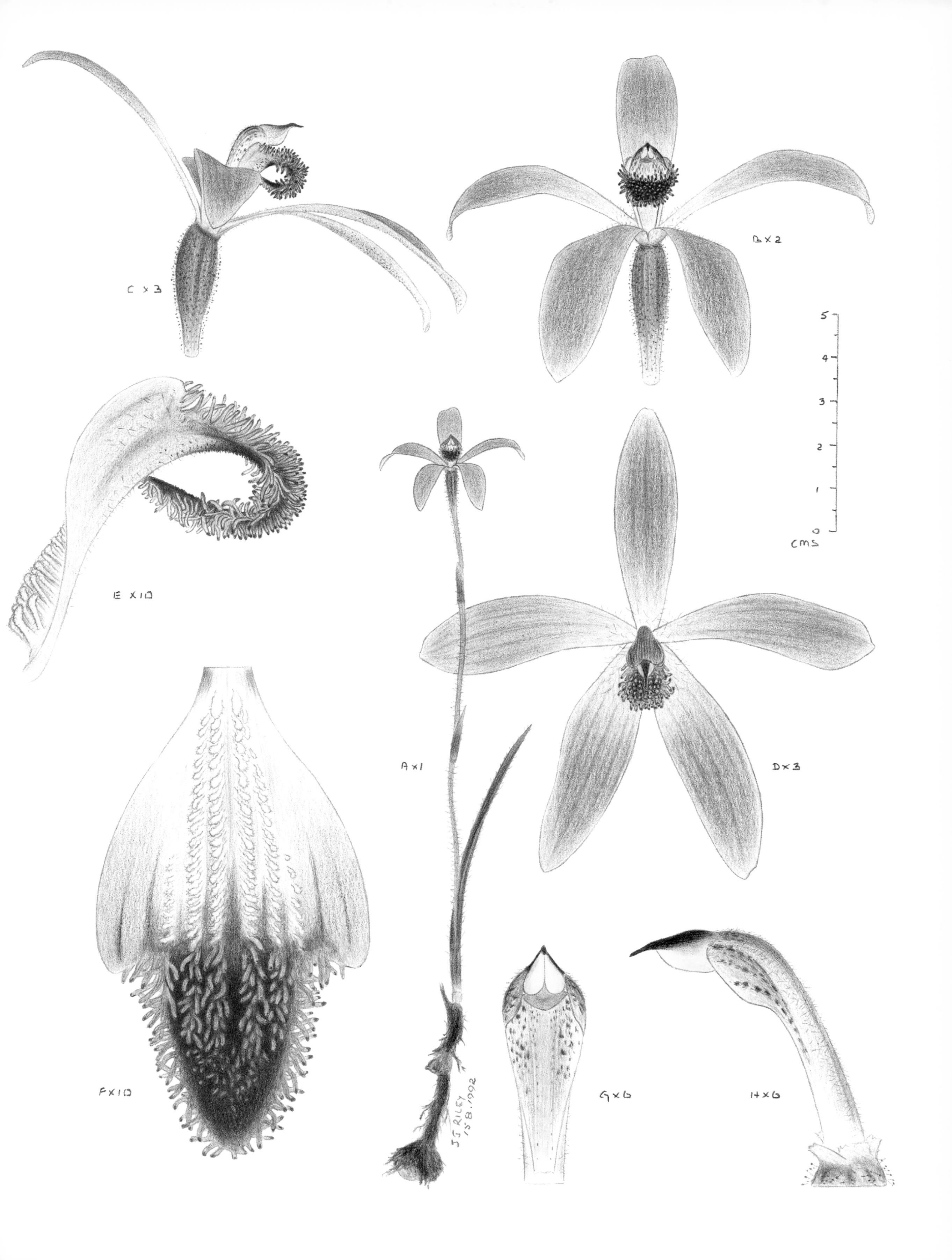
C x 3
B x 2
5
4
3
2
1
0
CMS
E x 10
A x 1
D x 3
F x 10
G x 6
H x 6
J.J. RILEY
15.8.1992

Caladenia cucullata Fitzg.

1876 | *Australian Orchids* 1(2)

TYPE LOCALITY Boorowa, New South Wales

ETYMOLOGY *cucullata* — hooded

FLOWERING TIME September to November

DISTRIBUTION South from the Central Tablelands and slopes of New South Wales to the central Goldfields of Victoria and the extreme south-eastern corner of South Australia.

ALTITUDINAL RANGE 100 m to 900 m

DISTINGUISHING FEATURES *C. cucullata* is a multiflowered species. It has small lemon-scented blooms with a distinctly hooded dorsal sepal. There are four rows of clubbed, purple calli on the midlobe of the labellum.

HABITAT *C. cucullata* prefers drier open forest and scrubland with a grassy understorey. It is most often found on well drained slopes and ridges in these areas.

CONSERVATION STATUS Common. Represented in national parks and nature reserves. Secure.

DISCUSSION This is a very attractive, small multiflowered *Caladenia*. It sometimes grows hidden among shrubs or with grasses in open ground. *C. cucullata* does not require a fire to stimulate blooming. This is a solitary species that occurs as scattered individuals and reproduces from seed. Was recently reclassified as *Stegostyla cucullata* (Fitzg.) D.L. Jones & M.A. Clem.

Caladenia cucullata
Kandos, New South Wales

15 October 1991

A plant
B flower from front
C flower from side
D flower from back
E flower from above
F labellum from side
G longitudinal section of labellum
H labellum flattened
J column from side
K column from front

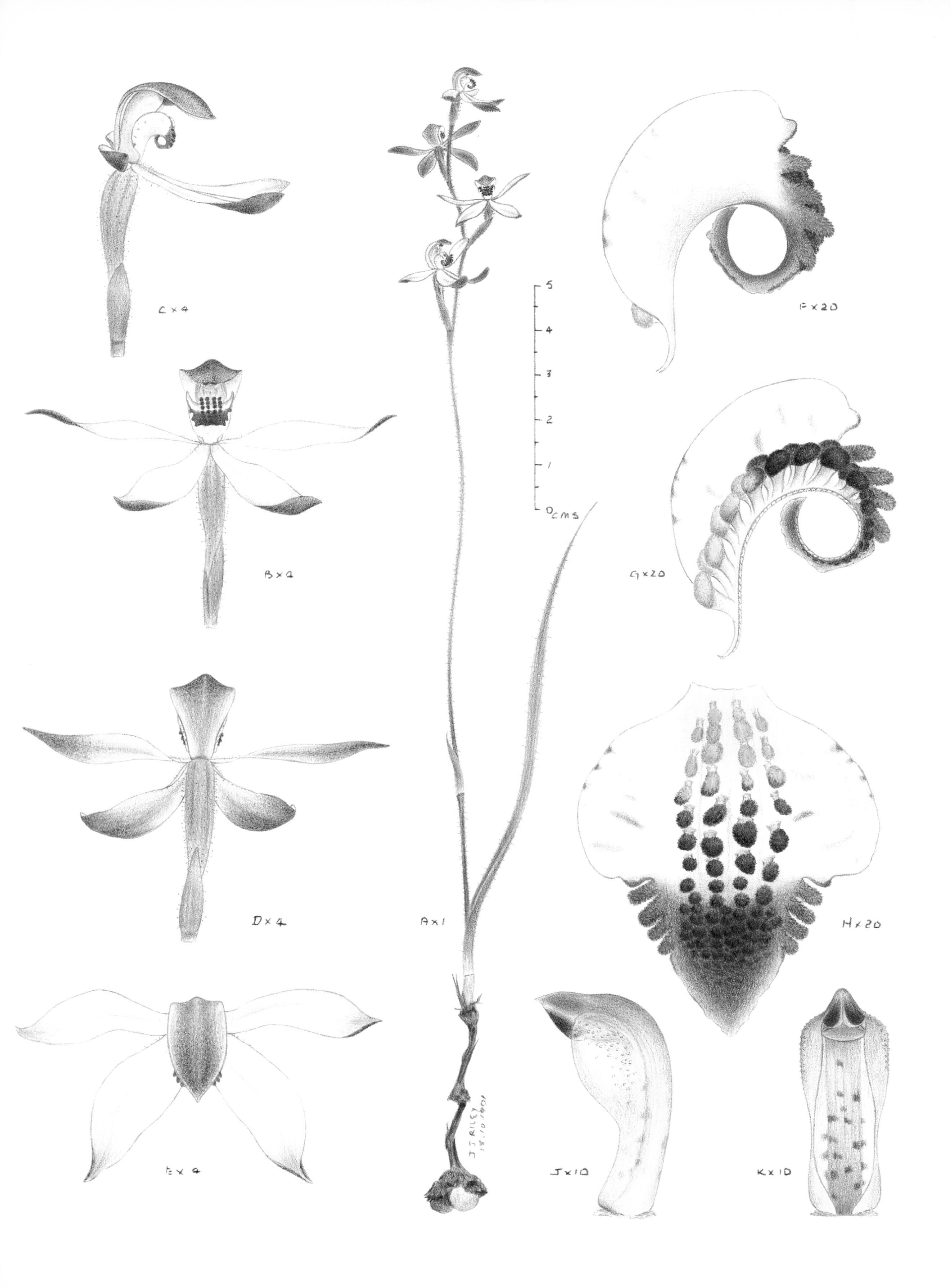

C×4
B×4
D×4
E×4
A×1
5
4
3
2
1
0 CMS
F×20
G×20
H×20
J×10
K×10
J J RILEY

Caladenia carnea R. Br.

1810 | *Prodromus Florae Novae Hollandiae*: 324

TYPE LOCALITY Port Jackson, New South Wales

ETYMOLOGY *carneus* — flesh-coloured

FLOWERING TIME September to November

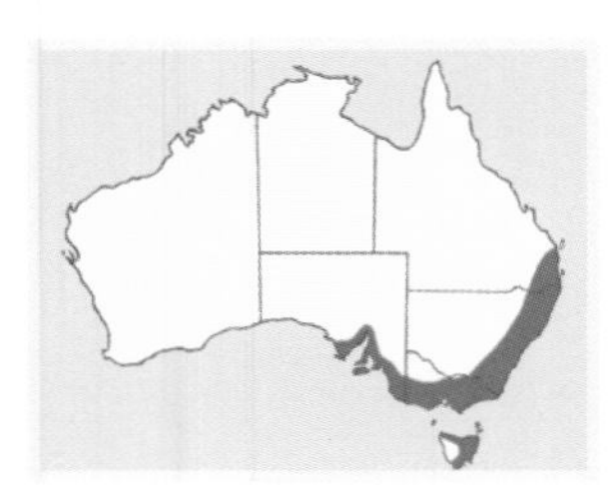

DISTRIBUTION South Australia, Victoria, Tasmania, New South Wales, Queensland.

ALTITUDINAL RANGE 10 m to 1000 m

DISTINGUISHING FEATURES *C. carnea* has a labellum with erect, rounded sidelobes. The labellum when flattened is three lobed with a triangular-shaped midlobe, pale yellow and fringed with marginal calli. Column and labellum heavily barred in reddish-maroon.

HABITAT Heath, woodland and open forest growing on a variety of soil types from sand, sandy clay loams to heavy clay loams.

CONSERVATION STATUS Widespread. Conserved in national parks and nature reserves. Secure.

DISCUSSION *Caladenia carnea* is part of a complex of similar species. Some of these have recently been named. Research being carried out may recognise and describe other taxa in this complex. The distribution given, therefore, includes *C. carnea* in the strict sense, plus unrecognised or, as yet, unnamed taxa. The flowers vary in colour from pale pinkish-white to deep pink. The illustrated plant is in the more common colour form. Flowers are mostly produced singly, sometimes with two blooms and rarely up to four on robust specimens, and are usually observed growing as scattered individuals or in small colonies. When growing along tracks or in clearings, where they receive more light, groups of numerous plants, with their pink flowers are quite a spectacle. Reproduces from seed. Was recently reclassified as *Petalochilus carneus* (R. Br.) D.L. Jones & M.A. Clem.

Caladenia carnea
Picnic Point, New South Wales

10 September 2001

A plant
B flower from front
C flower from rear
D labellum from side
E longitudinal section of labellum
F labellum flattened
G column from side
H column from front
J dorsal sepal
K lateral sepal
L petal

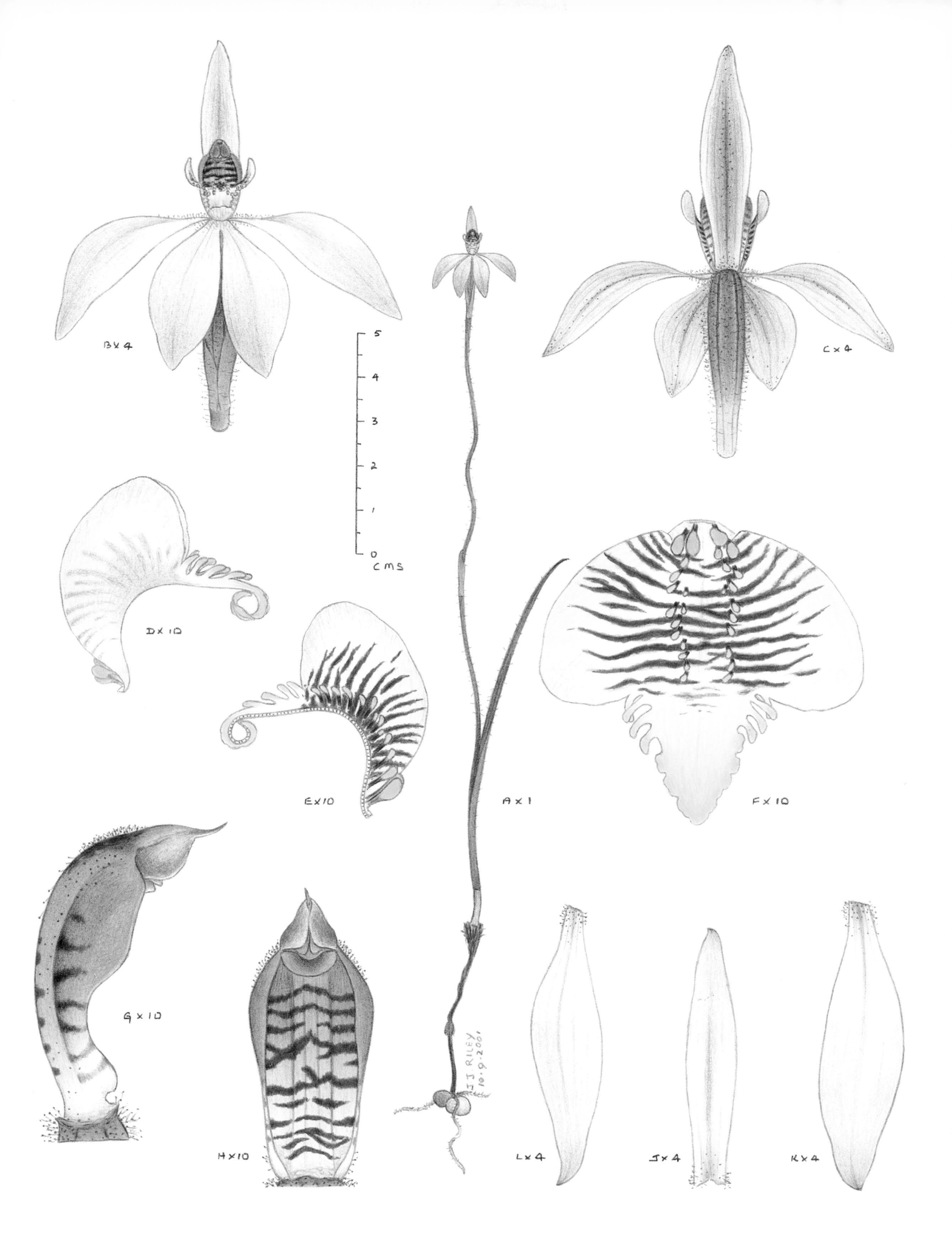
B×4
C×4
5
4
3
2
1
0
CMS
D×10
E×10
A×1
F×10
G×10
H×10
L×4
J×4
K×4
J J RILEY
10·9·2001

Caladenia coactilis D.L. Jones

1991 | *Australian Orchid Research* 2:20

TYPE LOCALITY Telowie Gorge, South Australia

ETYMOLOGY *coactilis* — thick

FLOWERING TIME August to September

DISTRIBUTION Endemic from the far northern portion of the Mount Lofty Ranges to the southern Flinders Ranges, South Australia.

ALTITUDINAL RANGE 300 m to 900 m

DISTINGUISHING FEATURES *C. coactilis* is a robust plant which is thick and fleshy in all its parts. The leaf and bracts are large, well-developed and very hairy. The flowers are usually produced in pairs and are pink on the inside and brownish on the outside. The column is wide at the base and narrow at the apex.

HABITAT Open, dry *Eucalyptus* and *Callitris* forest, growing in clay loams on the stony ridges of the Mount Lofty and Flinders Ranges, South Australia.

CONSERVATION STATUS Locally common. Conserved in national parks and conservation parks. Secure.

DISCUSSION *C. coactilis* is a striking orchid with a fairly narrow distribution, preferring the partly sheltered areas of dry ridges. While the plants are generally scattered, it can be locally common. Reproduces from seed. Was recently reclassified as *Petalochilus coactilis* (D.L. Jones) D.L. Jones & M.A. Clem.

Caladenia coactilis
Flinders Ranges, South Australia

1 September 1997

A plant
B flower from front
C flower from side
D flower from rear
E labellum flattened
F longitudinal section of labellum
G column from side
H column from front
J dorsal sepal
K lateral sepal
L petal

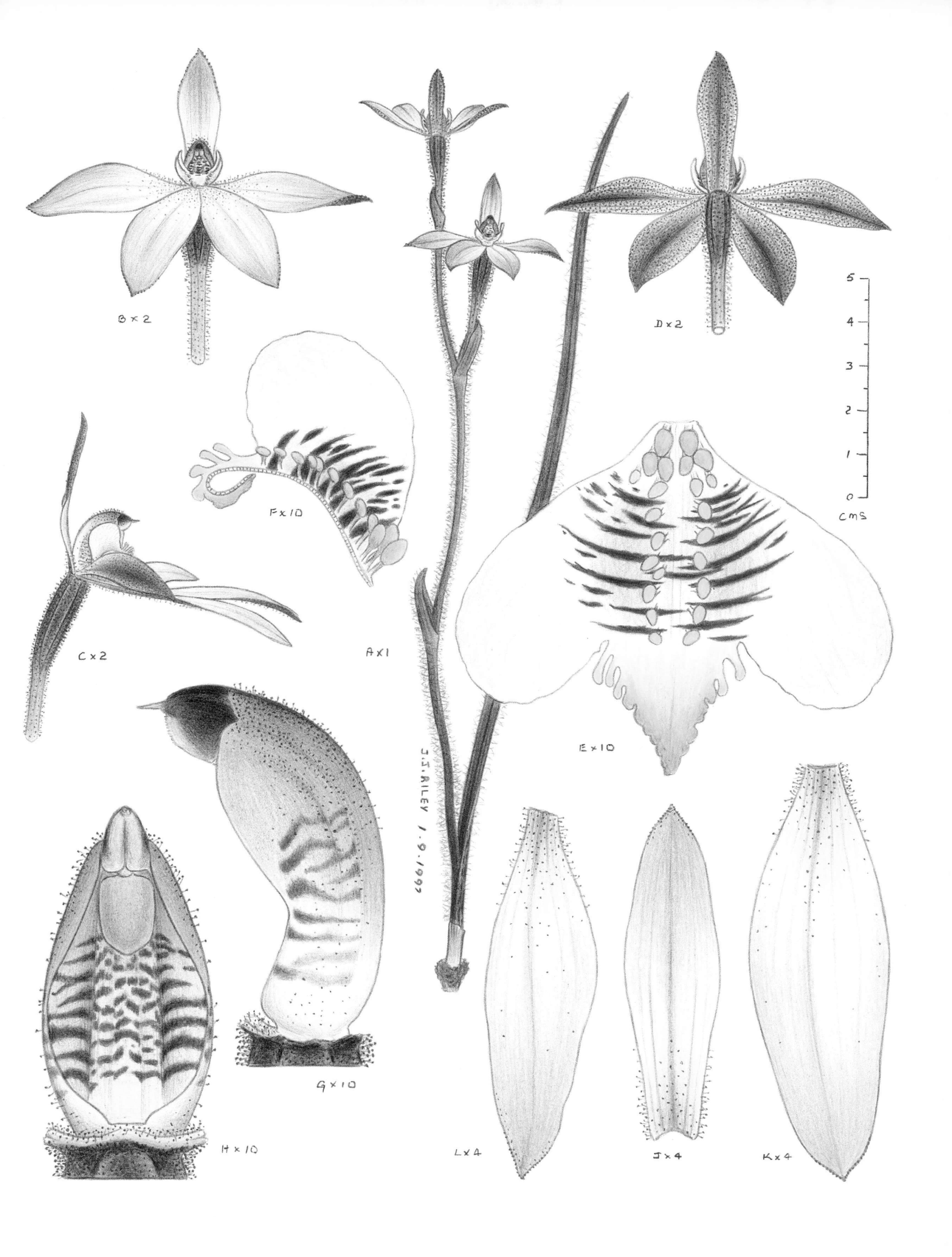
B × 2
D × 2
5
4
3
2
1
0
CMS
F × 10
C × 2
A × 1
E × 10
J.I. RILEY 1.9.1997
G × 10
H × 10
L × 4
J × 4
K × 4

Caladenia picta (Nicholls) M.A. Clem. & D.L. Jones

1989 | *Australian Orchid Research* 1:29

TYPE LOCALITY Waterfall Creek, New South Wales

RECENT SYNONYMS *Caladenia alba* R. Br. var. *picta* Nicholls

ETYMOLOGY *picta* — coloured

FLOWERING TIME April to June

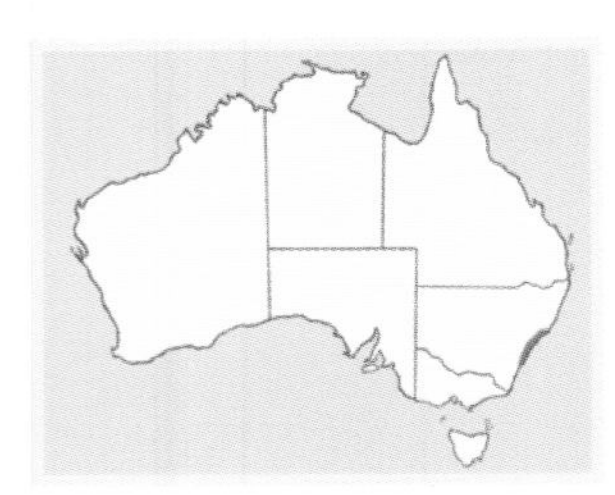

DISTRIBUTION South Coast to Newcastle, on the Central Coast of New South Wales. Mostly coastal, extending in parts to the nearby ranges.

ALTITUDINAL RANGE Up to 1000 m

DISTINGUISHING FEATURES *C. picta* is an autumn to early winter-flowering white *Caladenia*, with a column banded white, red and green.

HABITAT Grows in open forest with a grass and/or shrubby understorey in clay loams or sandy loams. Also grows in heathland.

CONSERVATION STATUS Common. Conserved in national parks and reserves. Secure.

DISCUSSION *C. picta* is often locally common. Mostly the blooms are white, but occasionally individuals are seen with the floral segments pinkish-red. With its white flowers, this species stands out in the bush and large groups of individuals make an impressive sight. Reproduces from seed. Was recently reclassified as *Petalochilus pictus* (Nicholls) D.L. Jones & M.A. Clem.

Caladenia picta
Nattai, New South Wales

10 May 1995

A plant
B flower from front
C flower from side
D flower from rear
E labellum flattened
F longitudinal section of labellum
G column from side
H column from front
J dorsal sepal
K lateral sepal
L petal

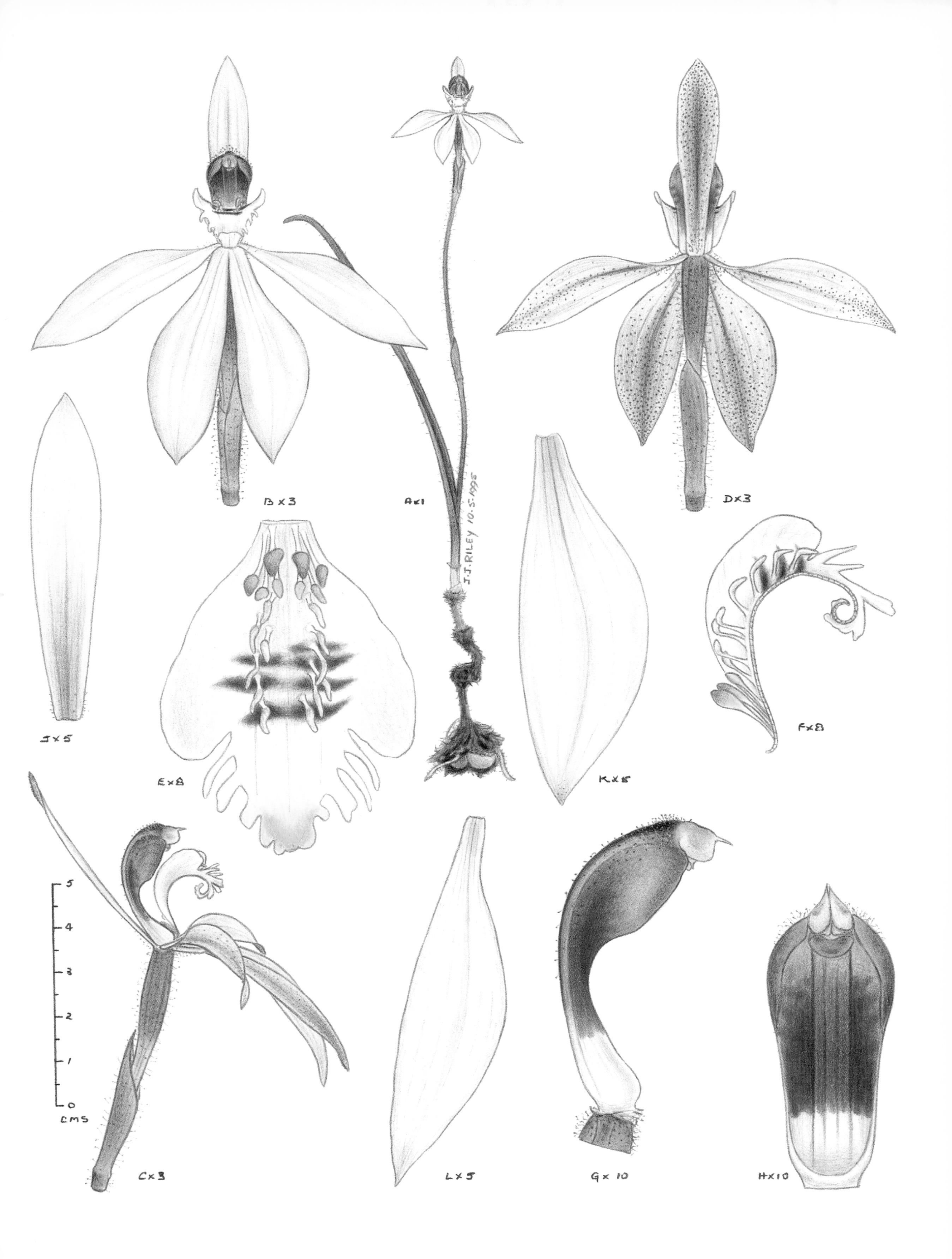
B×3
A×1
D×3
J.J. RILEY 10.5.1995
J×5
E×8
K×5
F×8
5
4
3
2
1
0
CMS
C×3
L×5
G×10
H×10

Caladenia porphyrea D.L. Jones

1999 | *Orchadian* 13:16

TYPE LOCALITY Norah Head, New South Wales

ETYMOLOGY *porphyreus* — purple

FLOWERING TIME August to October

DISTRIBUTION From Wyong on the Central Coast to the North Coast, New South Wales.

ALTITUDINAL RANGE Up to 50 m

DISTINGUISHING FEATURES *C. porphyrea* has highly coloured, pink to purple-red flowers. The labellum and column are heavily striated with reddish-maroon. The midlobe of the labellum has flat marginal teeth and an orange apex.

HABITAT Grows close to the coast in open heath and heathy woodland on grey/white sands and peaty sandy loams.

CONSERVATION STATUS Common and conserved at the type-site. Secure.

DISCUSSION *C. porphyrea* is a poorly known, yet very striking, orchid, the vibrant colour standing out among the surrounding vegetation. The intensity of colour varies through a colony, some being lighter and others much brighter. The illustrated specimen represents the most common hue. Reproduces from seed. Was recently reclassified as *Petalochilus porphyreus* (D.L. Jones) D.L. Jones & M.A. Clem.

Caladenia porphyrea
Norah Head, New South Wales

25 August 1999

A plant
B flower from front
C flower from rear
D labellum from side
E labellum flattened
F longitudinal section of labellum
G column from side
H column from front
J dorsal sepal
K lateral sepal
L petal

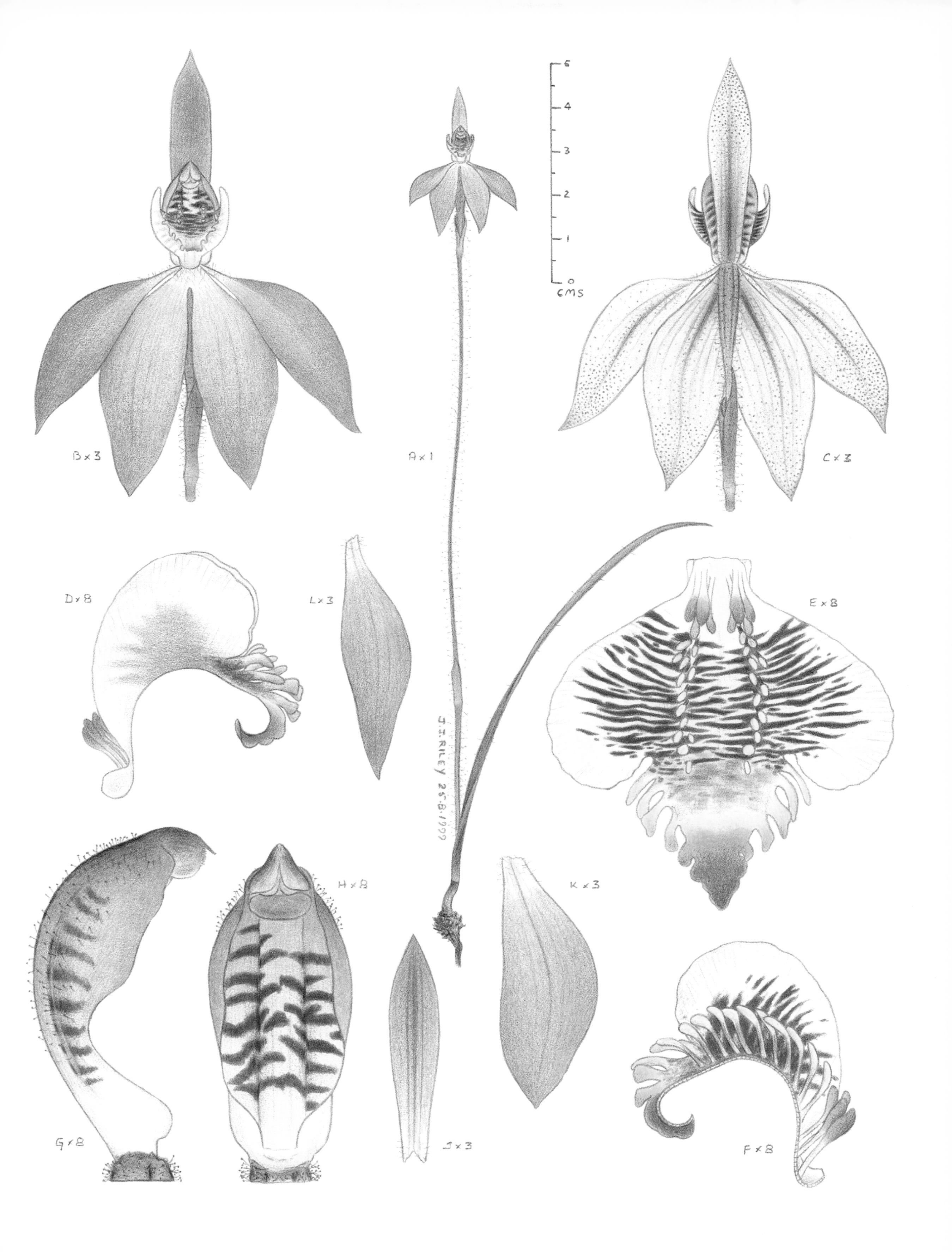

B × 3
A × 1
C × 3
5
4
3
2
1
0
CMS
D × 8
L × 3
E × 8
J.J. RILEY 25-8-1999
G × 8
H × 8
K × 3
J × 3
F × 8

Caladenia dilatata R. Br.

1810 | *Prodromus Florae Novae Hollandiae*: 325

TYPE LOCALITY Port Dalrymple, Tasmania

ETYMOLOGY *dilatata* — fingered, hand-shaped

FLOWERING TIME November to January

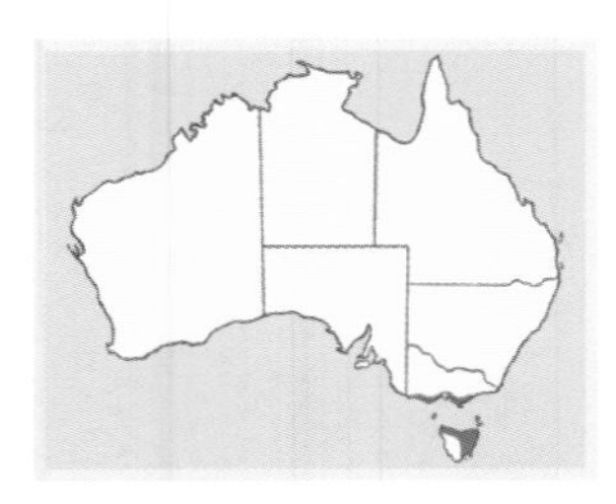

DISTRIBUTION Northern and eastern Tasmania, the Bass Strait islands and along the southern coastline of Victoria.

ALTITUDINAL RANGE Up to 150 m

DISTINGUISHING FEATURES The *Caladenia dilatata* complex has petaline osmophores. The *Caladenia tentaculata* complex lacks these petaline osmophores. True *C. dilatata* has pale green-yellow, narrow flowers with reddish-maroon markings. The labellum, when flattened, has a long midlobe and narrow sidelobes. It is also the most southerly occurring species in the complex.

HABITAT Open woodland and forest with a grass to heathy understorey. Open areas in heathland on sandy and clay loams.

CONSERVATION STATUS Common. Conserved in national parks and nature reserves. Secure.

DISCUSSION *C. dilatata* is a very attractive species in the green comb group of spider caladenias, so-named because of the prominent sidelobes to the labellum. While most flowers have prominent osmophores on the petals, some specimens appear to lack them. A closer examination reveals these are reduced to a collection of glands on the petal tips. Pollination is by pseudocopulation involving males of a species of thynnid wasp. Reproduces from seed. *C. dilatata* is free-flowering and under favourable conditions appears in large numbers. Many taxa from mainland Australia, previously thought to be this species, have been given individual specific status. Was recently reclassified as *Arachnorchis dilatata* (R. Br.) D.L. Jones & M.A. Clem.

Caladenia dilatata
Bullock Hill, Tasmania

16 December 1994

A plant
B flower from front
C labellum flattened
D longitudinal section of labellum
E column from front
F column from side
G dorsal sepal
H petal
J lateral sepal
K glands at base of column

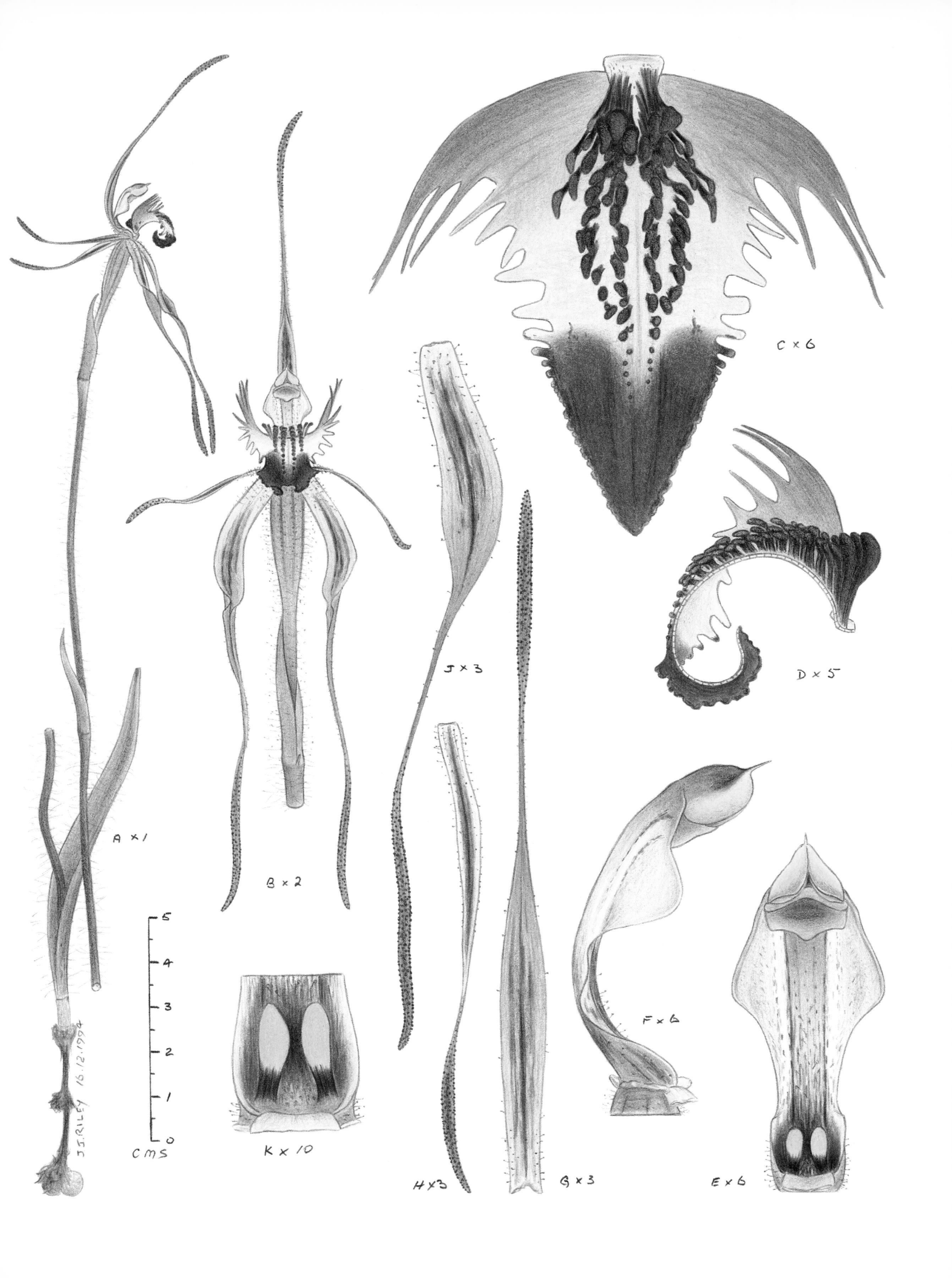

C × 6
D × 5
J × 3
A × 1
B × 2
F × 6
5
4
3
2
1
0
CMS
J.J. RILEY 16.12.1994
K × 10
H × 3
G × 3
E × 6

Caladenia amnicola D.L. Jones

1997 | *Orchadian* 12:166

TYPE LOCALITY Tea Tree Creek, New South Wales

ETYMOLOGY *amnicola* — living by a stream

FLOWERING TIME November to January

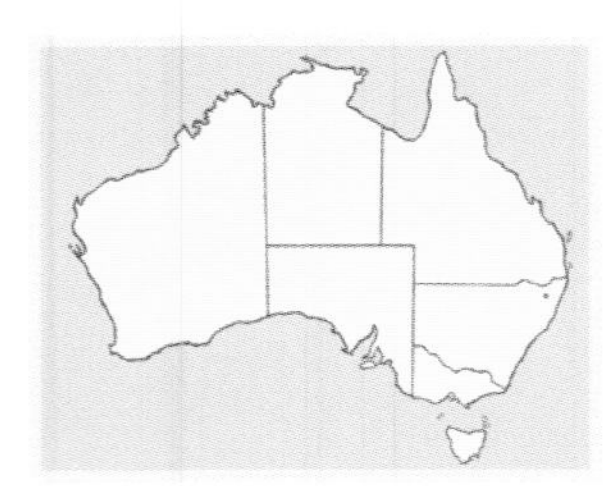

DISTRIBUTION Endemic to near Armidale on the New England Tableland, New South Wales.

ALTITUDINAL RANGE 800 m to 900 m

DISTINGUISHING FEATURES *C. amnicola* is the most northern species of the *C. dilatata* complex. It has greenish and maroon flowers, with stiffly held petals and sepals, and petaline osmophores. The labellum, when flattened, has a relatively short midlobe and wide, prominent sidelobes.

HABITAT Grows in gravelly loams among rocks and heathy shrubs and under *Leptospermum* and *Eucalyptus* spp. close to streams.

CONSERVATION STATUS Very rare. Known only from the type-site, and not conserved. Extremely vulnerable.

DISCUSSION *C. amnicola*, due to its rarity, is poorly understood. It has attractive blooms, which grow up through the low shrubs. Pollination is by pseudocopulation involving males of a species of thynnid wasp. Reproduces from seed. Was recently reclassified as *Arachnorchis amnicola* (D.L. Jones) D.L. Jones & M.A. Clem.

Caladenia amnicola
near Armidale, New South Wales

3 December 1993

A plant
B flower from front
C flower from side
D labellum flattened
E longitudinal section of labellum
F glands at base of column
G column from side
H column from front
J lateral sepal
K petal
L dorsal sepal

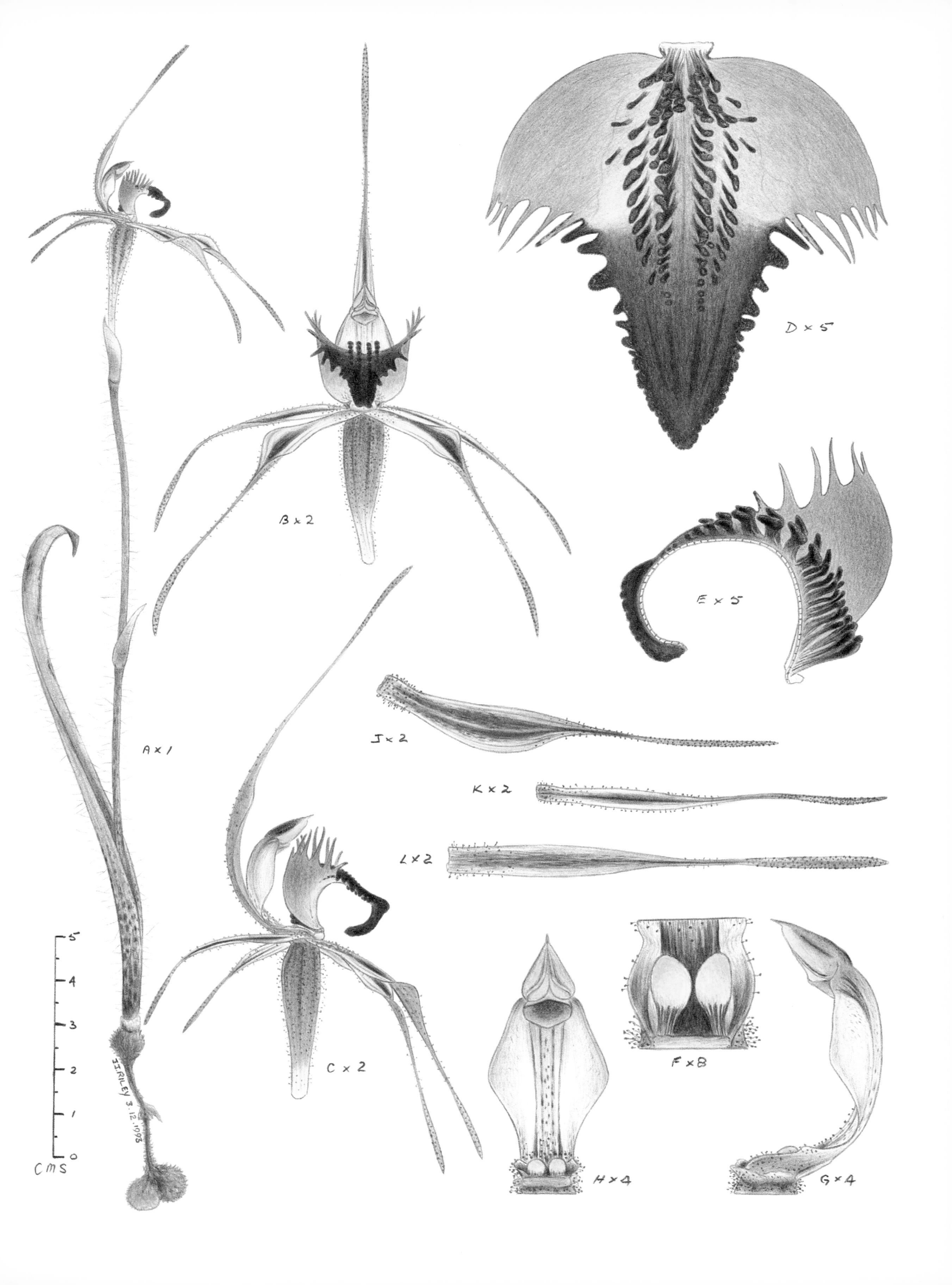

A × 1
B × 2
C × 2
D × 5
E × 5
F × 8
G × 4
H × 4
J × 2
K × 2
L × 2
5
4
3
2
1
0
CMS
J.J. RILEY 3.12.1993

Caladenia necrophylla D.L. Jones

1997 | *Orchadian* 12:162

TYPE LOCALITY Robe, South Australia

ETYMOLOGY *necrophylla* — dead leaf

FLOWERING TIME November to January

DISTRIBUTION South-eastern South Australia

ALTITUDINAL RANGE Up to 50 m

DISTINGUISHING FEATURES *C. necrophylla* has olive green flowers with reddish-maroon stripes and the petals have osmophores. The labellum, when flattened, has a long, broad pointed midlobe with narrow, poorly developed lateral lobes. The column leans backwards in *C. necrophylla* while it is erect in *C. dilatata*. The leaf is withered either before or at flowering.

HABITAT Grows in heathy woodland and open forest on sandy loams.

CONSERVATION STATUS Represented in conservation parks. Secure.

DISCUSSION *C. necrophylla* is the most westerly occurring member of the *C. dilatata* complex. All other species of the green comb spider orchids native to the south-east of South Australia have the leaf present at blooming and lack petaline osmophores. Pollination is by pseudocopulation involving males of a species of thynnid wasp. Reproduces from seed. Was recently reclassified as *Arachnorchis necrophylla* (D.L. Jones) D.L. Jones & M.A. Clem.

Caladenia necrophylla
Mary Seymour Conservation Park, South Australia

10 December 1996

A plant
B flower from front
C labellum flattened
D longitudinal section of labellum
E column from front
F column from side
G glands at base of column
H dorsal sepal
J petal
K lateral sepal

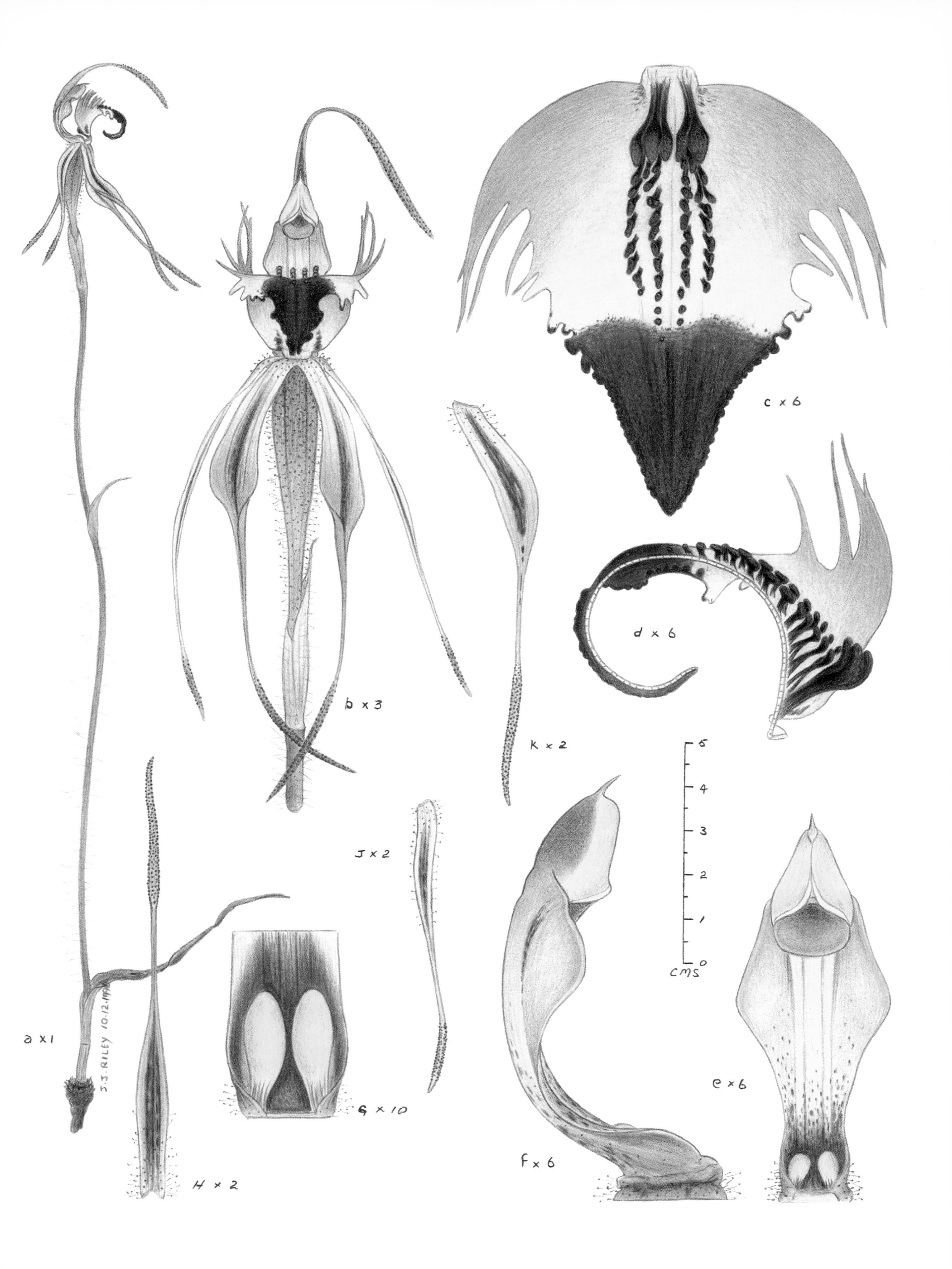

a x 1
b x 3
c x 6
d x 6
e x 6
f x 6
G x 10
H x 2
J x 2
K x 2
0
1
2
3
4
5
CMS

Caladenia rileyi D.L. Jones

1997 | *Orchadian* 12:164

TYPE LOCALITY Gillenbah State Forest, New South Wales

ETYMOLOGY After illustrator John James Riley

FLOWERING TIME September to October

DISTRIBUTION The Riverina district of New South Wales between Narranderra and Jerilderie.

ALTITUDINAL RANGE 150 m to 250 m

DISTINGUISHING FEATURES *C. rileyi* has yellow-green flowers with reddish-maroon stripes. All sepals and petals have very prominent osmophores. The labellum, when flattened, is wide with a long, rounded midlobe and prominent sidelobes. It is the earliest of the *C. dilatata* complex to bloom.

HABITAT Occurs on flat, low red sand ridges that support a woodland mostly of the native white pine (*Callitris glaucophylla*), with a sparse understorey of grasses and small herbaceous plants and coral lichen.

CONSERVATION STATUS Restricted occurrence. Not represented in a national park. Vulnerable.

DISCUSSION *C. rileyi* is an impressive species that stands out when in bloom. The colour of the blooms can be variable, with some individuals having faint reddish-maroon markings or lacking them entirely. These flowers are an attractive pale yellowish-green. Occasional hybrids occur where related species are growing in close proximity to each other. Pollination is by pseudocopulation involving males of a species of thynnid wasp. Reproduces from seed. Was recently reclassified as *Arachnorchis rileyi* (D.L. Jones) D.L. Jones & M.A. Clem.

Caladenia rileyi
Narrandera, New South Wales

6 October 1992

A plant
B flower from front
C flower from rear
D labellum flattened
E longitudinal section of labellum
F glands at base of column
G column from side
H column from front
J lateral sepal
K petal
L dorsal sepal

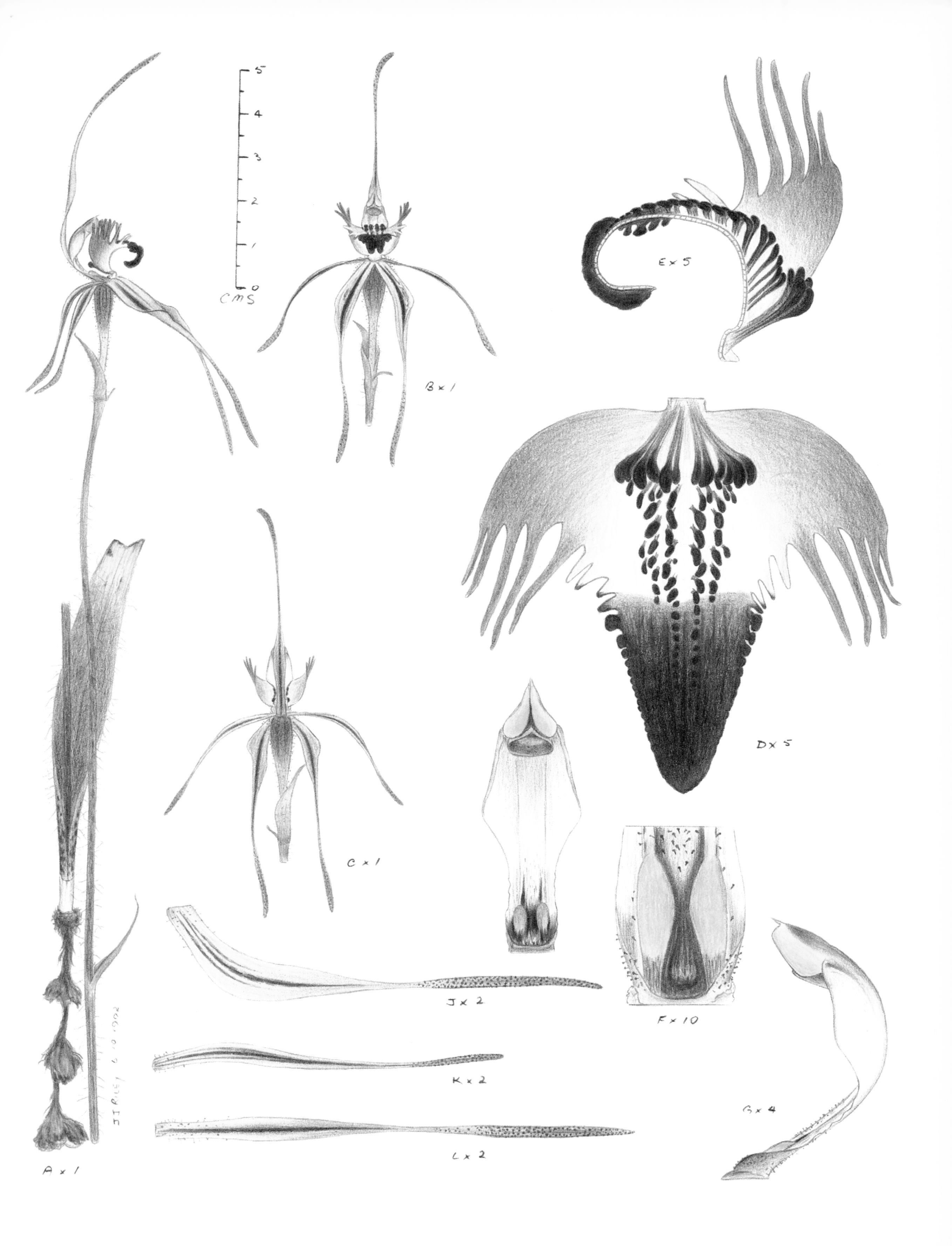

5
4
3
2
1
0
CMS
A x 1
B x 1
C x 1
D x 5
E x 5
F x 10
G x 4
J x 2
K x 2
L x 2
JJ RILEY 6.10.1992

Caladenia tentaculata Schldl.

1847 | *Linnaea* 20:571

TYPE LOCALITY Mount Lofty Ranges, South Australia

ETYMOLOGY *tentaculata* — tendril-like

FLOWERING TIME September to November

DISTRIBUTION Common in the Adelaide Hills, Barossa Valley and lower Flinders Ranges to the south-eastern corner of South Australia, also in southern Victoria.

ALTITUDINAL RANGE 50 m to 1000 m

DISTINGUISHING FEATURES *C. tentaculata* has a large flower lacking osmophores on the petals. The floral segments are more or less lax and the lateral sepals are often upsweeping. The labellum, which is very mobile, is wide with very long marginal combs and has a white centre. When flattened, the labellum is wider than long. The combs on the fringe of the labellum exceed the top of the column in height.

HABITAT Grows in a wide range of habitats, from heathland to open forest and woodland on stony ridges. Soils can be sandy loams to clay loams.

CONSERVATION STATUS Widespread and common. Conserved in national parks and nature reserves. Secure.

DISCUSSION The *C. tentaculata* complex contains a number of species, some of which have been formally named. This is a large, showy terrestrial orchid often found in large colonies. Robust specimens can carry two blooms at the same time. Pollination is by pseudocopulation involving males of a species of thynnid wasp. Reproduces from seed. Was recently reclassified as *Arachnorchis tentaculata* (Schldl.) D.L. Jones & M.A. Clem.

Caladenia tentaculata
Ironbank, South Australia

16 October 1994

A plant
B flower from front
C labellum flattened
D longitudinal section of labellum
E column from side
F column from front
G glands at base of column
H dorsal sepal
J lateral sepal
K petal

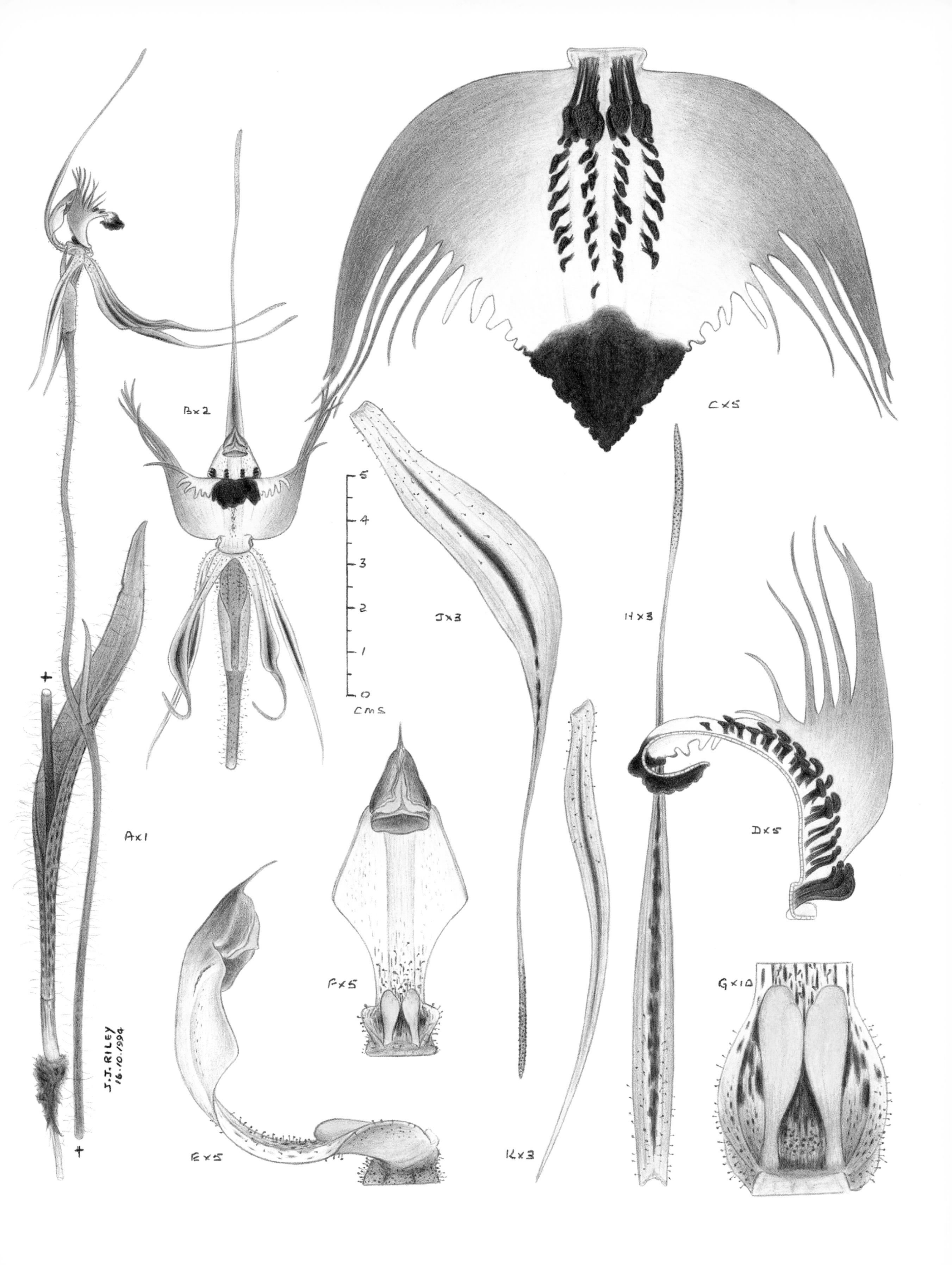

B×2
C×5
5
4
3
2
1
0
CMS
J×3
H×3
D×5
A×1
F×5
G×10
J.J. RILEY
16.10.1994
E×5
K×3

Caladenia tensa G.W. Carr

1991 | *Indigenous Flora & Fauna Association, Miscellaneous Paper* 1:15

TYPE LOCALITY Kiata, Victoria

ETYMOLOGY *tensa* — stiff

FLOWERING TIME September to November

DISTRIBUTION Restricted to far western Victoria in the Little Desert area, and South Australia.

ALTITUDINAL RANGE 50 m to 150 m

DISTINGUISHING FEATURES *C. tensa* lacks osmophores on the petals. The perianth segments are held rigidly. The labellum, when flattened, is longer than it is wide, with a noticeable white centre. The combs on the fringe of the labellum are shorter than those of *C. tentaculata*, only being level with the top of the column.

HABITAT Open woodland on sandy loamy soils growing under *Callitris preissii* and *Eucalyptus leucoxylon*.

CONSERVATION STATUS Uncommon in Victoria, common in South Australia. Represented in a national park and reserves. Considered endangered in Victoria.

DISCUSSION *C. tensa* is a poorly known representative of the *C. tentaculata* complex. Most of the *Callitris* and *Eucalyptus* woodland of western Victoria has been cleared and very little suitable habitat remains. *C. tensa* has moderately large flowers and to the untrained eye may be mistaken for *C. tentaculata*. Pollination is by pseudocopulation involving males of a species of thynnid wasp. Reproduces from seed. Was recently reclassified as *Arachnorchis tensa* (G.W. Carr) D.L. Jones & M.A. Clem.

Caladenia tensa
Kiata, Victoria

10 October 1995

A plant
B flower from front
C labellum flattened
D longitudinal section of labellum
E column from side
F column from front
G glands at base of column
H dorsal sepal
J lateral sepal
K petal

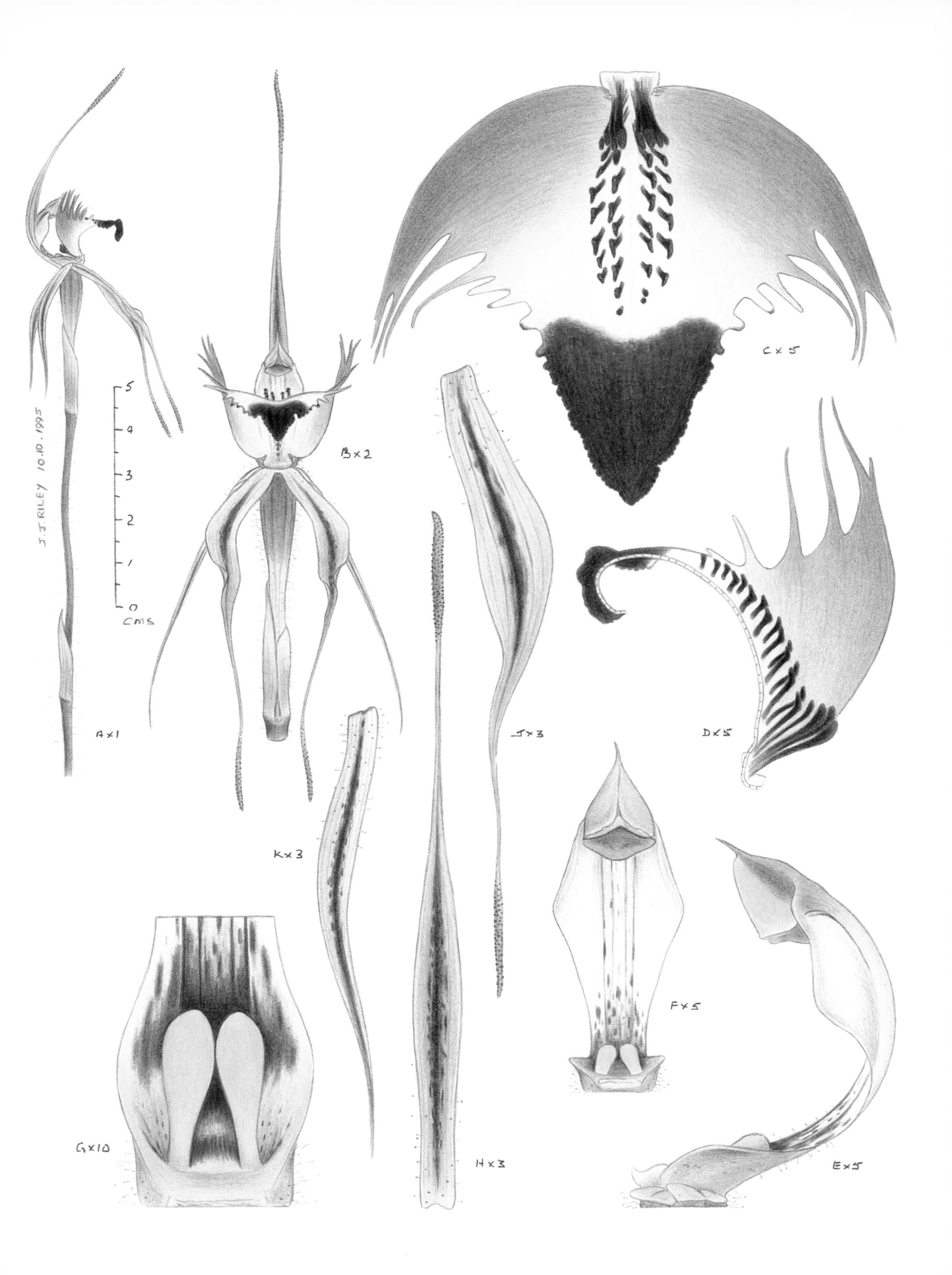
J.J. RILEY 10.10.1995
5
4
3
2
1
0
CMS
A×1
B×2
C×5
D×5
E×5
F×5
G×10
H×3
J×3
K×3

Caladenia corynephora A.S. George

1971 | *Nuytsia* 1(2):158

TYPE LOCALITY Pemberton, Western Australia

ETYMOLOGY *corynephora* — bearing clubs

FLOWERING TIME November to January

DISTRIBUTION Endemic to south-western Western Australia, scattered between Albany and Augusta.

ALTITUDINAL RANGE Up to 250 m

DISTINGUISHING FEATURES *C. corynephora* is a late-flowering, distinctive species with an unusual labellum. It has noticeable clubbing on the perianth segments. The labellum, when flattened, is long and narrow with heavy, prominent clubbed calli. The lateral lobes are fringed with numerous, reddish-maroon, filiform teeth. The midlobe has large, odd-shaped marginal teeth and a clubbed apex.

HABITAT This colourful species is found in forest and woodland, often on the slopes of streams and swamps and drainage areas, which are damp in winter.

CONSERVATION STATUS Sporadic. Conserved in reserves. Secure

DISCUSSION This is one of the last of the *Caladenia* species to bloom in Western Australia. In common with most spider caladenias, flowering is further stimulated by fires the previous year. Pollination is by pseudocopulation involving males of a species of thynnid wasp. Reproduces from seed. Was recently reclassified as *Arachnorchis corynephora* (A.S. George) D.L. Jones & M.A. Clem.

Caladenia corynephora
Mount Frankland, Western Australia

7 January 1997

A plant
B flower from front
C labellum flattened
D longitudinal section of labellum
E column from side
F column from front
G glands at base of column
H dorsal sepal
J lateral sepal
K petal

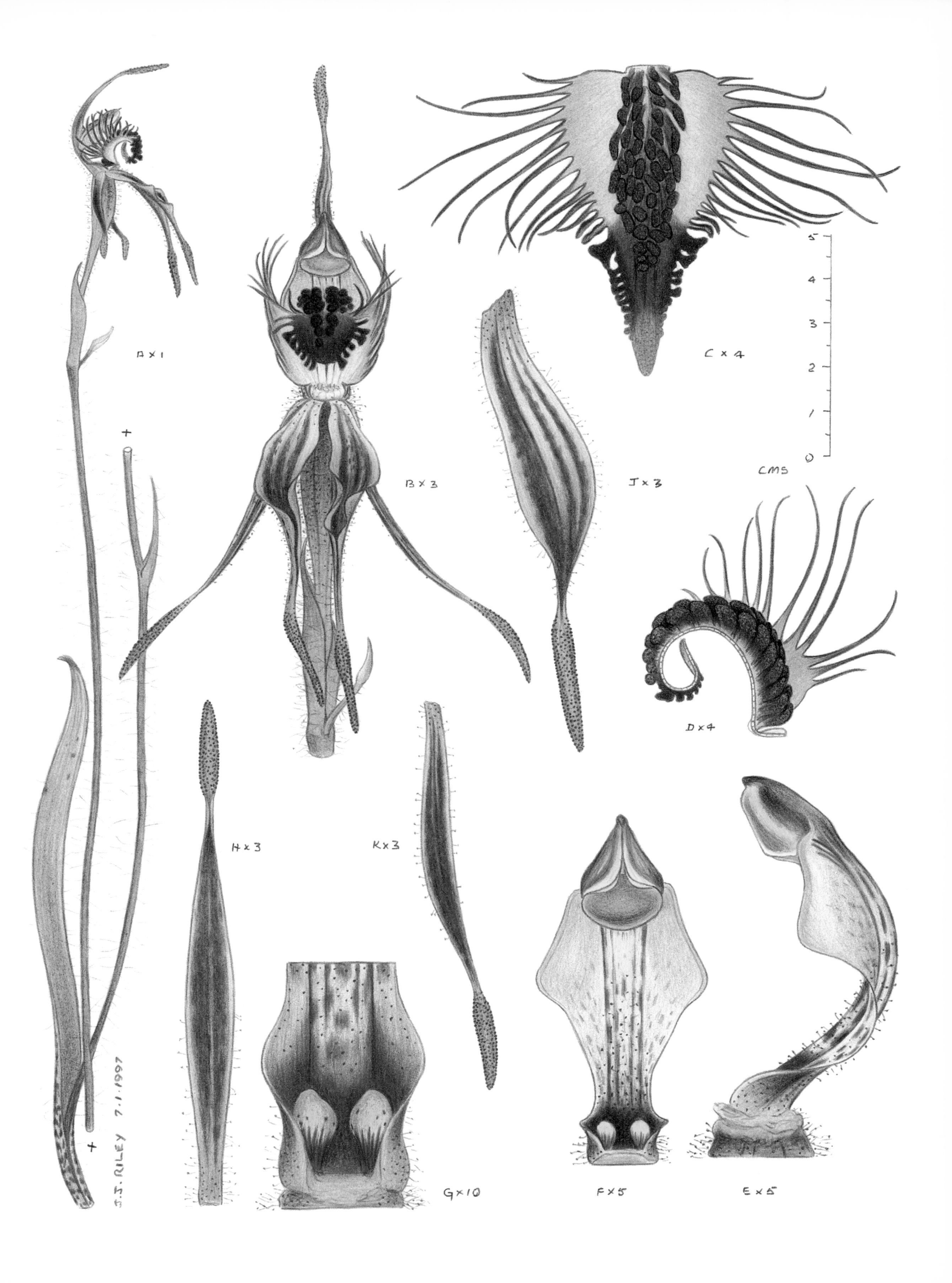
A x 1
B x 3
C x 4
D x 4
E x 5
F x 5
G x 10
H x 3
J x 3
K x 3
0
1
2
3
4
5
CMS
J.J. RILEY 7.1.1997

Caladenia gladiolata R.S. Rogers

1907 | *Transactions and Proceedings of the Royal Society of South Australia* 31:210

TYPE LOCALITY Hornsdale, South Australia

ETYMOLOGY *gladiolata* — like a small sword

FLOWERING TIME August to October

DISTRIBUTION Restricted to the lower Flinders Ranges and Mount Lofty Ranges, South Australia.

ALTITUDINAL RANGE 400 m to 900 m

DISTINGUISHING FEATURES *C. gladiolata* has very prominent, wide, flat osmophores that occupy half the length of the perianth segments. The labellum, when flattened, is obscurely three-lobed. All three lobes are fringed with short marginal teeth.

HABITAT This species is found in forest and woodland with an open, grassy and shrubby understorey, on flat or slightly sloping ridge tops. Soils are clay loams.

CONSERVATION STATUS Rare. Represented in a national park. Vulnerable.

DISCUSSION *C. gladiolata* with its prominent osmophores is a distinctive and impressive, small spider *Caladenia*. The flowers are noticeably scented. Once it was more widespread, but grazing and development have destroyed most of its former habitat. It was thought to be extinct in the Mount Lofty Ranges near Adelaide, but was fortunately rediscovered at one locality. Pollination is by pseudocopulation involving males of a species of thynnid wasp. Reproduces from seed. Sometimes hybridises with other spider caladenias. Was recently reclassified as *Arachnorchis gladiolata* (R.S. Rogers) D.L. Jones & M.A. Clem.

Caladenia gladiolata
Flinders Ranges, South Australia

14 September 1997

A plant
B flower from front
C flower from side
D labellum flattened
E longitudinal section of labellum
F column from side
G column from front
H glands at base of column
J dorsal sepal
K lateral sepal
L petal

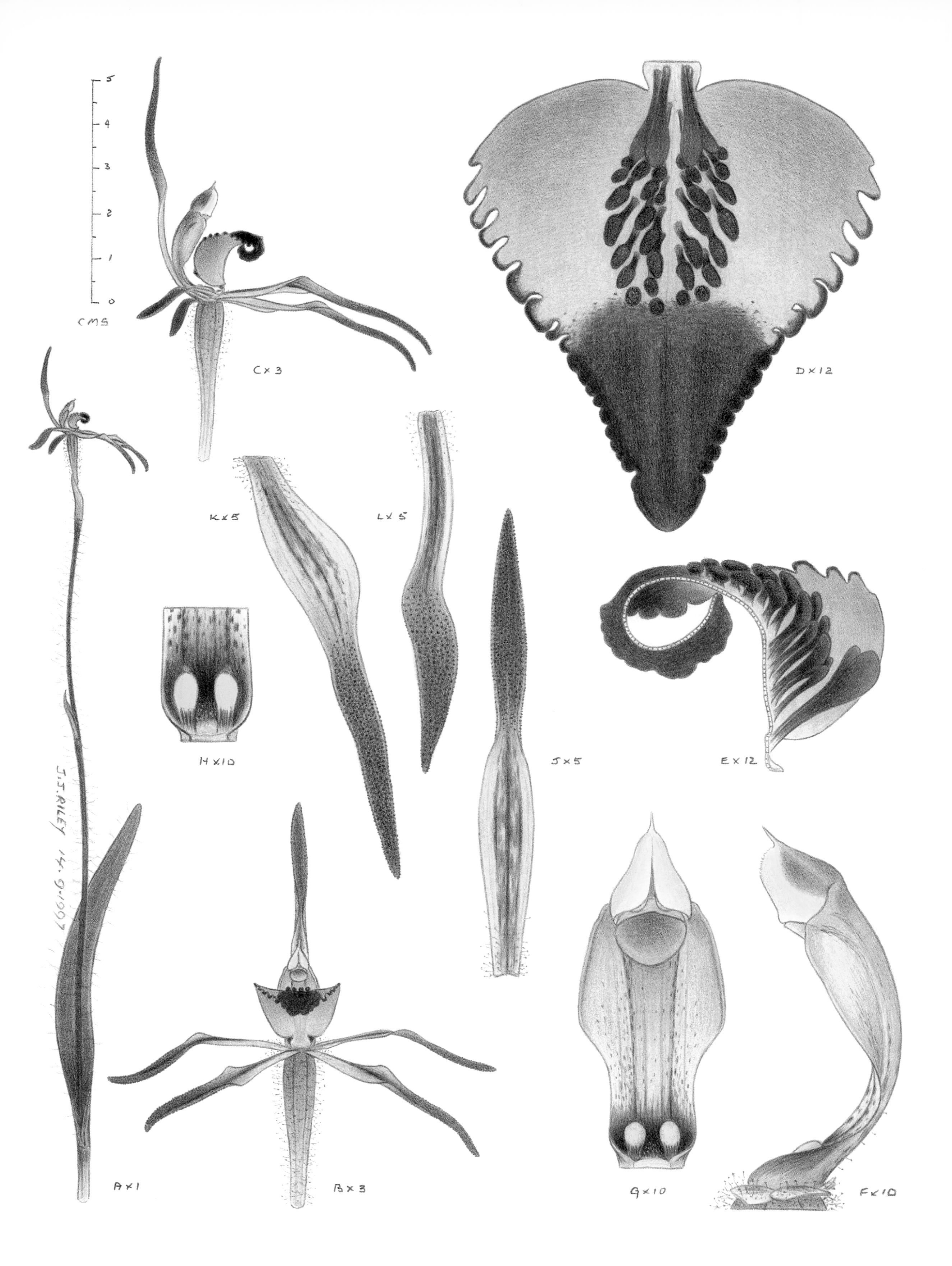

CMS
C×3
D×12
K×5
L×5
H×10
J×5
E×12
J.J. RILEY 14.9.1997
A×1
B×3
G×10
F×10

Caladenia lobata Fitzg.

1882 | *Gardener's Chronicle* 17:461

TYPE LOCALITY Upper Hay River, Western Australia

ETYMOLOGY *lobata* — lobed

FLOWERING TIME September to November

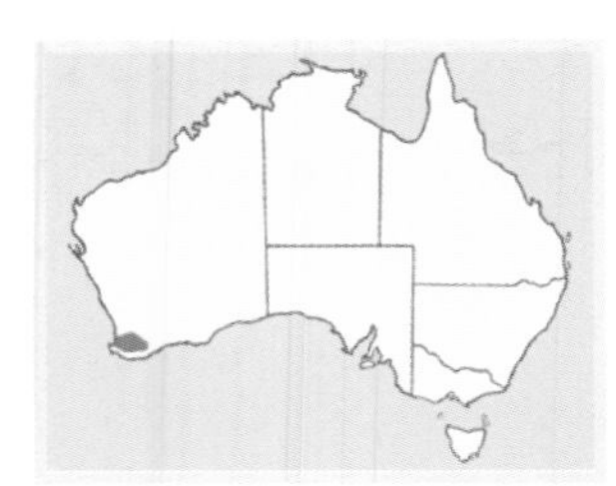

DISTRIBUTION Endemic to south-western Western Australia, being distributed from Bunbury to Mount Barker.

ALTITUDINAL RANGE 200 m to 600 m

DISTINGUISHING FEATURES *C. lobata* is a distinctive species that has a very wide labellum which is hinged at the base. The column, when viewed from the front, is cross-shaped. There is some variation in colour of the floral segments, with variations from a dark to yellowish-green through to blood red.

HABITAT Open forest with a shrubby understorey, growing on slopes above creeks.

CONSERVATION STATUS Uncommon, but conserved in reserves. Secure.

DISCUSSION *C. lobata* is a striking, colourful and large-flowered spider *Caladenia*. It is a tall-growing species that sometimes produces two blooms. Although well known, it is uncommon, occurring in small isolated pockets. Sometimes numerous individuals are encountered growing among shrubs. Pollination is by pseudocopulation involving males of a species of thynnid wasp. Reproduces from seed. It occasionally hybridises with other *Caladenia* species. Was recently reclassified as *Arachnorchis lobata* (Fitzg.) D.L. Jones & M.A. Clem.

Caladenia lobata
Mount Barker, Western Australia

8 November 1995

A plant
B flower from front
C labellum flattened
D longitudinal section of labellum
E column from side
F column from front
G glands at base of column
H dorsal sepal
J lateral sepal
K petal

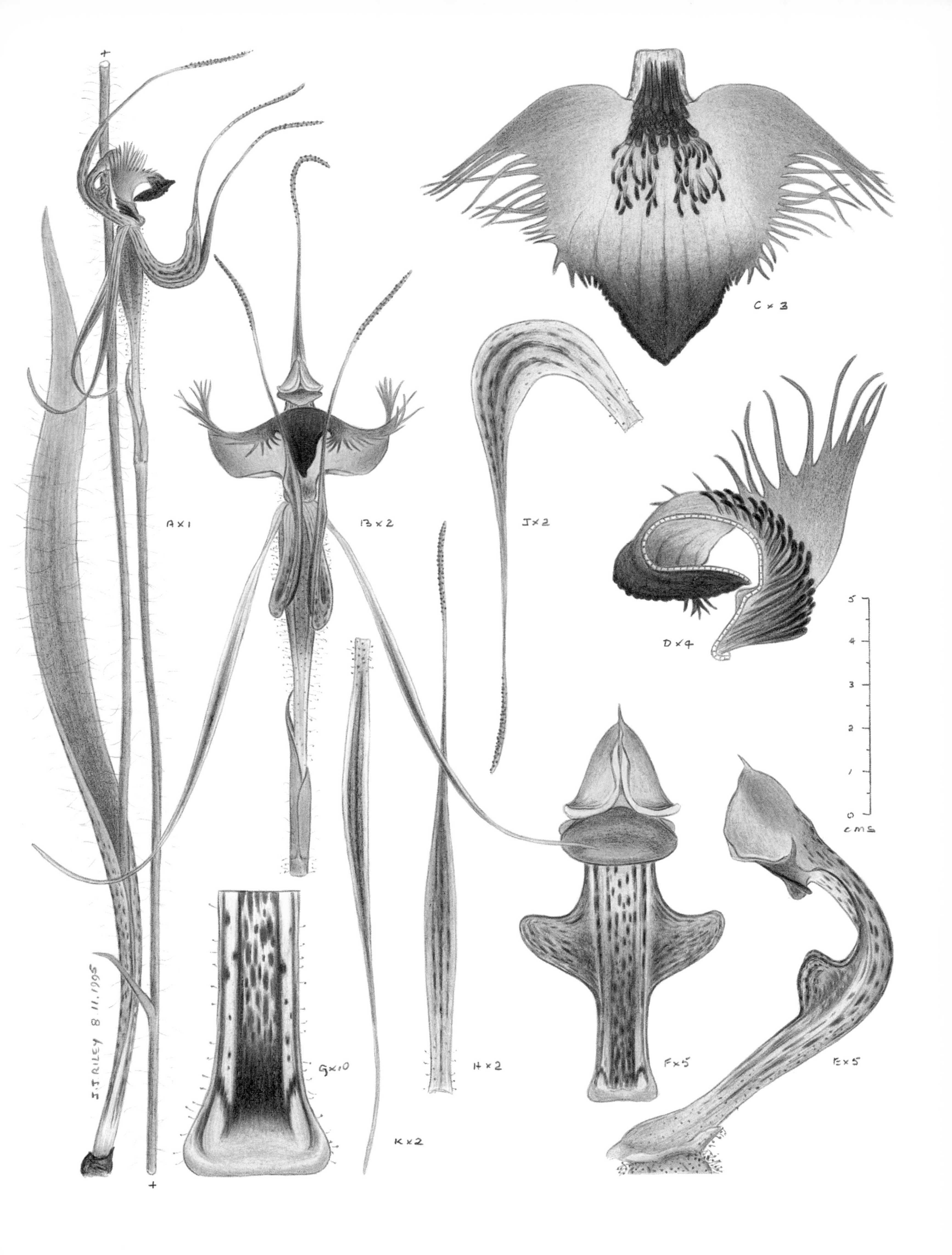

C x 3
A x 1
B x 2
J x 2
D x 4
5
4
3
2
1
0
cms
J.J. RILEY 8.11.1995
G x 10
H x 2
F x 5
E x 5
K x 2

Caladenia radiata Nicholls

1949 | *Victorian Naturalist* 65:267

TYPE LOCALITY Yarloop, Western Australia

ETYMOLOGY *radiata* — radiating outwards

FLOWERING TIME October to December

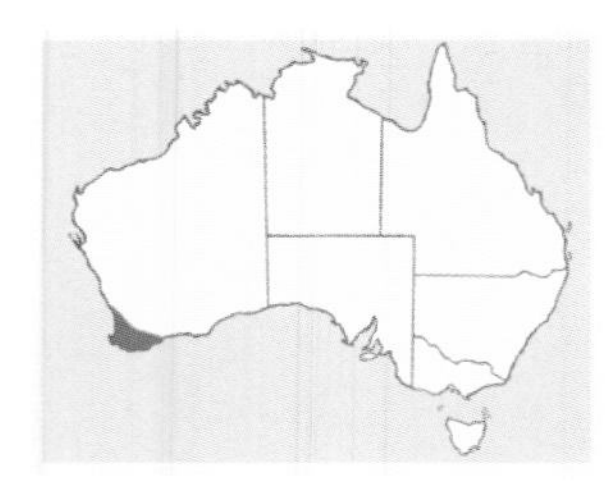

DISTRIBUTION Found from around Perth to Albany in south-western Western Australia.

ALTITUDINAL RANGE Up to 200 m

DISTINGUISHING FEATURES *C. radiata* is a tall, slender species with its dorsal and lateral sepals having long, fine osmophores. The labellum is long and narrow, indistinctly three-lobed and is fringed with slender marginal teeth extending onto the midlobe.

HABITAT Low lying, seasonally wet areas and around the margins of swamps.

CONSERVATION STATUS Locally common. Conserved in national parks and reserves. Secure.

DISCUSSION *C. radiata* grows in a much wetter habitat than other species in this group. It is not uncommon to encounter flowering plants standing in shallow water. Fire promotes flowering and large numbers bloom after a burn the previous year. Pollination is by pseudocopulation involving males of a species of thynnid wasp. Reproduces from seed. Was recently reclassified as *Arachnorchis radiata* (Nicholls) D.L. Jones & M.A. Clem.

Caladenia radiata
Lake Muir, Western Australia

16 November 1995

A plant
B flower from front
C labellum flattened
D longitudinal section of labellum
E column from side
F column from front
G glands at base of column
H dorsal sepal
J lateral sepal
K petal

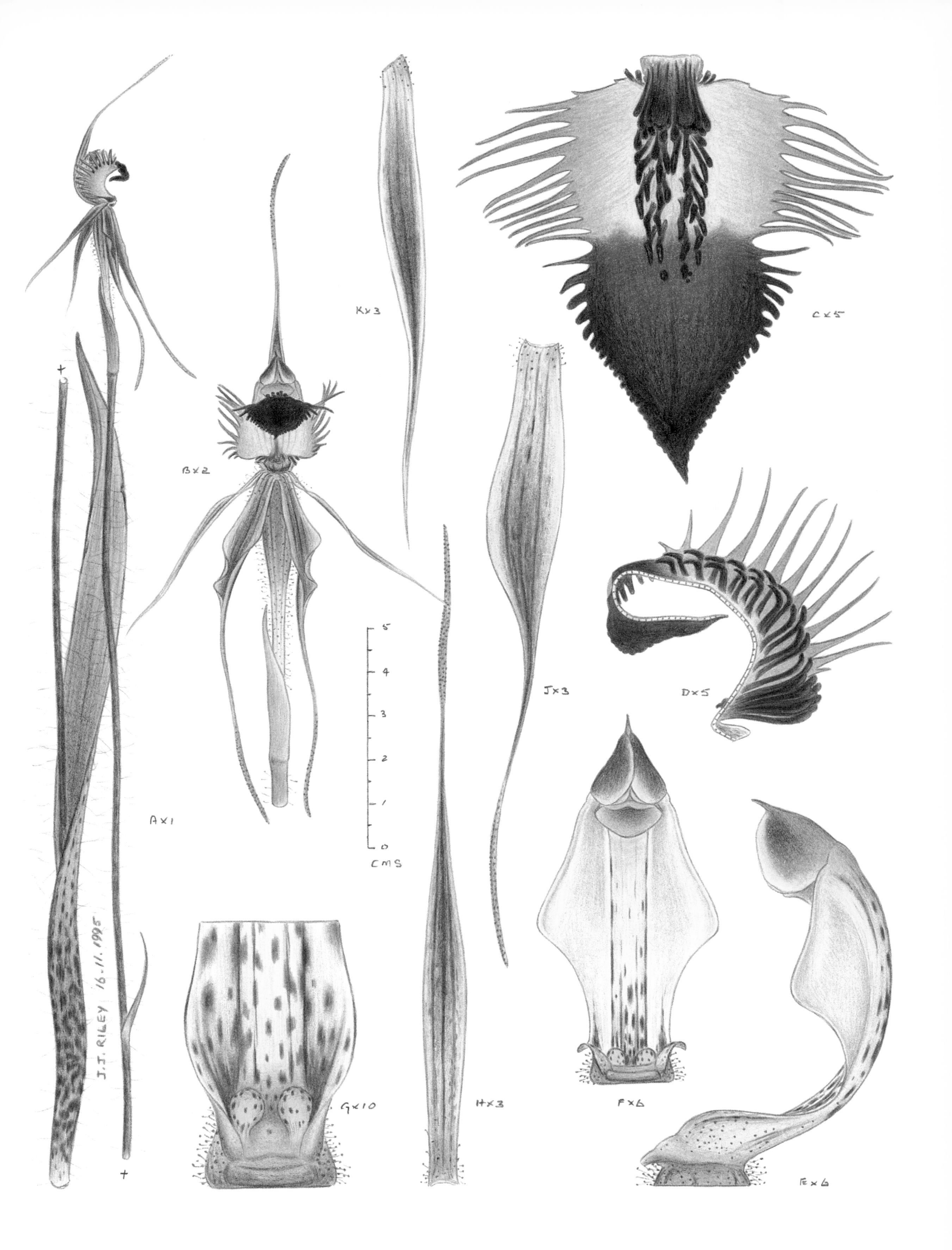

A×1
B×2
C×5
D×5
E×6
F×6
G×10
H×3
J×3
K×3
5
4
3
2
1
0
CMS
J.J. RILEY 16.11.1995

Caladenia pilotensis D.L. Jones

1999 | *Orchadian* 13:15

TYPE LOCALITY Mount Pilot, Victoria

ETYMOLOGY From Mount Pilot

FLOWERING TIME September and October

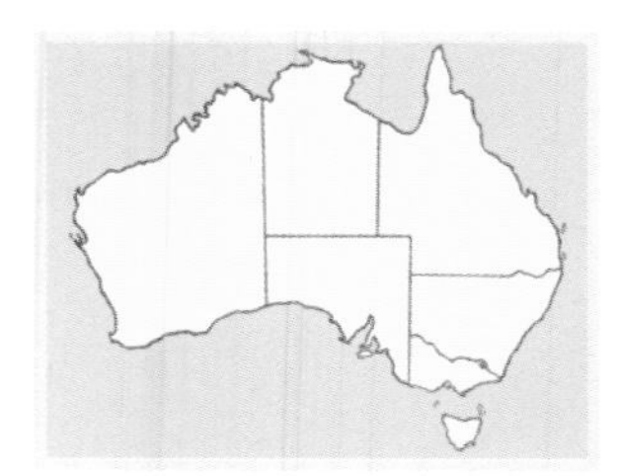

DISTRIBUTION Only known from Mount Pilot and Beechworth, Victoria.

ALTITUDINAL RANGE 400 m to 600 m

DISTINGUISHING FEATURES *C. pilotensis* is a recently described species with a narrow distribution, that has pale, whitish yellow-green blooms with a faint scent. The labellum is fringed with clubbed marginal teeth.

HABITAT Grows in open woodland of *Eucalyptus* and *Callitris* species, with a very sparse understorey of some grass and small herbs, plus extensive areas of coral lichens. Soils are a gravelly loam derived from weathered granite.

CONSERVATION STATUS Rare. Not represented in national parks. Vulnerable.

DISCUSSION *C. pilotensis* is a rare orchid with attractive blooms. It is most often seen at the base of granite boulders and outcropping granite sheets, and flowering plants stand out among the surrounding low vegetation. Members of the *C. tentaculata* complex also occur in the same habitat, but no hybrids have been seen to date. Pollination is by pseudocopulation involving males of a species of thynnid wasp. Reproduces from seed. Was recently reclassified as *Arachnorchis pilotensis* (D.L. Jones) D.L. Jones & M.A. Clem.

Caladenia pilotensis
Mount Pilot, Victoria

13 September 1993

A plant
B flower from front
C labellum from side
D labellum flattened
E longitudinal section of labellum
F glands at base of column
G column from front
H column from side
J lateral sepal
K petal
L dorsal sepal

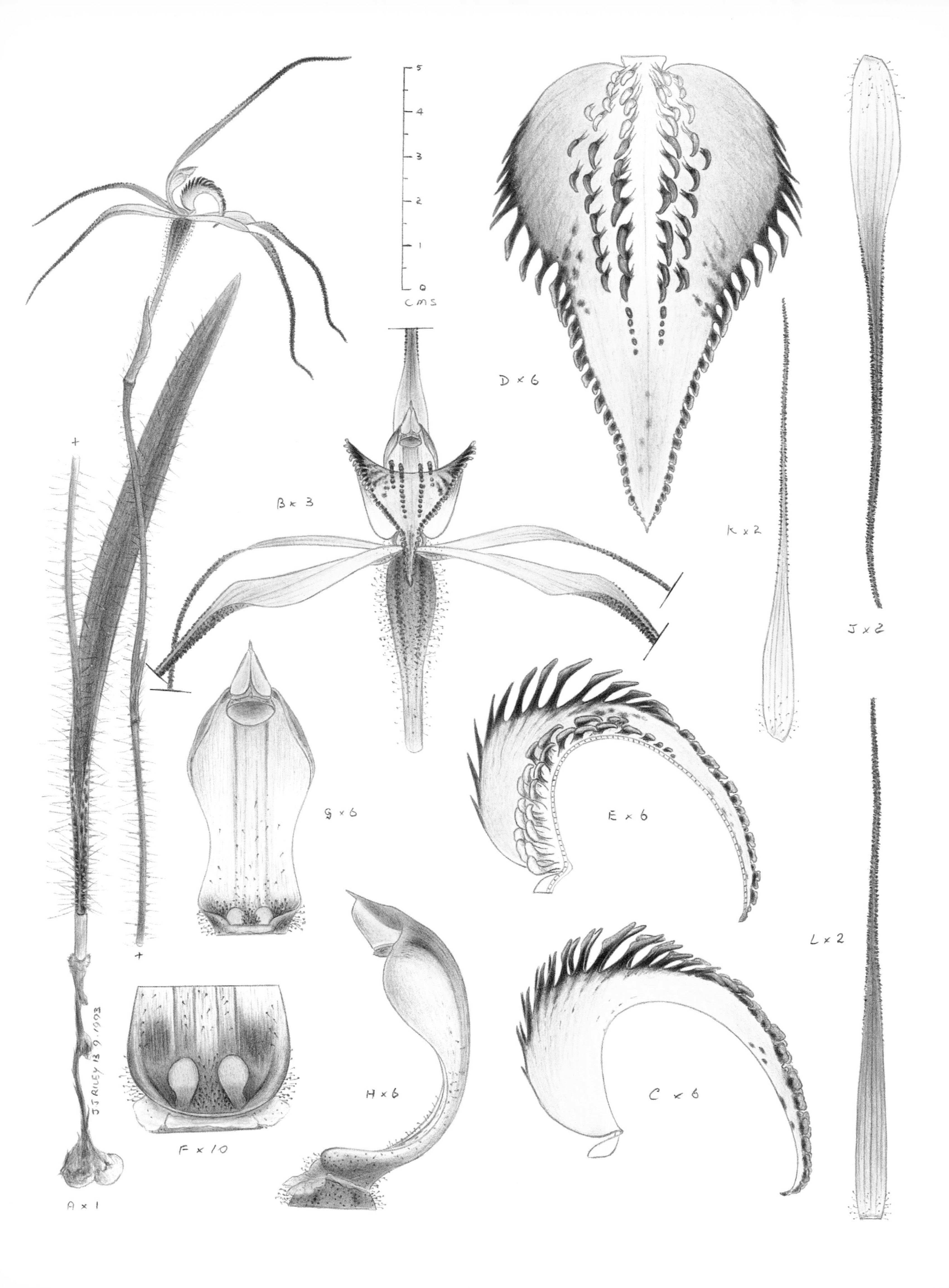

5
4
3
2
1
0
CMS
D x 6
B x 3
K x 2
J x 2
G x 6
E x 6
L x 2
H x 6
C x 6
F x 10
A x 1
J J RILEY 13 9 1993

Caladenia cardiochila Tate

1887 | *Transactions and Proceedings of the Royal Society of South Australia* 9:60

TYPE LOCALITY Golden Grove, South Australia

ETYMOLOGY *cardiochila* — heart-shaped labellum

FLOWERING TIME August to October

DISTRIBUTION Found in Victoria west from Port Phillip Bay, extending inland to the south-eastern corner of South Australia and the Mount Lofty Ranges, the York and Eyre Peninsulas and Kangaroo Island.

ALTITUDINAL RANGE 20 m to 1000 m

DISTINGUISHING FEATURES *C. cardiochila* is a distinctive species that is rarely confused with other caladenias. It lacks osmophores on the petals and sepals. The labellum has entire margins with a congested cluster of calli that do not extend to the midlobe.

HABITAT Grows in a range of vegetation communities, from coastal heath, heathy woodland, heathy mallee to open forest with a grassy and shrubby understorey. Soils are sandy clay loams and white to grey sand.

CONSERVATION STATUS Common. Well represented in national parks and nature reserves. Secure.

DISCUSSION *C. cardiochila* is a colourful, small-flowered spider *Caladenia*. The intensity of colour is very variable. The specimen illustrated is of one of the paler forms. On some flowers the labellum is almost entirely reddish-maroon and the perianth segments are heavily striated with red markings. Usually found as individuals or small groups. Pollination is by pseudocopulation involving males of a species of thynnid wasp. Reproduces from seed. *C. cardiochila* hybridises with other related caladenias. Was recently reclassified as *Arachnorchis cardiochila* (Tate) D.L. Jones & M.A. Clem.

Caladenia cardiochila
Anglesea, Victoria

26 September 1993

A plant
B flower from front
C flower from side
D flower from above
E labellum from side
F labellum flattened
G longitudinal section of labellum
H column from side
J column from front
K glands at base of column
L dorsal sepal
M lateral sepal
N petal

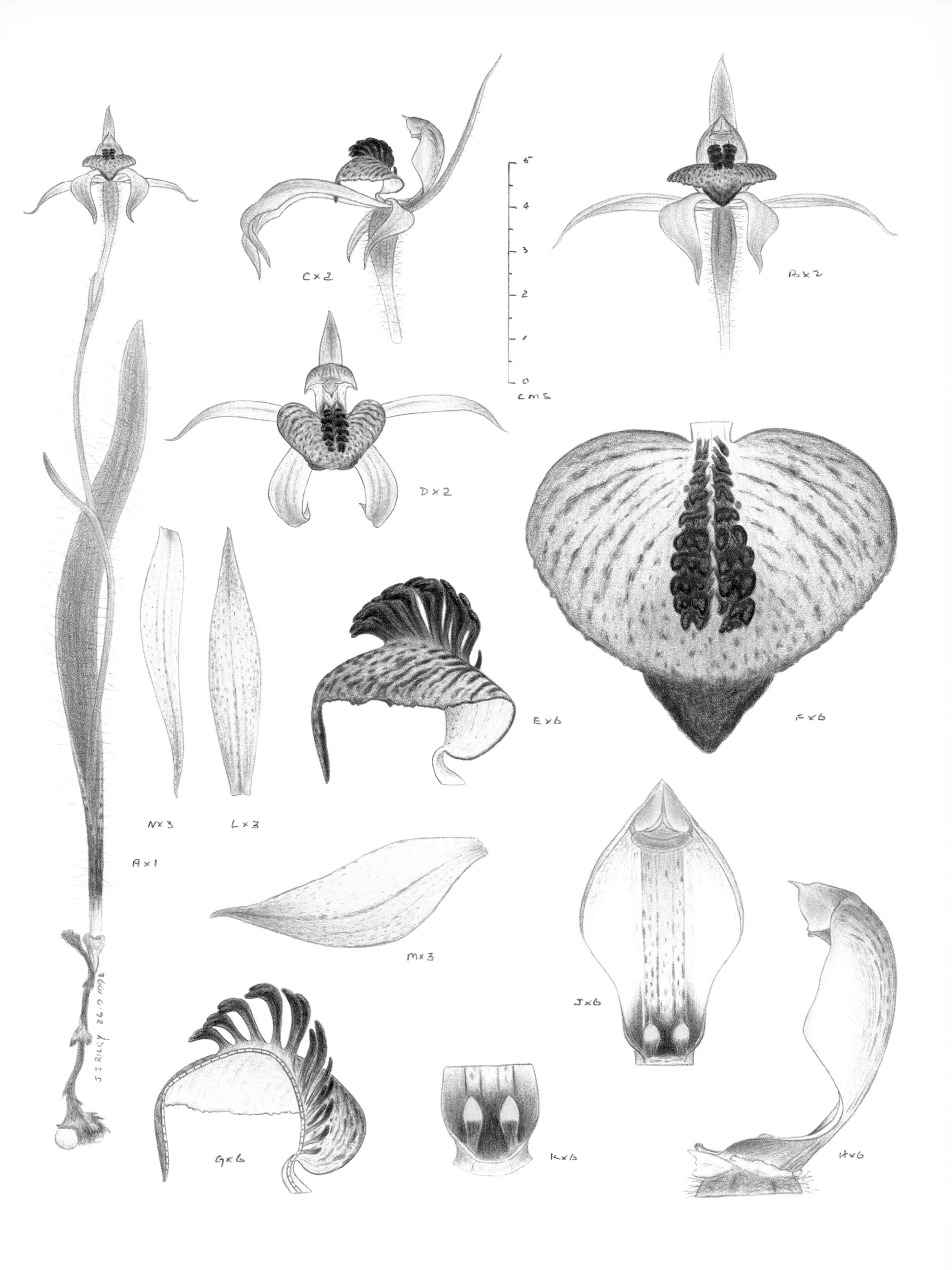
C x 2
B x 2
5
4
3
2
1
0
cms
D x 2
E x 6
F x 6
N x 3
L x 3
A x 1
M x 3
J x 6
J J Riley 26.9.1998
G x 6
K x 6
H x 6

Caladenia fitzgeraldii Rupp

1942 | *Victorian Naturalist* 58:199

TYPE LOCALITY Mudgee and Bathurst, New South Wales

ETYMOLOGY After Robert David Fitzgerald

FLOWERING TIME September to November

DISTRIBUTION Found on the Central Tablelands of New South Wales, from near Kandos to Bathurst, and the Mullion Range in the Orange–Wellington area.

ALTITUDINAL RANGE 500 m to 1200 m

DISTINGUISHING FEATURES *C. fitzgeraldii* has sepals with long, dark, prominent clubs. The labellum, when flattened, is not obviously three-lobed. The base of the labellum is yellowish-green, partly veined and fringed with maroon teeth. The apex is maroon and entire.

HABITAT Found in open *Eucalyptus* forest with a grass and heath understorey. Terrain is often moderately rugged. Grows in clay loams between rocks and near shrubs and grass tussocks. Is usually seen growing as individuals or small colonies in shallow soils among rocks on the sides of hills and in their drainage systems. Rain falls sporadically throughout the year and water drains rapidly from the slopes, however, the area, while generally dry, supports a healthy and varied orchid flora.

CONSERVATION STATUS Moderately rare. Occurs in national parks and nature reserves. Secure.

DISCUSSION *C. fitzgeraldii* is a poorly known species with large, attractive flowers. Pollination is by pseudocopulation involving males of a species of thynnid wasp. Reproduces from seed. Was recently reclassified as *Arachnorchis fitzgeraldii* (Rupp) D.L. Jones & M.A. Clem.

Caladenia fitzgeraldii
Kandos, New South Wales

16 October 1999

A plant
B flower from front
C labellum flattened
D longitudinal section of labellum
E column from front
F column from side
G glands at base of column
H dorsal sepal
J lateral sepal
K petal

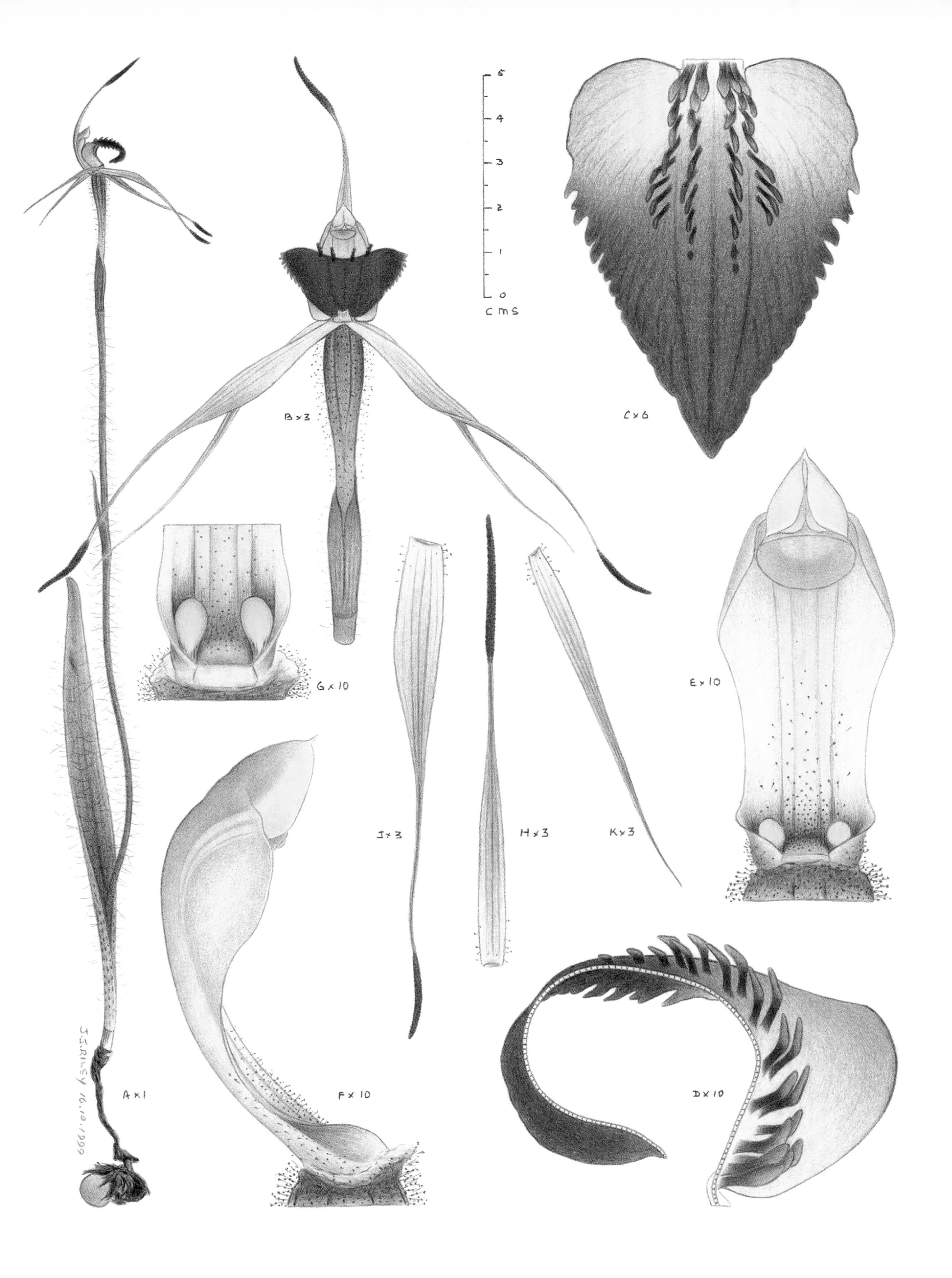

A × 1
B × 3
C × 6
D × 10
E × 10
F × 10
G × 10
H × 3
J × 3
K × 3
cms
0
1
2
3
4
5
J.J. RILEY 16.10.1990

Caladenia callitrophila D.L. Jones

1999 | *Orchadian* 13:5

TYPE LOCALITY Berrigan, New South Wales

ETYMOLOGY *callitrophila* — growing with *Callitris*

FLOWERING TIME September to October

DISTRIBUTION Endemic to the south-eastern Riverina district of New South Wales, from Savernake to Balldale.

ALTITUDINAL RANGE 90 m to 140 m

DISTINGUISHING FEATURES *C. callitrophila* has colourful flowers with stiff segments. It exhibits well developed, dark reddish-maroon, sepaline osmophores. The labellum is heavily marked with reddish-maroon veins and, when flattened, is indistinctly three-lobed with round sidelobes.

HABITAT Grows in open woodland of *Callitris glaucophylla*, with a very sparse understorey of grass and other small herbs. Usually found on flat ground or gently undulating slopes. Soils are fine, reddish sandy loams.

CONSERVATION STATUS Rare. The known populations are not contained in a national park or nature reserve. Vulnerable.

DISCUSSION *C. callitrophila* grows as scattered individuals or sometimes in moderately large groups. The specimen illustrated represents the most common colour, with other clones much darker or lighter. Robust specimens often have two flowers. Pollination is by pseudocopulation involving males of a species of thynnid wasp. Reproduces from seed. Very little of its former habitat remains, most having been cleared for agriculture. Recently conducted professional searches for new populations failed to turn up any new recordings. Was recently reclassified as *Arachnorchis callitrophila* (D.L. Jones) D.L. Jones & M.A. Clem.

Caladenia callitrophila
Balldale, New South Wales

24 September 1999

A plant
B flower from front
C labellum flattened
D longitudinal section of labellum
E column from side
F column from front
G glands at base of column
H dorsal sepal
J lateral sepal
K petal

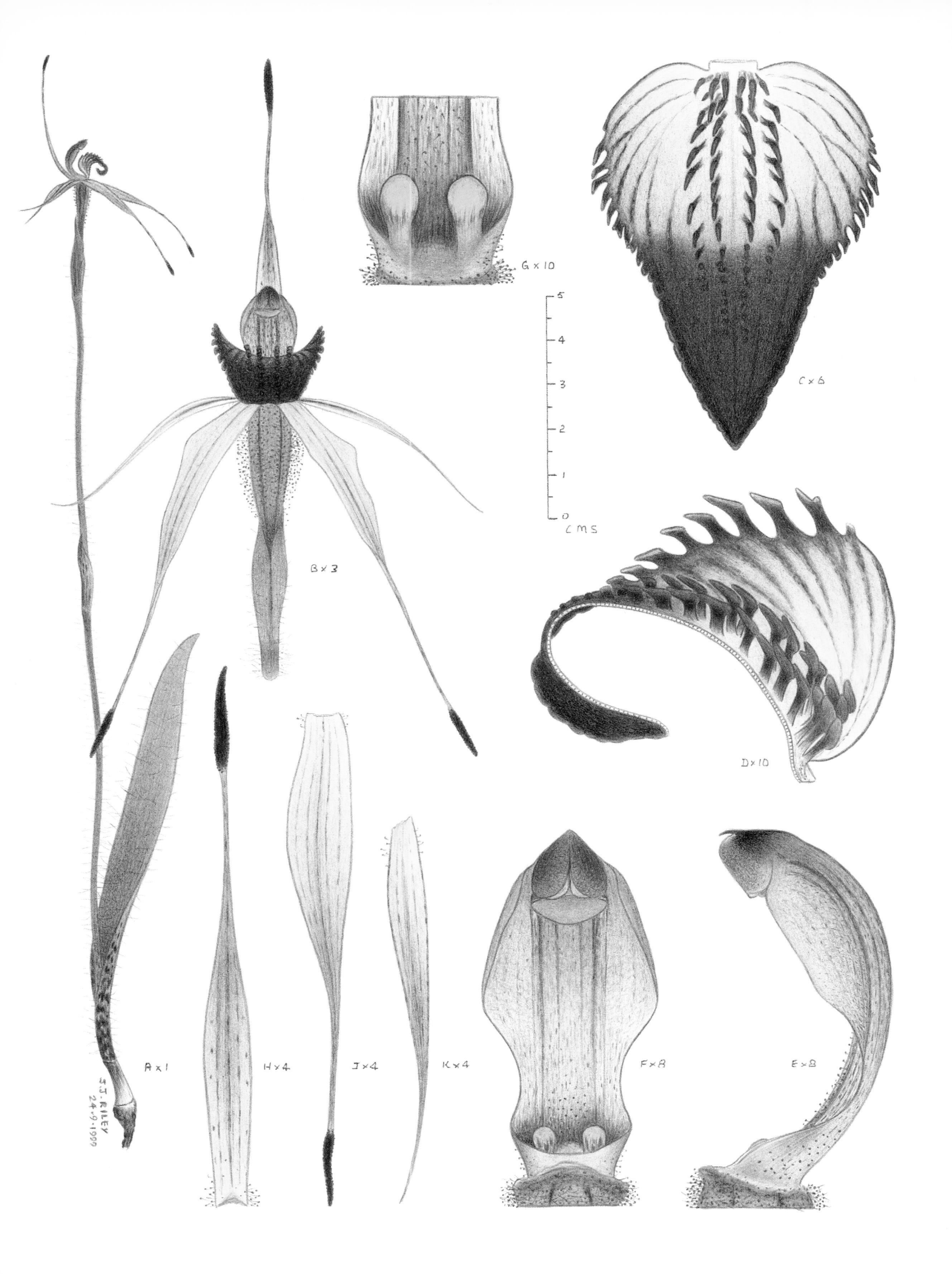

G × 10
C × 6
5
4
3
2
1
0
CMS
B × 3
D × 10
A × 1
H × 4
J × 4
K × 4
F × 8
E × 8
S.J. RILEY
24.9.1999

Caladenia subtilis D.L. Jones

1999 | *Orchadian* 13:21

TYPE LOCALITY Giro State Forest, New South Wales

ETYMOLOGY *subtilis* — fine, delicate, nice

FLOWERING TIME October to November

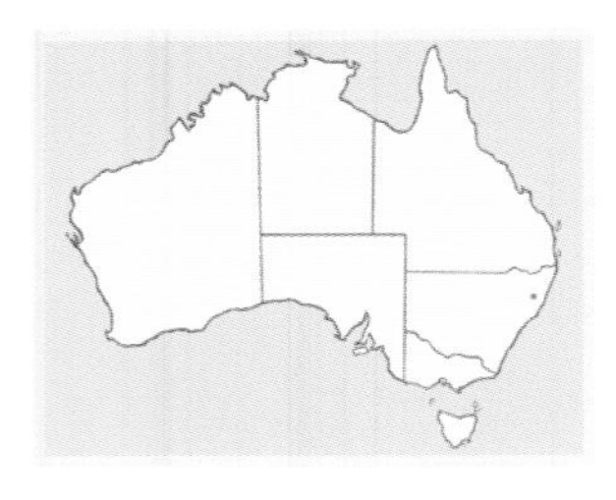

DISTRIBUTION On the ranges bordering the south-eastern edge of the New England Tableland in New South Wales. Known from two locations, one near Nowendoc, the other between Nundle and Woolomin.

ALTITUDINAL RANGE 600 m to 850 m

DISTINGUISHING FEATURES *C. subtilis* has rigidly held flowers. The labellum, when flattened, is not obviously three-lobed. Marginal teeth are often curved and extending more or less to the base of the labellum.

HABITAT This species grows on moderately steep slopes and ridges in open forest with a mostly grassy understorey on gravelly clay loams.

CONSERVATION STATUS Rare. Known only from two small areas, which are not in a national park or nature reserve. Vulnerable.

DISCUSSION *C. subtilis* is a colourful and striking species with a most attractive labellum contrasting with the rest of the flower. While restricted in distribution, small colonies of numerous plants can be observed growing among clumps of tussocky grass. It is often accompanied by other *Caladenia* species together with *Thelymitra* spp. and *Diuris* spp. *C. subtilis* may be more widespread, however, the rugged terrain, limited access and remoteness restrict searches. Pollination is by pseudocopulation involving males of a species of thynnid wasp. Reproduces from seed. Was recently reclassified as *Arachnorchis subtilis* (D.L. Jones) D.L. Jones & M.A. Clem.

Caladenia subtilis
Duncans Creek, New South Wales

30 September 1995

A plant
B flower from front
C labellum flattened
D longitudinal section of labellum
E column from side
F column from front
G glands at base of column
H dorsal sepal
J lateral sepal
K petal

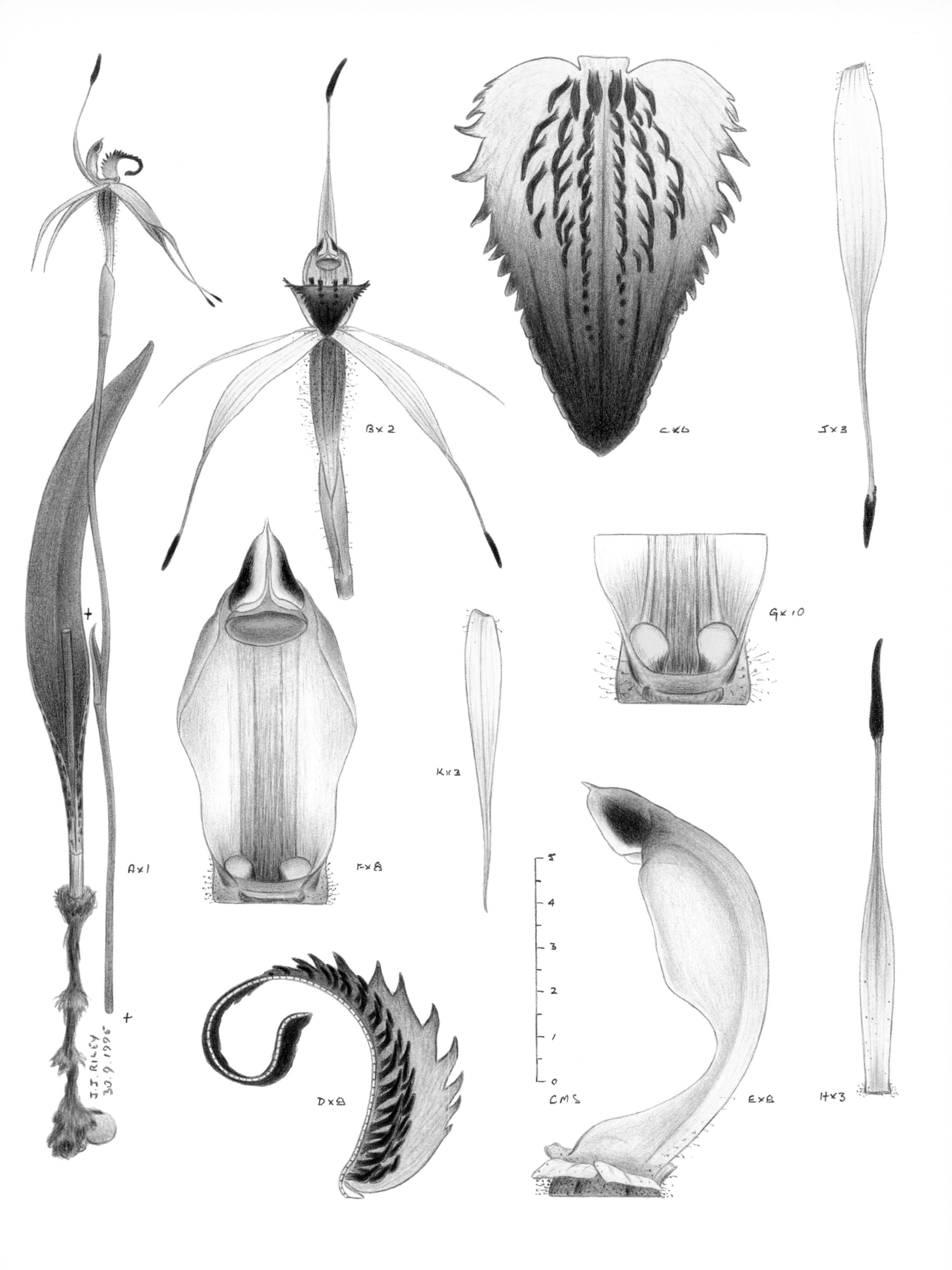
A×1
B×2
C×6
J×3
F×8
G×10
K×3
D×8
E×8
H×3
CMS
0
1
2
3
4
5
J.J. RILEY
30.9.1996

Caladenia stellata D.L. Jones

1991 | *Australian Orchid Research* 2:33

TYPE LOCALITY Sims Gap, New South Wales

ETYMOLOGY *stellata* — starry

FLOWERING TIME September to October

DISTRIBUTION Found from Rankin Springs to the Urana and Narrandera districts, New South Wales.

ALTITUDINAL RANGE 100 m to 250 m

DISTINGUISHING FEATURES *C. stellata* has star-shaped, greenish-yellow flowers whose segments are held outstretched and stiff. The labellum is three-lobed and yellowish-green with a prominent maroon midlobe. The marginal teeth are straight with rounded ends.

HABITAT This species grows in open *Callitris glaucophylla* woodland with a sparse grassy understorey. Occurs on flat, or slightly undulating, red sandy loams and also on stony ridges on grey, sandy clay loams.

CONSERVATION STATUS Uncommon. Not known to occur in a national park or nature reserve. Unsecure and not conserved.

DISCUSSION *C. stellata* may be locally common in some areas but it is poorly known. Clearing of native woodland for agriculture has reduced the orchid to a few populations. While the flower is usually greenish-yellow with maroon markings, in all populations albino plants may occur, with pale yellowish-green blooms. Pollination is by pseudocopulation involving males of a species of thynnid wasp. Reproduces from seed. *C. stellata* forms hybrids with *C. arenaria*, *C. rileyi*, *C.* sp. aff. *tentaculata* and *C. concinna* where the species grow and bloom together. Was recently reclassified as *Arachnorchis stellata* (D.L. Jones) D.L. Jones & M.A. Clem.

Caladenia stellata
Sims Gap, New South Wales

3 September 1992

A plant
B flower from side
C flower from above
D labellum from side
E labellum flattened
F longitudinal section of labellum
G column from side
H column from front

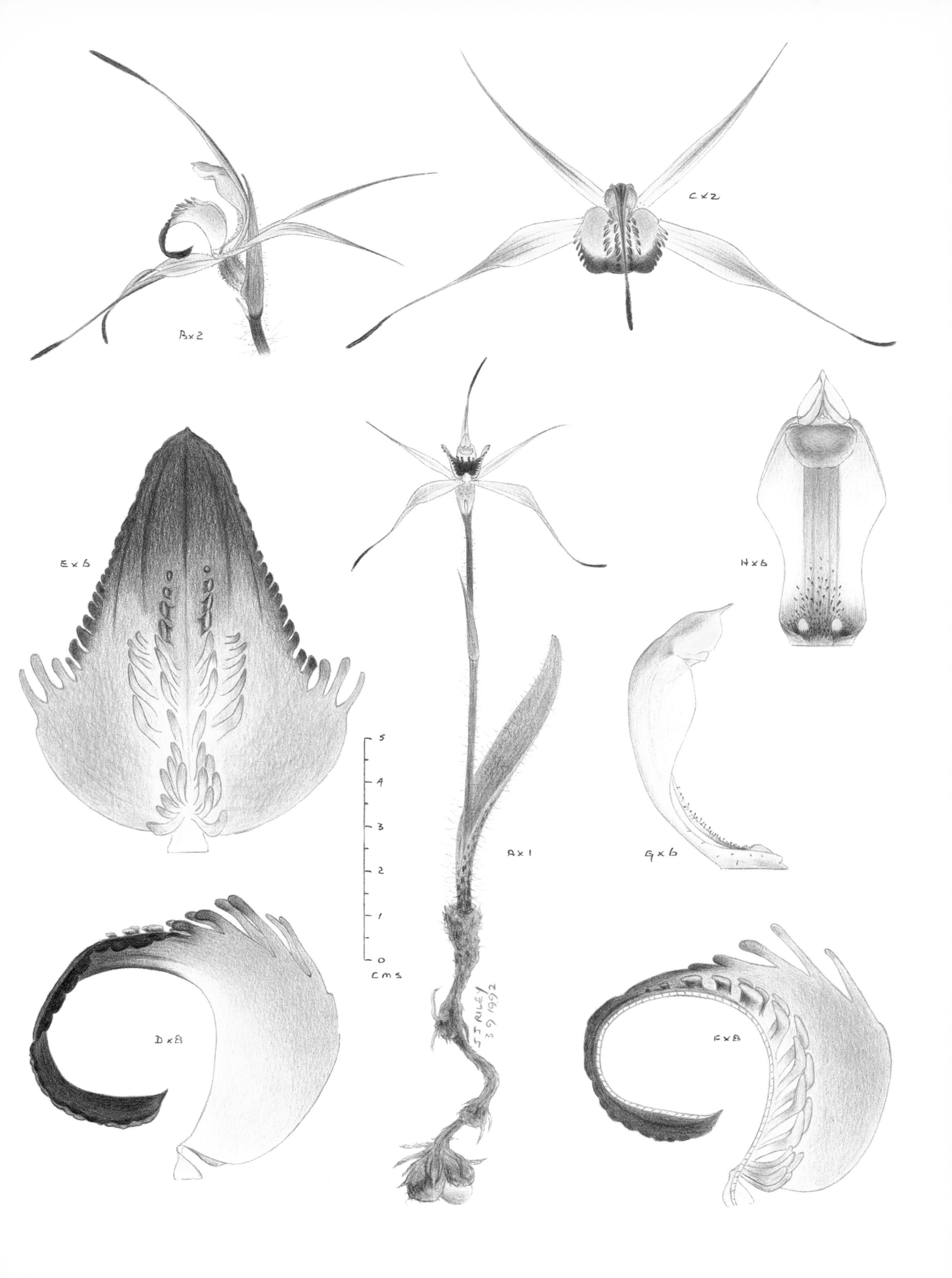
B×2
C×2
E×6
H×6
A×1
G×6
5
4
3
2
1
0
cms
D×8
F×8
J I RILEY
3 9 1992

Caladenia capillata D.L. Jones

2000 | *Orchadian* 13:255

TYPE LOCALITY Caroona Hill, South Australia

RECENT SYNONYMS *Caladenia filamentosa* R. Br. var. *tentaculata* R.S. Rogers

ETYMOLOGY *capillata* — hair-like

FLOWERING TIME July to October

DISTRIBUTION South-western New South Wales, Victoria, south-east South Australia to the Eyre Peninsula and Kangaroo Island.

ALTITUDINAL RANGE 20 m to 500 m

DISTINGUISHING FEATURES *C. capillata* has spidery, creamy-grey flowers, and this group of caladenias is often referred to by the common name of Daddy Long Legs, in apt reference to the common spider, *Pholcus phalangioides*. The white labellum, when flattened is not three-lobed and is triangular in shape. The teeth on the labellum and margins are narrow. There are also no glands at the base of the column, another feature that separates this group from other spider caladenias.

HABITAT Grows in scrubby woodland and open mallee woodland with a scrubby understorey. Also found on shrubby stabilised dunes. Soils are sand, sandy loams and red soils over limestone.

CONSERVATION STATUS Common and conserved in national parks and nature reserves. Secure.

DISCUSSION *C. capillata* is part of the *C. filamentosa* complex. Although common, it is easily overlooked in the field when growing with grasses due to its pale flowers with very narrow tepals. Reproduces mainly from seed but can also reproduce vegetatively, occasionally forming small tufts.

Caladenia capillata
Dimboola, Victoria

28 September 1990

A plant
B flower from front
C flower from rear
D flower from above
E labellum from side
F labellum flattened
G column from side
H column from front

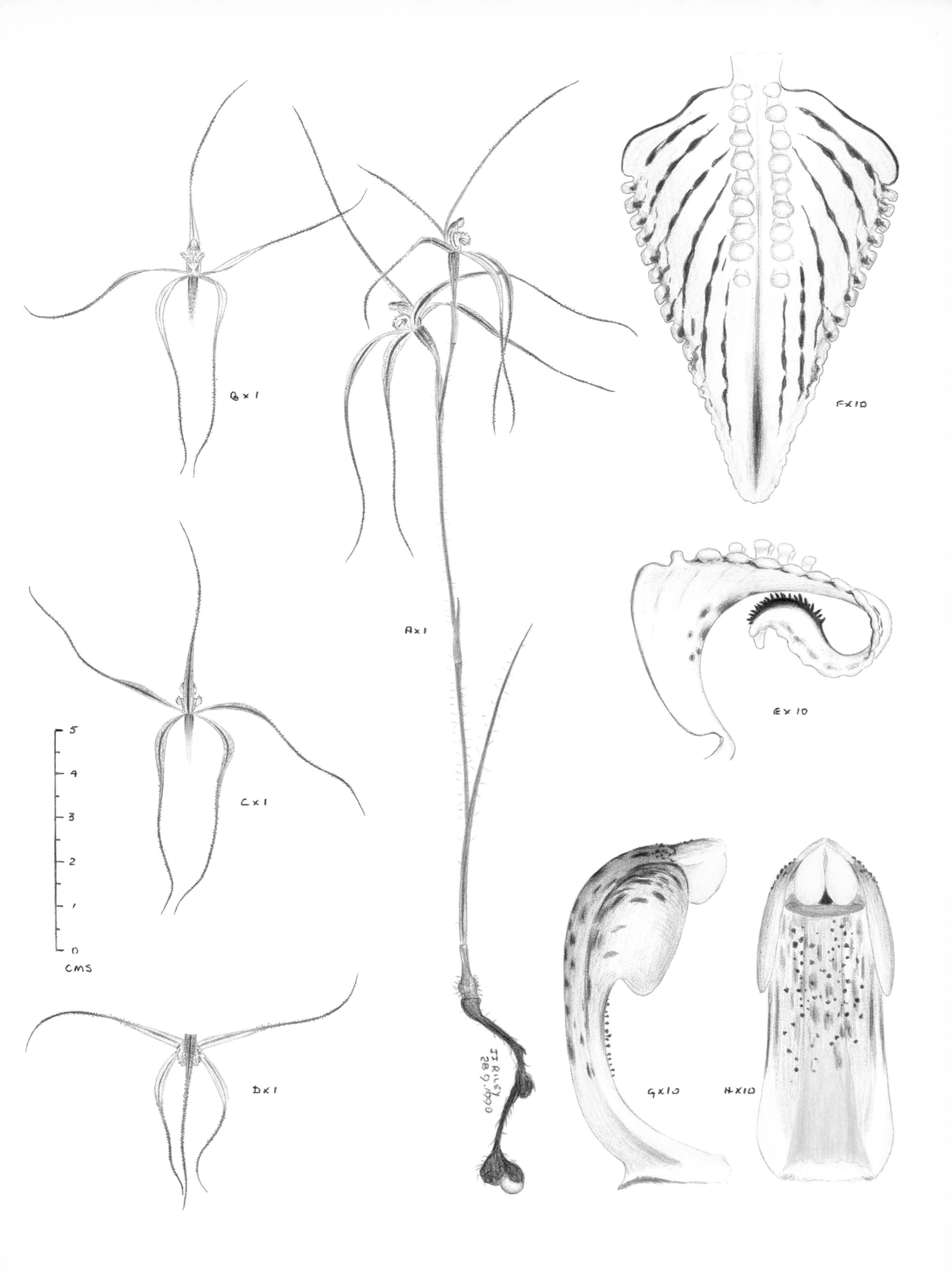
B x 1
A x 1
F x 10
E x 10
C x 1
5
4
3
2
1
0
CMS
D x 1
J J RILEY
28.9.1990
G x 10
H x 10

Caladenia flaccida D.L. Jones

1991 | *Australian Orchid Research* 2:24

TYPE LOCALITY Sims Gap, New South Wales

ETYMOLOGY *flaccida* — weak, drooping

FLOWERING TIME August to October

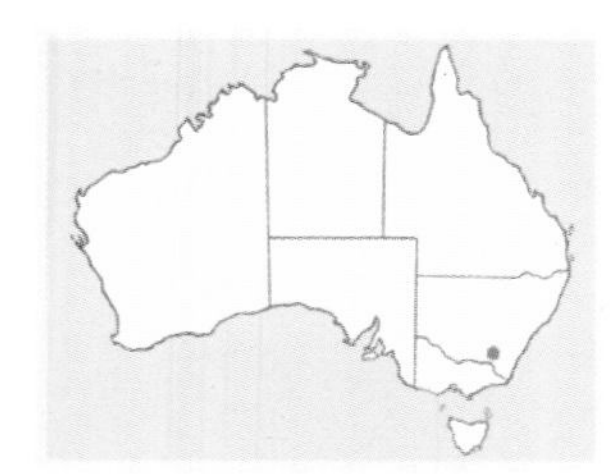

DISTRIBUTION Found on the western slopes and plains of New South Wales in the Narrandera to Griffith area.

ALTITUDINAL RANGE 150 m to 500 m

DISTINGUISHING FEATURES *C. flaccida* has cream to pinkish-red flowers, with drooping petals and sepals. The labellum, when flattened, is distinctly three-lobed with broad lamina calli and prominent, broad marginal teeth.

HABITAT Prefers open *Callitris glaucophylla* woodland with a sparse grassy understorey on flat, or slightly undulating, red sandy loams. May also be seen in open *Eucalyptus* forest and woodland on ridges and slopes with a sparse grass or shrub understorey on sandy and clay loams.

CONSERVATION STATUS Uncommon. Occurs in national parks and nature reserves. Not threatened. Secure.

DISCUSSION *C. flaccida* is part of the *C. filamentosa* complex. Can have from one to three flowers on the erect inflorescence. Grows as isolated individuals or small tufts and groups, often in association with other spider caladenias. Reproduces mainly from seed but, as with other members of the *C. filamentosa* complex, can also reproduce vegetatively, forming small tight clumps.

Caladenia flaccida
Sims Gap, New South Wales

27 July 1993

A plant
B labellum from side
C labellum flattened
D longitudinal section of labellum
E column from side
F column from front
G dorsal sepal
H lateral sepal
J petal

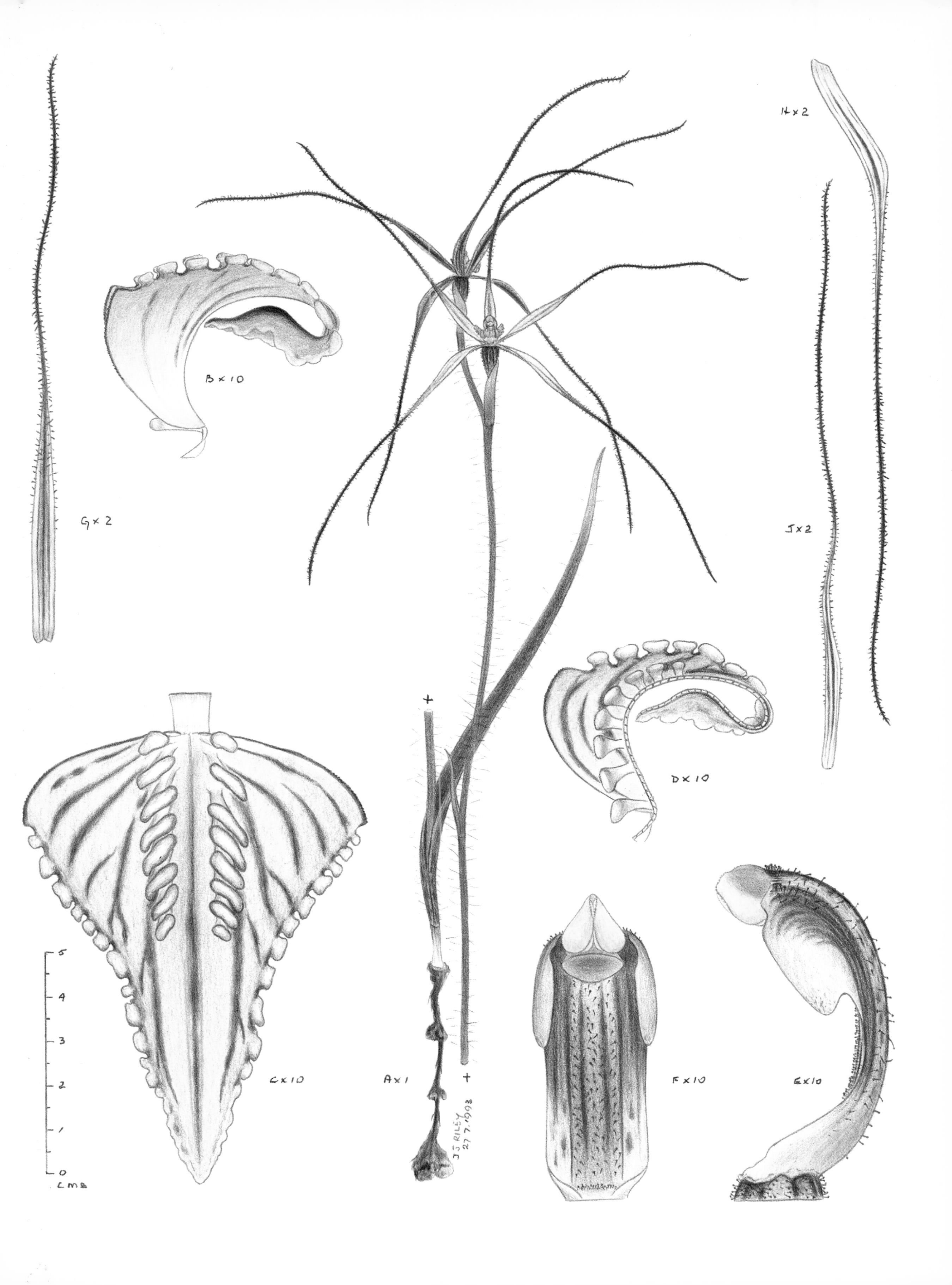

B × 10
G × 2
H × 2
J × 2
D × 10
C × 10
A × 1
F × 10
E × 10
5
4
3
2
1
0
cms
J J RILEY
27.7.1993

Caleana minor R. Br.

1810 | *Prodromus Florae Novae Hollandiae*: 329

TYPE LOCALITY Port Jackson, New South Wales

RECENT SYNONYMS *Paracaleana minor* (R. Br.) Blaxell

ETYMOLOGY *minor* — small

FLOWERING TIME November to January

DISTRIBUTION Widespread but sporadic in eastern Australia, from the Blackdown Tableland and Fraser Island in Queensland, south through eastern New South Wales, southern Victoria and west to the south-eastern corner of South Australia, and Tasmania.

ALTITUDINAL RANGE 20 m to 1000 m

DISTINGUISHING FEATURES The small flower resembles a duck in flight when viewed from the side. The duck-head shaped labellum is entirely covered with dark wart-like calli. The labellum is held above the rest of the flower and snaps shut once touched, a mechanism used to secure pollination.

HABITAT Found growing in a variety of habitats, from coastal, heathy dunes, in moss gardens over shallow soils on rock outcrops, to slopes and ridges in heathy woodland and open forest. Soils are grey sand, sandy loams and clay loams.

CONSERVATION STATUS Common. Occurs throughout its range in national parks and nature reserves. Secure.

DISCUSSION *C. minor* is one of Australia's most fascinating and unusual terrestrial orchids. It never fails to captivate those seeing it for the first time, due to its uncanny resemblance to a flying duck. Due to its small stature and late flowering, it is often overlooked among the shrubs and grasses. *C. minor* is pollinated by males of a species of thynnid wasp during pseudocopulation. Reproduces from seed and also vegetatively from tubers formed on the ends of thread-like stolons. At the Grampians in Victoria, plants with deformed, self-pollinating flowers occur rarely and have been called *C. sullivanii*. Similar apomictic plants from the upper Blue Mountains were known as *C. nublingii*. Was recently reinstated as *Paracaleana minor* (R. Br.) Blaxell.

Caleana minor
Port Stephens, New South Wales

9 November 1994

A plant
B flower from front
C flower from side
D flower from above
E labellum from side
F labellum from above
G column from side
H column from front
J dorsal sepal
K lateral sepal
L petal

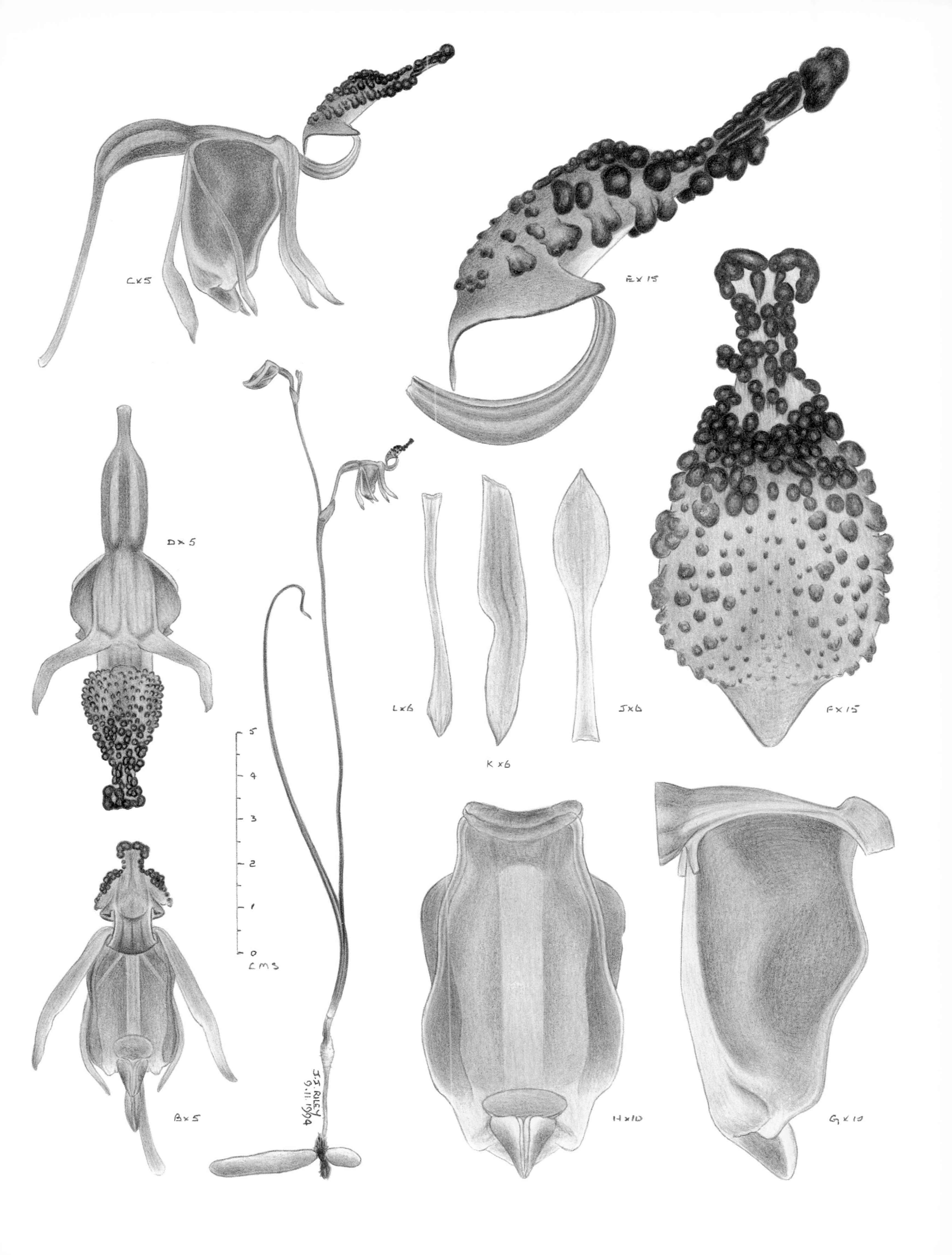
C×5
E×15
D×5
F×15
L×6
K×6
J×6
B×5
H×10
G×10
5
4
3
2
1
0
CMS
J.J. RILEY
9.11.1994

Chiloglottis reflexa (Labill.) Druce

1917 | *Botanical Exchange Club and Society of the British Isles Report for 1916*, Supp. 2:614

TYPE LOCALITY Tasmania

RECENT SYNONYMS *Chiloglottis trilabra* Fitzg.

ETYMOLOGY *reflexa* — bent backwards

FLOWERING TIME January to April

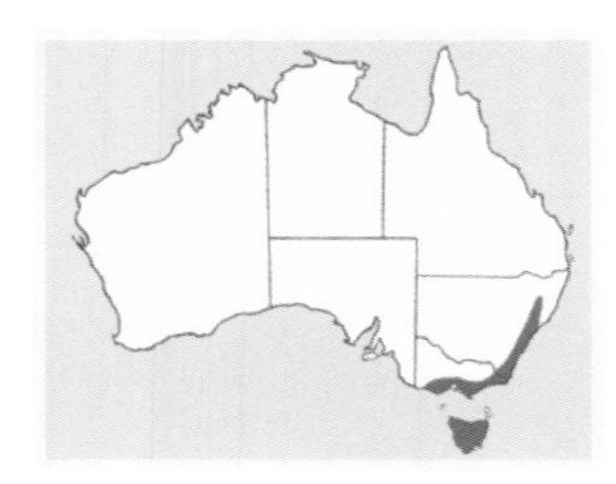

DISTRIBUTION Tasmania, eastern Victoria and north to at least Point Lookout on the New England Tableland of New South Wales.

ALTITUDINAL RANGE 20 m to 1550 m

DISTINGUISHING FEATURES *C. reflexa* has green flowers, heavily suffused with maroon-red. The labellum, when flattened, is wide and rounded, densely encrusted with dark maroon, sessile and stalked calli and glands that are wide and often slightly notched.

HABITAT Moist heathy woodland and open forest with an understorey of ferns and shrubs. In New South Wales, *C. reflexa* grows on the ranges and tablelands among snow grasses.

CONSERVATION STATUS Common. Well represented in national parks and nature reserves. Secure.

DISCUSSION On the mainland of Australia there is great confusion regarding the true identity of *C. reflexa*. The orchid growing in the montane areas of New South Wales and north-eastern Victoria known as *C. trilabra* appears to be identical to *C. reflexa* from Tasmania. The coastal taxon from the mainland that has been known as *C. reflexa* is in fact an undescribed species. Because of this misidentification, the distribution above only covers areas where true *C. reflexa* is currently known to occur. It can form extensive colonies, reproducing vegetatively from daughter tubers formed on the end of stolonoid roots. Also reproduces from seed to form new colonies, as *C. reflexa* is pollinated by males of a species of thynnid wasp, the result of pseudocopulation. It can be a shy-flowering species. As with all *Chiloglottis* species, flowering is greatly stimulated by fires the previous year.

Chiloglottis reflexa
Blackmans Bay, Tasmania

26 February 1996

A plant
B flower from front
C flower from side
D labellum from side
E labellum flattened
F column from side
G column from front
H dorsal sepal
J lateral sepal
K petal

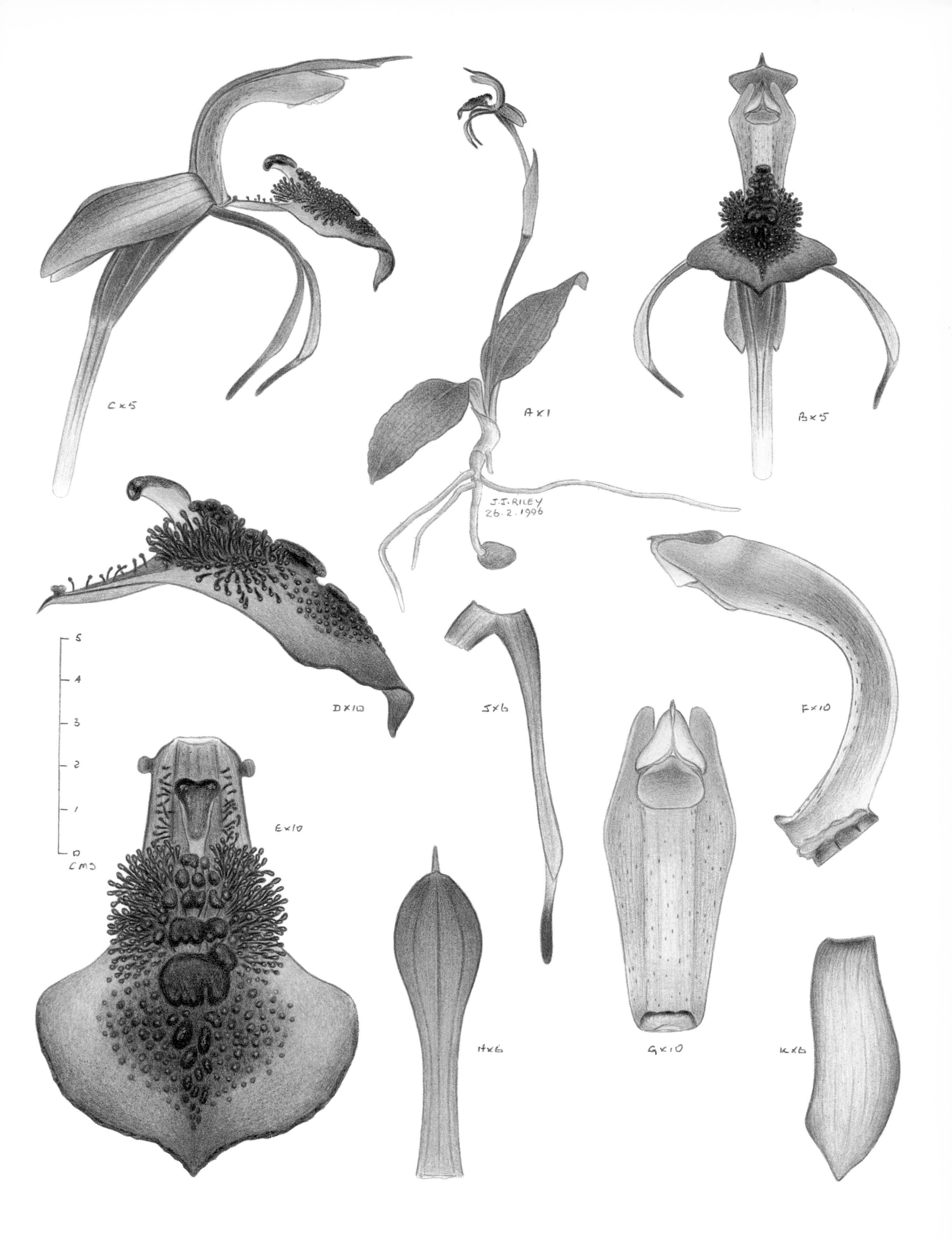
C×5
A×1
B×5
J.J. RILEY
26.2.1996
D×10
J×6
F×10
E×10
CMS
H×6
G×10
K×6

Chiloglottis seminuda D.L. Jones

1991 | *Australian Orchid Research* 2:41

TYPE LOCALITY Penrose State Forest, New South Wales

ETYMOLOGY *seminuda* — half naked

FLOWERING TIME January to May

DISTRIBUTION The Central Coast of New South Wales, from Gosford south to around Robertson and Penrose. Also extends onto the Blue Mountains.

ALTITUDINAL RANGE 50 m to 1100 m

DISTINGUISHING FEATURES *C. seminuda* lacks calli on the apical third or half of the labellum. The labellum, when flattened, is more or less rhomboidal in outline. The major stalked gland, when viewed from the side, resembles the head and neck of a turtle.

HABITAT This shy-flowering species grows in gullies and on slopes in moist *Eucalyptus* forest with a rich shrub and fern understorey. Soils are sandy loams and occasional peaty loams.

CONSERVATION STATUS Common. Widespread and conserved in national parks and nature reserves. Secure.

DISCUSSION At the type-site, *C. seminuda* grows within plantations of the introduced *Pinus radiata*. Usually seen as large colonies tucked in among shrubs and bracken ferns. Colonies often form in the rich organic soil at the base of large *Eucalyptus* trees. It can form extensive colonies, reproducing vegetatively from daughter tubers formed on the end of stolonoid roots. Also reproduces from seed to form new colonies, as *C. seminuda* is pollinated by males of a species of thynnid wasp. As with all *Chiloglottis*, the pedicel significantly elongates after pollination to aid seed dispersal.

Chiloglottis seminuda
Penrose State Forest, New South Wales

16 April 1994

A plant
B flower from front
C flower from side
D labellum from side
E labellum flattened
F column from side
G column from front
H dorsal sepal
J lateral sepal
K petal

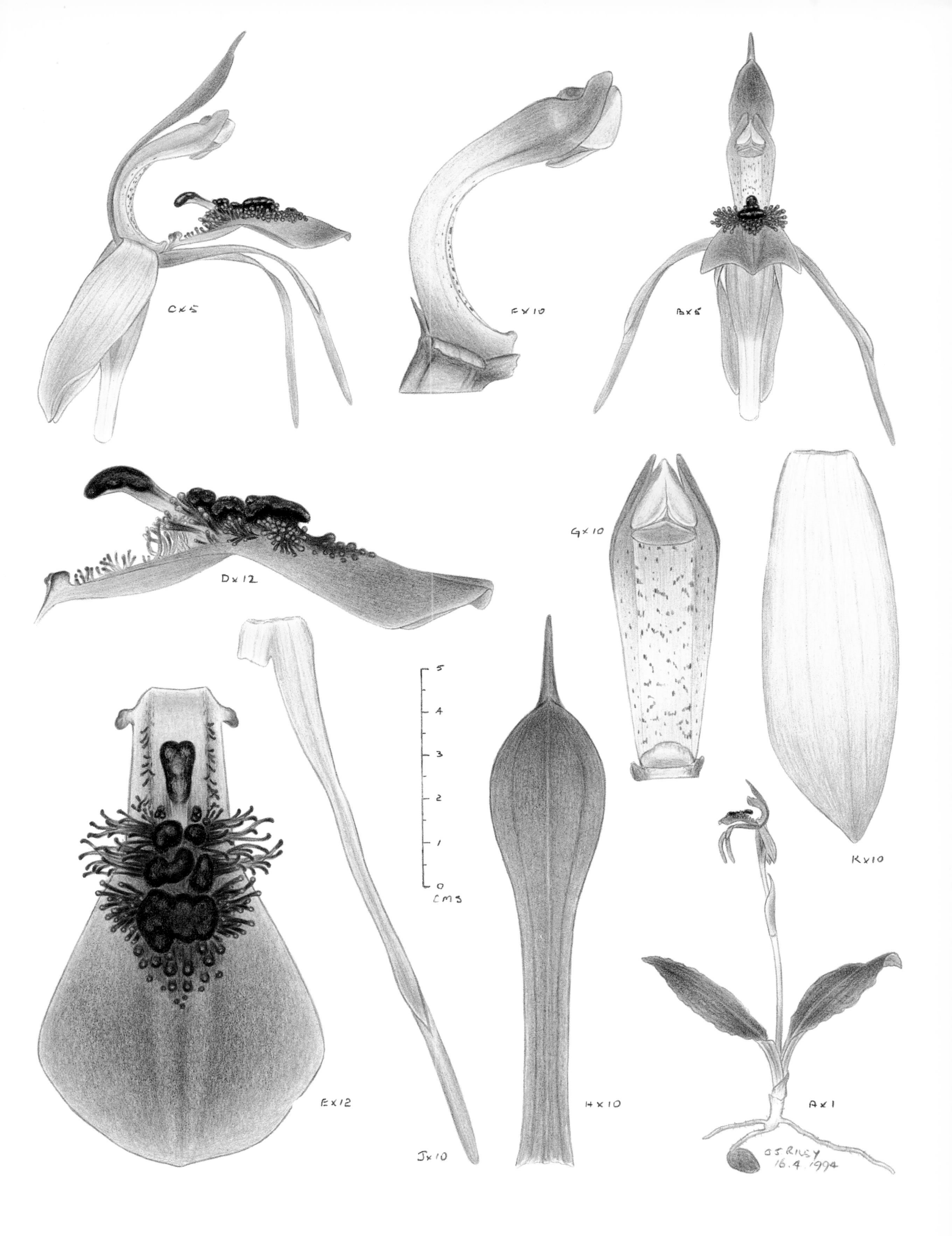

C×5
F×10
B×5
D×12
G×10
E×12
J×10
H×10
K×10
A×1
CMS
0
1
2
3
4
5
16.4.1994

Chiloglottis diphylla R. Br.

1810 | *Prodromus Florae Novae Hollandiae*: 323

TYPE LOCALITY Port Jackson, Sydney and Parramatta, New South Wales

ETYMOLOGY *diphylla* — two-leafed

FLOWERING TIME January to May

DISTRIBUTION Found from around Batemans Bay in southern New South Wales, northwards along the coast to Fraser Island and the adjacent mainland, Queensland. It is primarily a coastal species but it does extend in parts onto the nearby ranges.

ALTITUDINAL RANGE 20 m to 600 m

DISTINGUISHING FEATURES *C. diphylla* has dark flowers. The labellum is narrow with recurved sides, and, when flattened, is diamond-shaped. The apical half of the labellum is also heavily encrusted with stalked and sessile calli.

HABITAT Prefers moist heath, heathy woodland and forest. Soils variable, but often grows on sandy loams.

CONSERVATION STATUS Common. Widespread and conserved in national parks and nature reserves. Secure.

DISCUSSION In New South Wales, *C. diphylla* is the most frequently encountered of the bird orchids. Common on the coastal plains, it is still found in remnant bushland in towns and cities. It can form extensive colonies, reproducing vegetatively from daughter tubers formed on the end of stolonoid roots. Also reproduces from seed to form new colonies, as *C. diphylla* is pollinated by males of a species of thynnid wasp. There is some morphological variation throughout the range of this *Chiloglottis*. Further research is required to determine if these are simply local variations or part of a complex of similar species. *C. diphylla* is the type species of the genus *Chiloglottis*.

Chiloglottis diphylla
Oakdale, New South Wales

6 April 1992

A plant
B flower from front
C flower from side
D flower from above
E labellum from side
F labellum flattened
G column from side
H column from front
J dorsal sepal
K lateral sepal
L petal

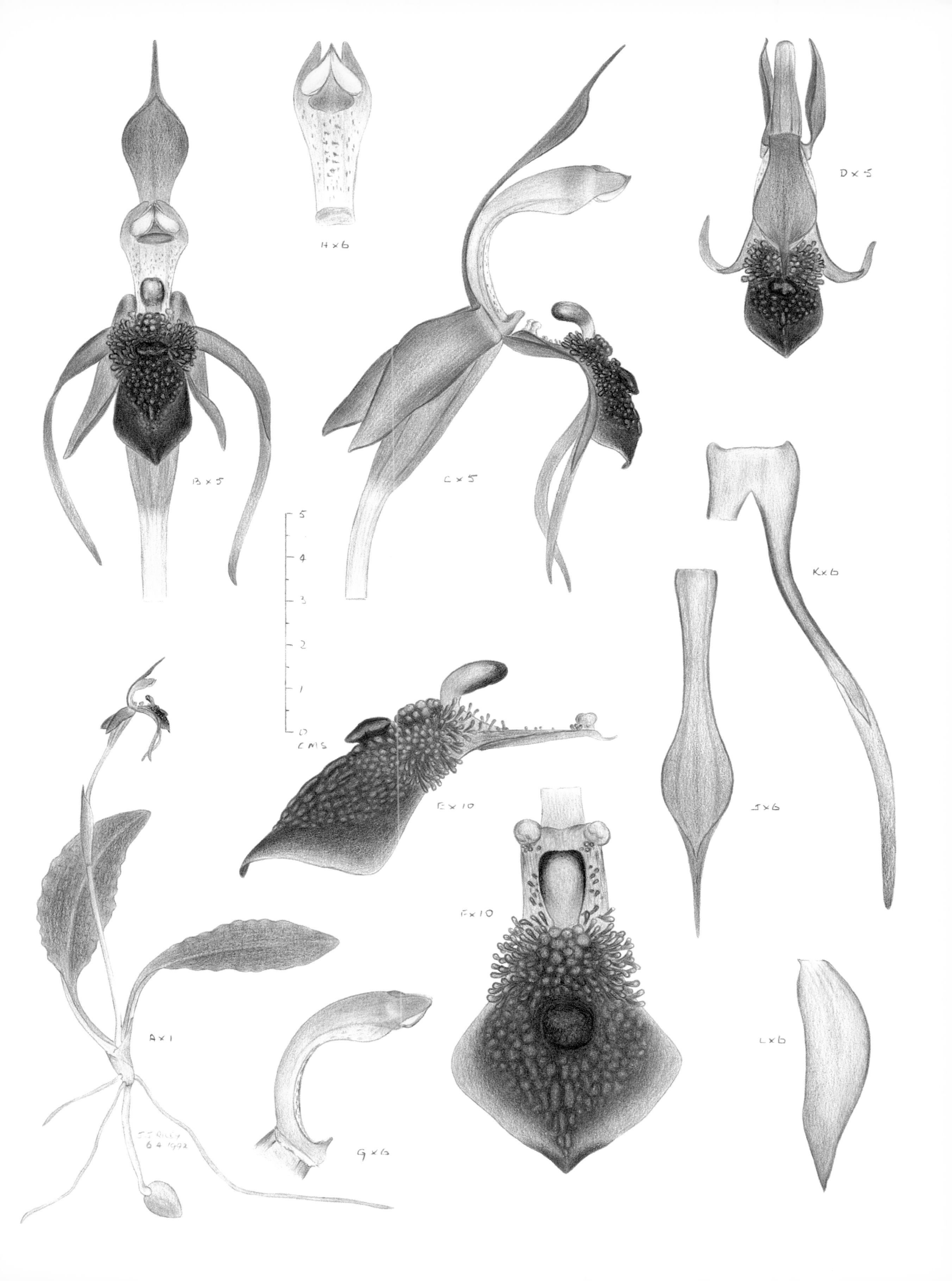
H x 6
B x 5
C x 5
D x 5
K x 6
5
4
3
2
1
0
CMS
E x 10
J x 6
F x 10
A x 1
G x 6
L x 6
6 4 1992

Chiloglottis longiclavata D.L. Jones

1991 | *Australian Orchid Research* 2:38

TYPE LOCALITY Herberton Range, Queensland

ETYMOLOGY *longiclavata* — long clubs

FLOWERING TIME April to June

DISTRIBUTION This is primarily a species from the ranges of North Queensland, being found from Cairns, the Atherton Tableland and Herberton Ranges, south to the Blackdown Tableland.

ALTITUDINAL RANGE 400 m to 700 m

DISTINGUISHING FEATURES Apart from its northern distribution, *C. longiclavata* can be recognised by its green flowers with maroon calli and the dorsal and lateral sepals terminating in long, pronounced osmophores.

HABITAT Found in moist, sheltered areas in open woodland and forest on gravelly clay loams.

CONSERVATION STATUS Uncommon. Conserved in national parks.

DISCUSSION *C. longiclavata* was previously included under *C. diphylla*, but both species are easily distinguished from each other. Plants from the Blackdown Tableland and Eungella differ slightly from those populations further north. Research is needed to determine if this is local variation or a complex of similar species. On Blackdown Tableland, *C. longiclavata* grows in sand often sheltering at the base of large, flat-topped sandstone boulders. This population occurs in a national park and is secure irrespective of its identity. It forms small, scattered colonies, and multiplies vegetatively from daughter tubers formed on the end of stolonoid roots. Also reproduces from seed to form new colonies, and is pollinated by males of a species of thynnid wasp, the result of pseudocopulation. This species is poorly known and rarely encountered as it occurs away from major towns.

Chiloglottis longiclavata
Herberton Range, Queensland

10 May 1994

A plant
B flower from front
C flower from side
D labellum from side
E labellum flattened
F column from side
G column from front
H dorsal sepal
J lateral sepal
K petal

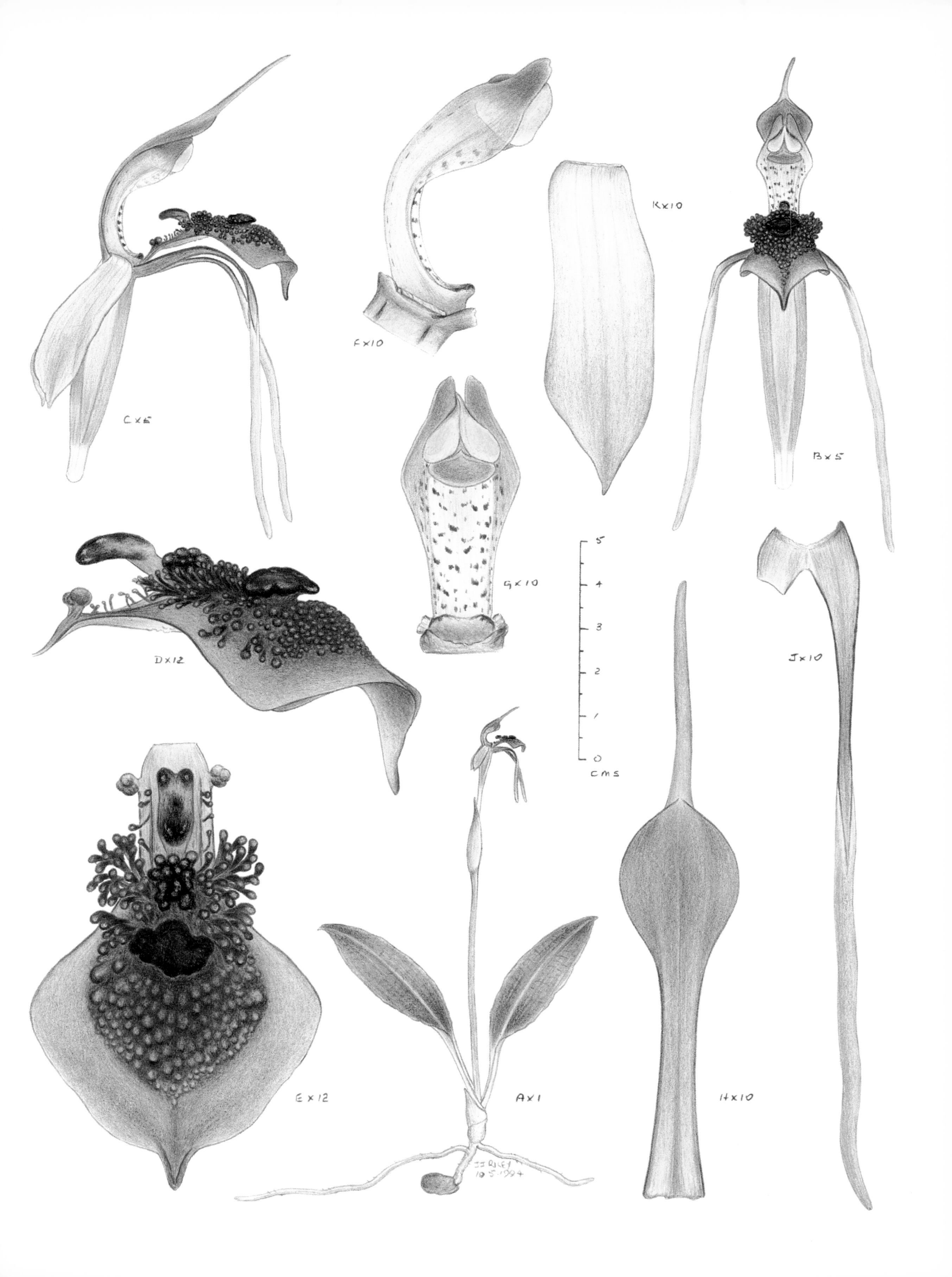
C x6
F x10
K x10
B x5
G x10
D x12
J x10
5
4
3
2
1
0
CMS
E x12
A x1
H x10
JJ RILEY
10.5.1994

Chiloglottis anaticeps D.L. Jones

1991 | *Australian Orchid Research* 2:37

TYPE LOCALITY Forbes River, New South Wales

ETYMOLOGY *anaticeps* — like a duck's head

FLOWERING TIME December to February

DISTRIBUTION Occurs in the mountains west of Wauchope and near Ebor on the New England Tableland, New South Wales.

ALTITUDINAL RANGE 1000 m to 1300 m

DISTINGUISHING FEATURES *C. anaticeps* has a distinctive labellum, with the major stalked calli shaped like a duck's head when viewed from the side. It also has a prominent, large and wide column.

HABITAT This poorly known species grows in open, sparse forest in gravelly soils on rocky slopes, often in moderately exposed situations.

CONSERVATION STATUS Rare. Grows in a national park only at the type-site.

DISCUSSION *C. anaticeps* has a very limited distribution and very few orchid enthusiasts have seen it growing in the wild. The high ranges and plateaus of the upper Forbes River section of the Great Dividing Range have very limited access, which hinders searches for this orchid. It is also rare near Ebor, where the population displays some morphological differences from the Forbes River plants. Further study will be needed to confirm their taxonomic status. It forms small, scattered, loose colonies, and multiplies vegetatively from daughter tubers formed on the end of stolonoid roots. Also reproduces from seed to form new colonies, and is pollinated by males of a species of thynnid wasp, the result of pseudocopulation.

Chiloglottis anaticeps
Werrikimbe National Park, New South Wales

28 February 1995

A plant
B flower from front
C flower from side
D labellum from side
E labellum flattened
F column from side
G column from front
H dorsal sepal
J lateral sepal
K petal

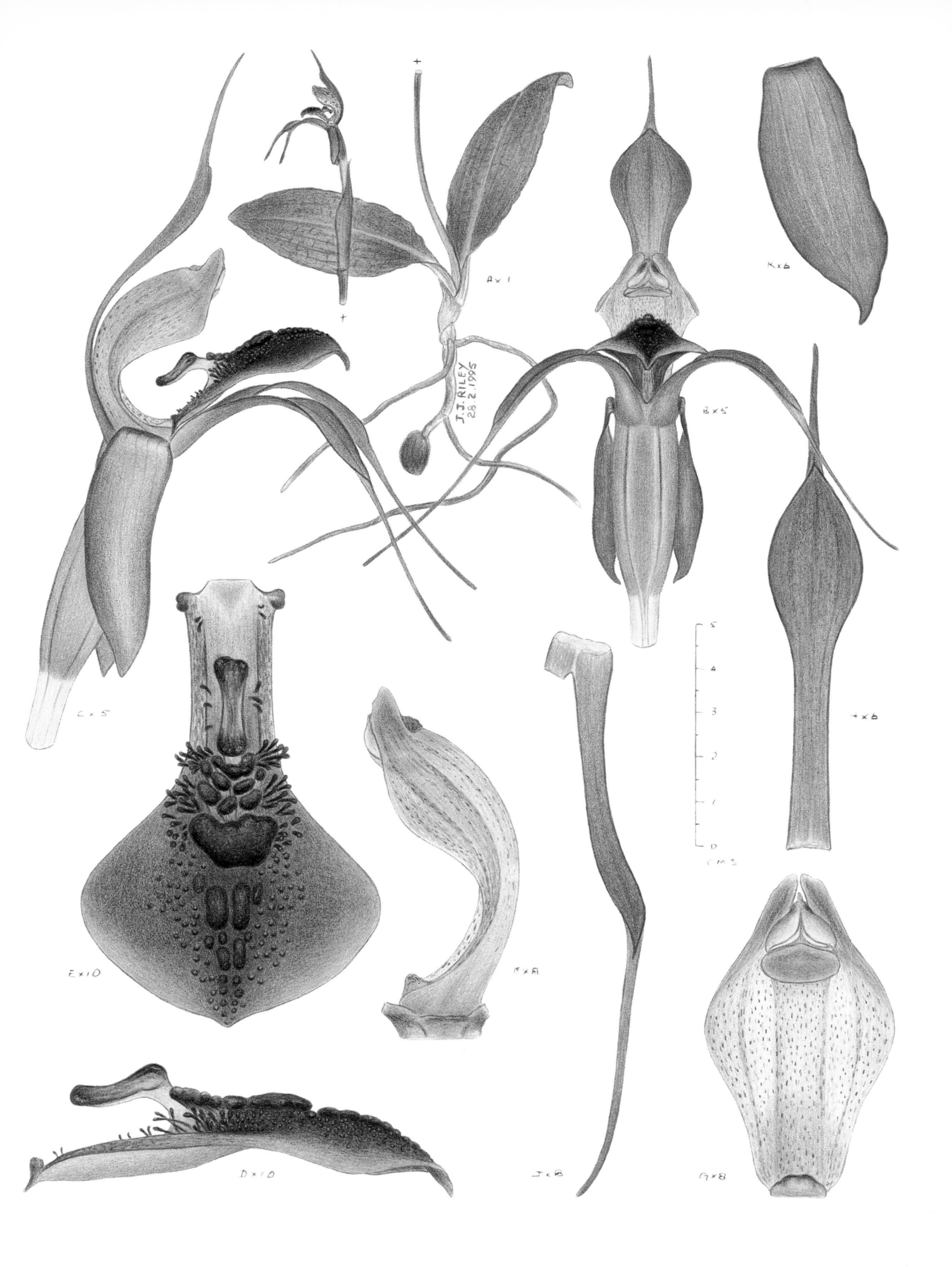

A x 1
J. J. RILEY
28.2.1995
B x 5
K x 6
C x 5
H x 6
CMS
E x 10
F x 8
D x 10
J x 8
G x 8

Chiloglottis sylvestris D.L. Jones & M.A. Clem.

1987 | *Proceedings of the Royal Society of Queensland* 98:123–34

TYPE LOCALITY Springbrook, Queensland

ETYMOLOGY *sylvestris* — dwelling in forest

FLOWERING TIME December to May

DISTRIBUTION South from the Moreton District of south-east Queensland, to around Robertson in New South Wales. It occurs along the coastal plain to the adjacent ranges and tablelands.

ALTITUDINAL RANGE 50 m to 1500 m

DISTINGUISHING FEATURES *C. sylvestris* has a labellum with only a few, poorly developed calli on the apical half. This labellum, when flattened, has a long, narrow distal half and a wide, rounded apical half. The major stalked calli are not notched.

HABITAT Has a preference for moist gullies and slopes in *Eucalyptus* woodland, forest with a ferny understorey, and the margins of rainforest. On the edges of rainforest it can be found on banks or along the edges of tracks or small clearings.

CONSERVATION STATUS Common. Widespread and conserved in national parks and nature reserves. Secure.

DISCUSSION *C. sylvestris* grows in small to large colonies in moist, sheltered areas tucked among ferns and shrubs. It multiplies vegetatively from daughter tubers formed on the end of stolonoid roots. Also reproduces from seed to form new colonies, and is pollinated by males of a species of thynnid wasp, the result of pseudocopulation. As with all other late summer to autumn blooming *Chiloglottis*, large numbers of plants are encountered with only very few flowering. The number of flowering plants increases spectacularly when fires occurred the previous year.

Chiloglottis sylvestris
Washpool National Park, New South Wales

20 February 1995

A plant
B flower from front
C flower from side
D labellum from side
E labellum flattened
F column from side
G column from front
H dorsal sepal
J lateral sepal
K petal

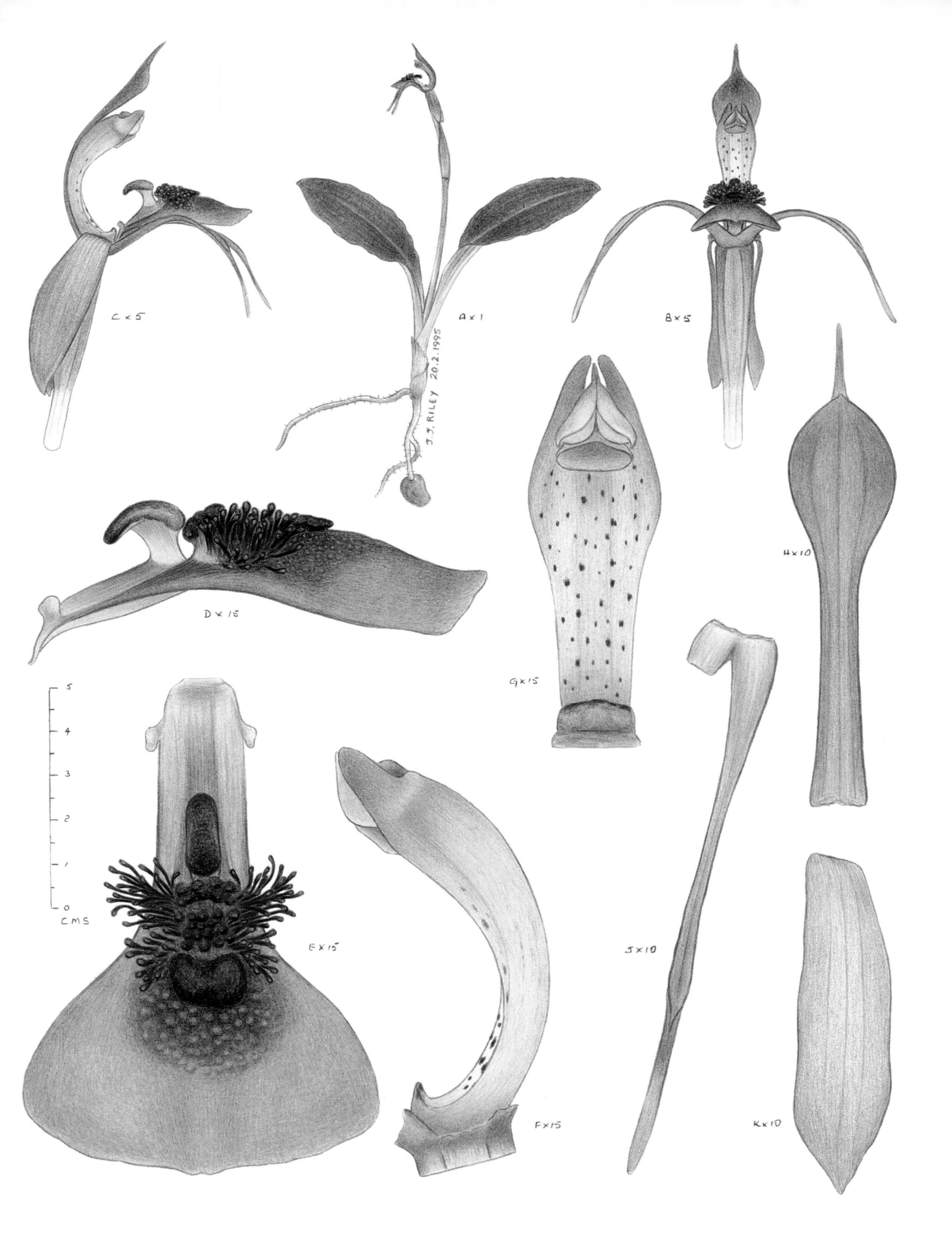
C×5
A×1
B×5
J.J. RILEY 20.2.1995
H×10
D×15
G×15
5
4
3
2
1
0
CMS
E×15
J×10
F×15
K×10

Chiloglottis gunnii Lindl.

1840 | *Genera and Species of Orchidaceous Plants*: 387

TYPE LOCALITY Circular Head, Tasmania

RECENT SYNONYMS *Chiloglottis platychila* G.W. Carr

ETYMOLOGY After Tasmanian naturalist Ronald Campbell Gunn

FLOWERING TIME November to January

DISTRIBUTION Tasmania, excluding the east and north-east.

ALTITUDINAL RANGE 10 m to 450 m

DISTINGUISHING FEATURES True *C. gunnii* has stalked labellum calli that are erect, terminating with a swollen, dark maroon head. The globose head of the main calli is distinctive.

HABITAT Prefers moist forest and woodland and the edges of temperate rainforest. Grows in sheltered positions among shrubs and ferns.

CONSERVATION STATUS Common. Grows in national parks and nature reserves. Secure.

DISCUSSION Until the early 1990s, all large-flowered *Chiloglottis* on the mainland, as well as Tasmania, were referred to as *C. gunnii*. Recent research has shown this species to be endemic to Tasmania. Even in that island state, *C. gunnii* was, and still is, a poorly known taxon, the name being misapplied to other recently named species. This confusion led to *C. gunnii* being described as a new species, *C. platychila*. As with almost all of the *Chiloglottis* species, pseudocopulation by males of a species of thynnid wasp is the means of pollination. Reproduces vegetatively from daughter tubers formed on stolonoid roots and also propagates from seed to form new colonies.

Chiloglottis gunnii
Mount Catamaran, Tasmania

12 January 1996

A plant
B flower from front
C flower from side
D flower from rear
E labellum flattened
F longitudinal section of labellum
G column from side
H column from front
J dorsal sepal
K lateral sepal
L petal

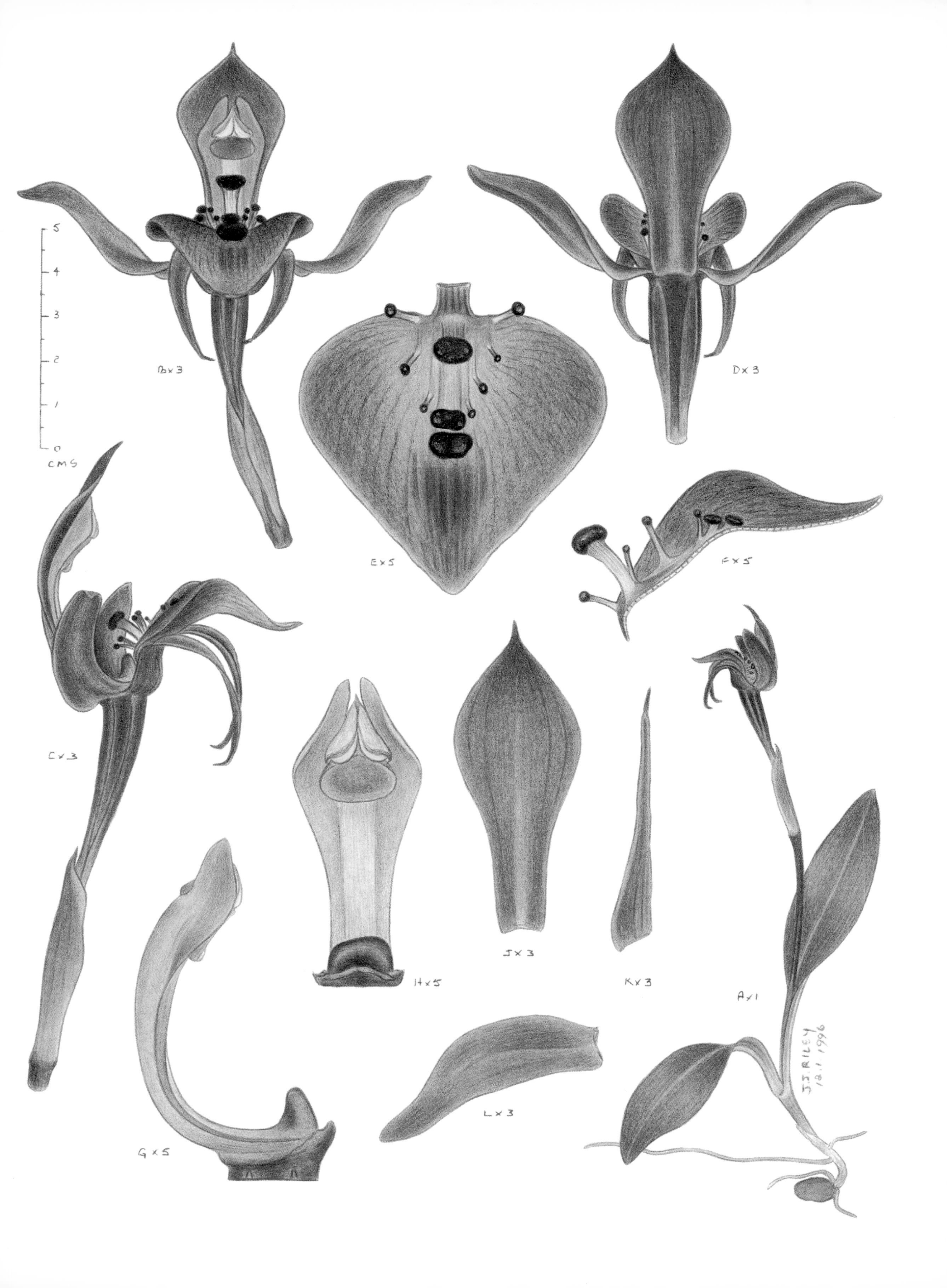

B×3
D×3
E×5
F×5
C×3
H×5
J×3
K×3
A×1
G×5
L×3
CMS
0
1
2
3
4
5
J.J. RILEY
18.1.1996

Chiloglottis grammata G.W. Carr

1991 | *Indigenous Flora & Fauna Association Miscellaneous Paper* 1:20

TYPE LOCALITY Jackeys Marsh, Tasmania

ETYMOLOGY *grammata* — having a labellum with marks like handwriting

FLOWERING TIME October to February

DISTRIBUTION Tasmania, excluding West Coast.

ALTITUDINAL RANGE 500 m to 1200

DISTINGUISHING FEATURES *C. grammata* has flowers that are greenish-maroon, with narrow petals that are spread widely. The labellum is dark with short, stubby, darker calli.

HABITAT Occurs in tall open forest, rainforest and open montane forest. Soils are moist, well drained and loamy.

CONSERVATION STATUS Common. Widespread and represented in national parks and reserves. Secure.

DISCUSSION *C. grammata* is a common species endemic to Tasmania that was mistakenly thought to be *C. gunnii*. It grows in colonies, which can be extensive, in moderately open vegetation and around the base of tall trees, often at high altitude. Sometimes true *C. gunnii* and *C. triceratops* can be found growing with *C. grammata*, but no natural hybrids have been recorded. *C. grammata* is probably the best known of the Tasmanian large bird orchids. Pollination results from pseudocopulation by male thynnid wasps. Also reproduces vegetatively from daughter tubers formed on stolonoid roots.

Chiloglottis grammata
Launceston, Tasmania

5 November 1998

A plant
B flower from front
C flower from side
D flower from rear
E labellum flattened
F longitudinal section of labellum
G column from side
H column from front
J dorsal sepal
K lateral sepal
L petal

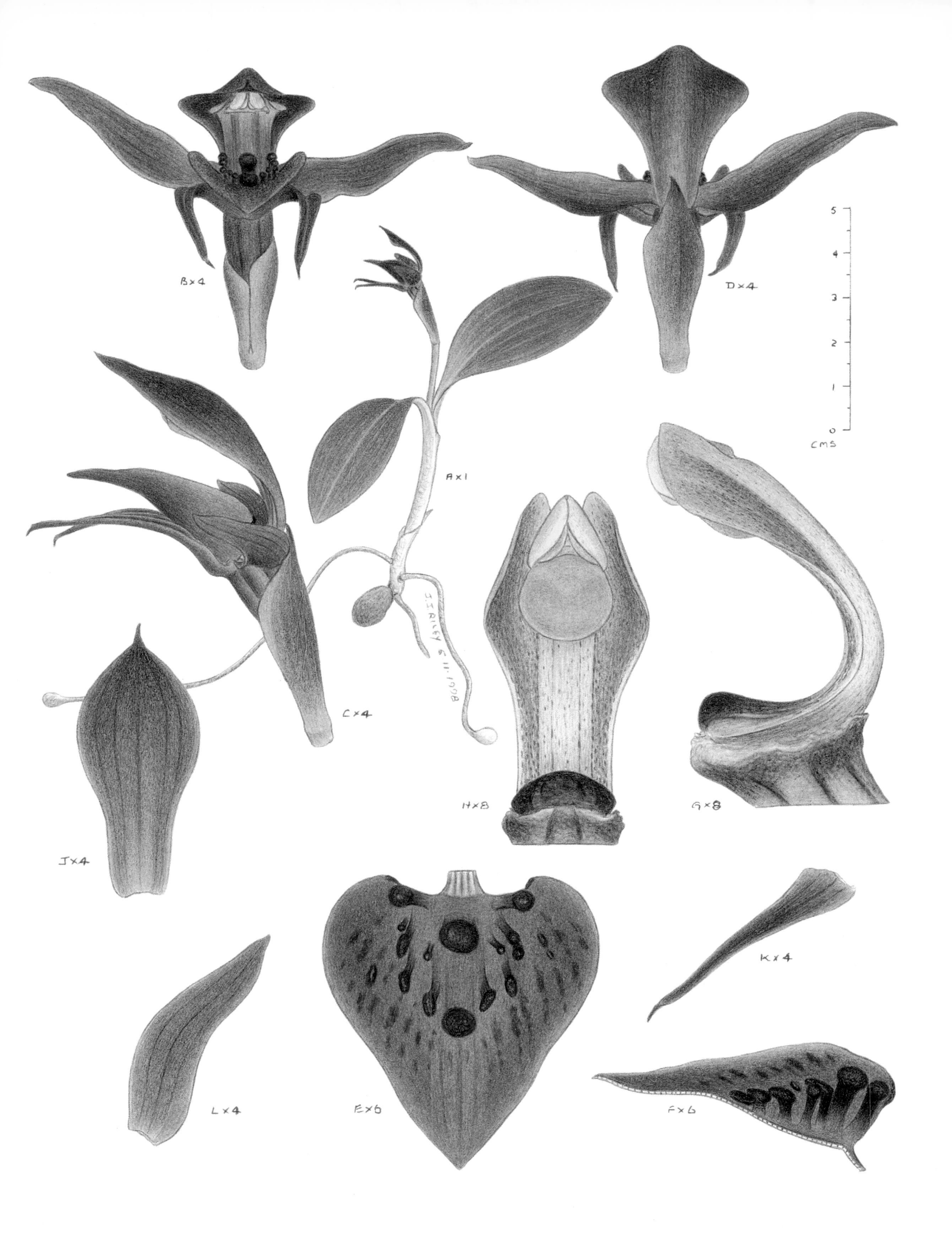
B×4
D×4
5
4
3
2
1
0
CMS
A×1
J.I. RILEY 6.11.1998
C×4
H×8
G×8
J×4
K×4
L×4
E×6
F×6

Chiloglottis valida D.L. Jones

1991 | *Australian Orchid Research* 2:43

TYPE LOCALITY Near Ginini Flats, Australian Capital Territory

ETYMOLOGY *validus* — strong, robust

FLOWERING TIME October to January

DISTRIBUTION Occurs in the Brindabella Range in the Australian Capital Territory, southern alps and surrounding mountains of New South Wales, and in Victoria on the higher eastern parts of the Great Dividing Range.

ALTITUDINAL RANGE 500 m to 1500 m

DISTINGUISHING FEATURES *C. valida* has a broad dorsal sepal and wide, incurved petals. It has a broad labellum with few stalked calli.

HABITAT Grows in moist, open montane forest, in patches of snow gums (*Eucalyptus pauciflora*) and surrounding frost hollows in subalpine areas. Also along creek banks, often in *Sphagnum* moss hummocks.

CONSERVATION STATUS Common. Widespread and represented in national parks and reserves. Secure.

DISCUSSION Up until 1991, this species was interpreted as *C. gunnii* (a Tasmanian endemic). Recent research has shown many populations known as *C. valida* are, in fact, unnamed species. Because of this, the exact distribution of the species is unclear. Our distribution notes cover areas known to contain *C. valida* in the strict sense. This species is common and forms extensive colonies in moist, open forest. It seems to flower readily, the number of blooming plants in large colonies being quite high. Pollination results from pseudocopulation by male thynnid wasps. Also reproduces vegetatively from daughter tubers formed on stolonoid roots.

Chiloglottis valida
Mount Selwyn, New South Wales

16 December 1993

A plant
B flower from front
C flower from side
D flower from rear
E labellum flattened
F longitudinal section of labellum
G column from side
H column from front
J dorsal sepal
K lateral sepal
L petal

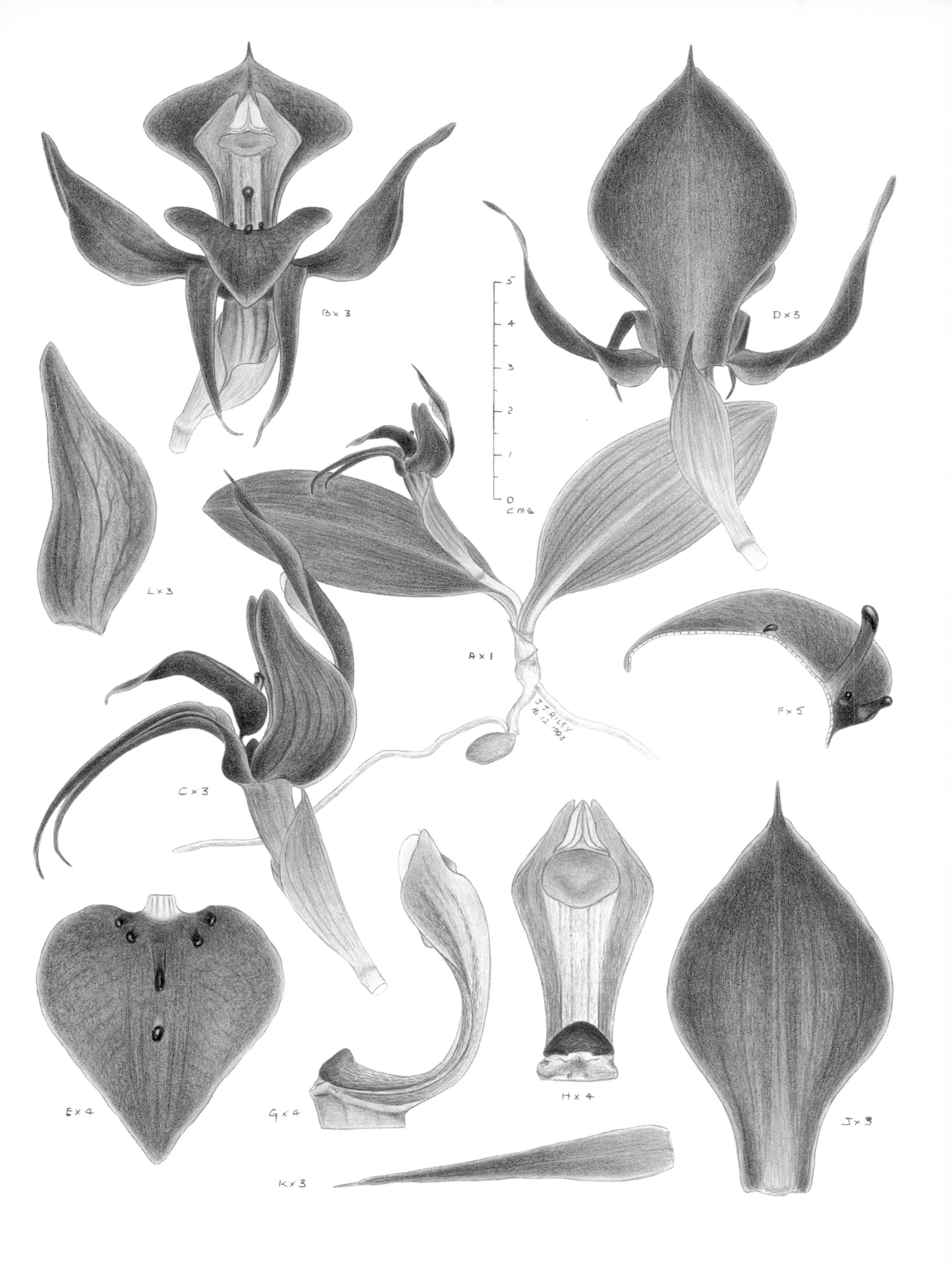
B x 3
D x 3
5
4
3
2
1
0
CMS
L x 3
A x 1
F x 5
J.J. RILEY
16.12.1993
C x 3
E x 4
G x 4
H x 4
J x 3
K x 3

Chiloglottis pluricallata D.L. Jones

1991 | *Australian Orchid Research* 2:40

TYPE LOCALITY Point Lookout, New South Wales

ETYMOLOGY *pluricallata* — many glands

FLOWERING TIME November to January

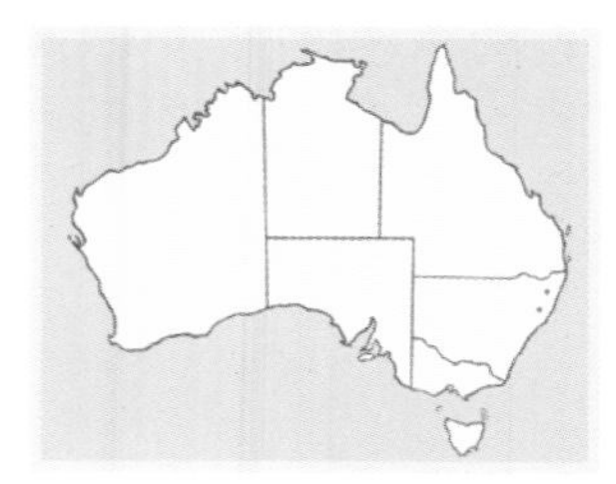

DISTRIBUTION *C. pluricallata* is endemic to the Barrington Tops and higher eastern parts of the New England Tableland, New South Wales.

ALTITUDINAL RANGE 900 m to 1560 m

DISTINGUISHING FEATURES *C. pluricallata* has a narrow dorsal sepal and widely spreading petals. The labellum, when flattened, is longer that wide. Calli are in two rows, the basal two flat-sided and distinctively bent.

HABITAT Grows in montane and sub-alpine, moist, tall forest with a grass, fern and shrub understorey. Soils are well structured loams.

CONSERVATION STATUS Represented in national parks. Secure.

DISCUSSION An unnamed species with affinities to *C. pluricallata* also grows on Barrington Tops. Another similar unnamed taxon occurs on the New England Tableland. Both of these yet to be named species differ from *C. pluricallata* in having more numerous multi-stalked calli in four rows. Both the unnamed taxa grow in, or very close to, the edge of swamps, creeks or very wet areas, often in *Sphagnum* moss hummocks. *C. pluricallata* grows in drier conditions among grasses, ferns and shrubs in open forest or grassland. This species flowers freely and can be locally common. Pollination results from pseudocopulation by male thynnid wasps. Also reproduces vegetatively from daughter tubers formed on stolonoid roots.

Chiloglottis pluricallata
Barrington Tops, New South Wales

8 December 1998

A plant
B flower from front
C flower from side
D flower from rear
E labellum flattened
F longitudinal section of labellum
G column from side
H column from front
J dorsal sepal
K lateral sepal
L petal

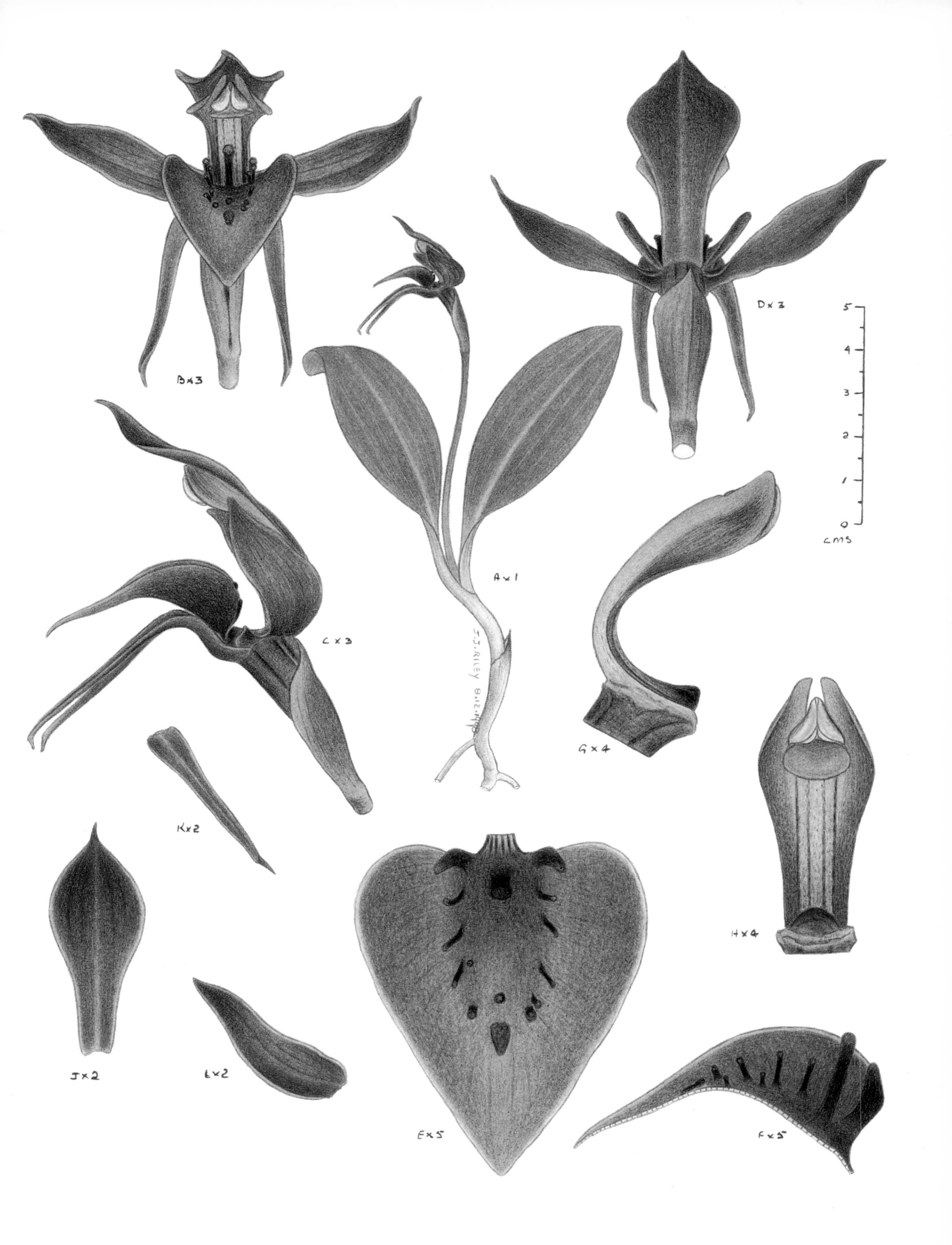

B×3
D×3
A×1
C×3
G×4
K×2
H×4
J×2
L×2
E×5
F×5
5
4
3
2
1
0
cms

Chiloglottis jeanesii D.L. Jones

1997 | *Orchadian* 12:233

TYPE LOCALITY Toorongo, Victoria

ETYMOLOGY After Jeffrey Alan Jeanes, co-author of *The Orchids of Victoria*

FLOWERING TIME December to February

DISTRIBUTION *C. jeanesii* is of restricted distribution in the Eastern Highlands of Victoria, found around Sherbrooke Forest to the Baw Baw plateau.

ALTITUDINAL RANGE 800 m to 1500 m

DISTINGUISHING FEATURES *C. jeanesii* has small flowers with a narrow dorsal sepal. The petals are very narrow and widely spreading. The labellum has sparse, narrow, dark calli.

HABITAT Found in montane, wet, tall mountain ash (*Eucalyptus regnans*) forest and forest dominated by beech trees, (*Nothofagus cunninghamii*).

CONSERVATION STATUS Uncommon. Of limited distribution in national parks and reserves. Secure.

DISCUSSION *C. jeanesii* is a poorly known species. As it becomes more familiar to botanists and orchid enthusiasts, its restricted distribution may be expanded. This is the smallest species in the *C. gunnii* complex of bird orchids. Forms small to medium colonies, growing in leaf litter among ferns and grasses. Pollination results from pseudocopulation by male thynnid wasps. Also reproduces vegetatively from daughter tubers formed on stolonoid roots.

Chiloglottis jeanesii
Toorongo, Victoria

18 January 1997

A plant
B flower from side
C flower from front
D flower from rear
E labellum flattened
F longitudinal section of labellum
G column from front
H column from side
J dorsal sepal
K lateral sepal
L petal

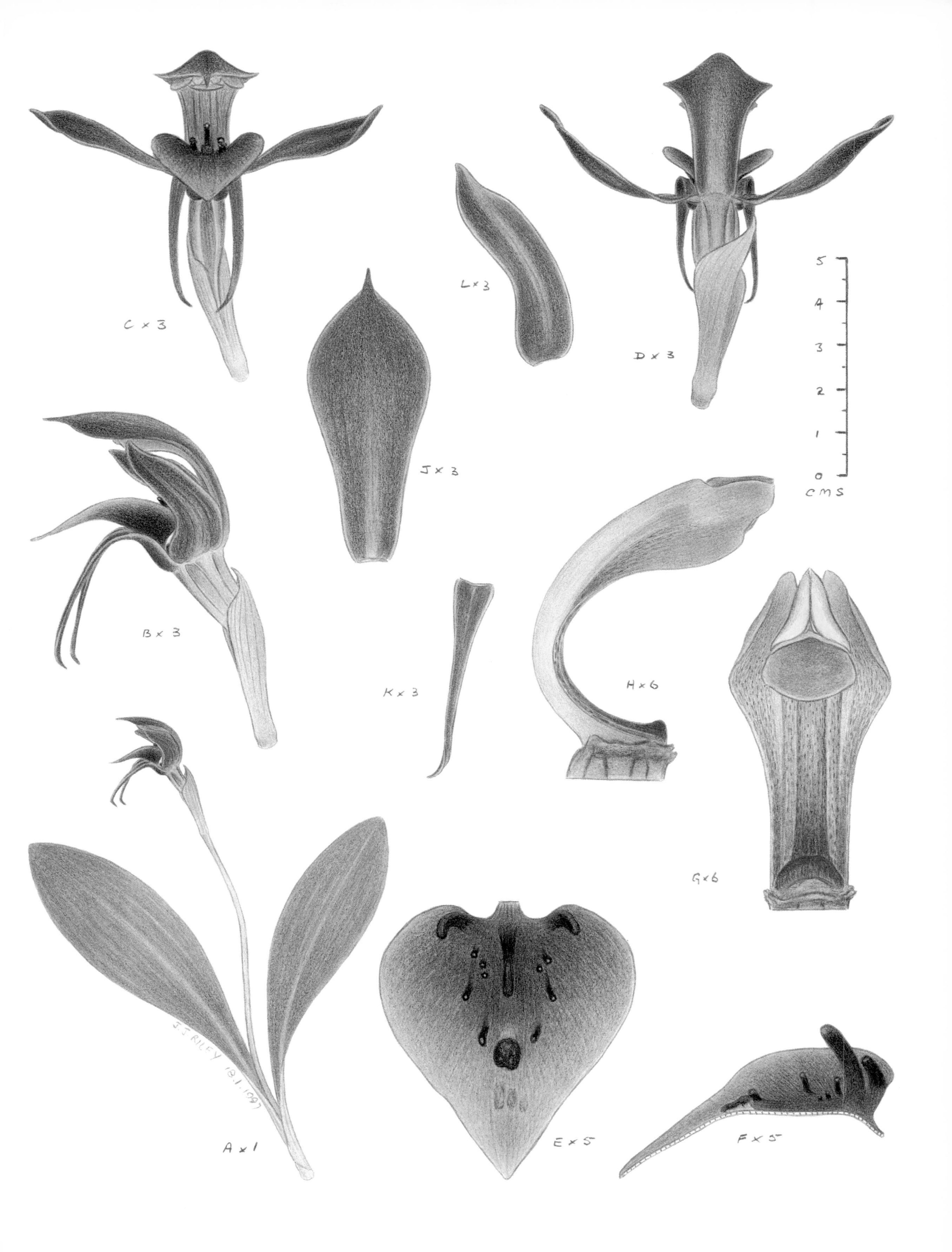
C × 3
L × 3
D × 3
J × 3
B × 3
K × 3
H × 6
G × 6
A × 1
E × 5
F × 5
5
4
3
2
1
0
cms
J.J. RILEY 18.1.1997

Chiloglottis triceratops D.L. Jones

1998 | *Australian Orchid Research* 3:66

TYPE LOCALITY Coquette Creek, Tasmania

ETYMOLOGY *triceratops* — three-horned

FLOWERING TIME August to December

DISTRIBUTION Widespread throughout Tasmania, but is not recorded from the West Coast.

ALTITUDINAL RANGE 10 m to 800 m

DISTINGUISHING FEATURES *C. triceratops* has greenish-brown to reddish-brown flowers with narrow spreading petals. The arrangement and shape of the three prominent basal calli is distinctive.

HABITAT Favours moist, moderately sheltered areas in open forest with an understorey of heath, small shrubs and grasses.

CONSERVATION STATUS Common. Occurs in national parks and nature reserves. Secure.

DISCUSSION *C. triceratops* is an early-flowering, widespread and generally lowland species that can be locally common. It grows close to the coast, mainly in the south-east. In more montane areas, it can be found with *C. gunnii* and *C. grammata*. It is reasonably free-flowering. In suitable conditions, it forms moderately large or scattered colonies vegetatively from daughter tubers formed on the end of stolonoid roots. As with most *Chiloglottis* species, pollination results from pseudocopulation by male thynnid wasps.

Chiloglottis triceratops
Tinderbox, Tasmania

30 September 1995

A plant
B flower from front
C flower from side
D flower from rear
E labellum flattened
F longitudinal section of labellum
G column from side
H column from front
J dorsal sepal
K lateral sepal
L petal

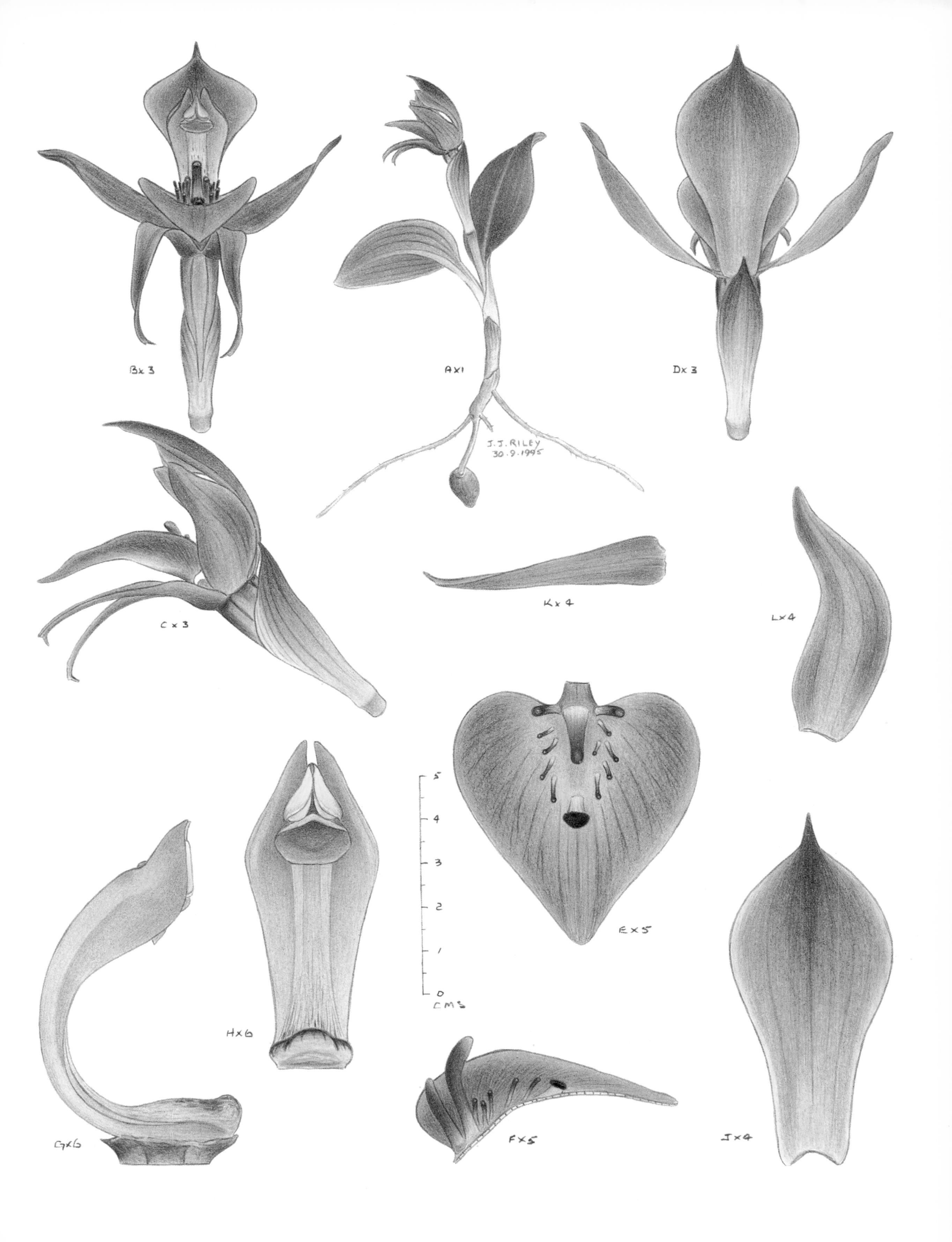
B x 3
A x1
D x 3
J. J. RILEY
30.9.1995
C x 3
K x 4
L x 4
E x 5
H x 6
G x 6
F x 5
J x 4
CMS
5
4
3
2
1
0

Chiloglottis turfosa D.L. Jones

1991 | *Australian Orchid Research* 2:43

TYPE LOCALITY Kosciuszko National Park, New South Wales

ETYMOLOGY *turfosa* — growing in peat bogs

FLOWERING TIME November to December

DISTRIBUTION This alpine species is found around Mount Selwyn, Kiandra and Tantangara in the Kosciuszko National Park and Snowy Mountains of New South Wales.

ALTITUDINAL RANGE 1300 m to 1600 m

DISTINGUISHING FEATURES *C. turfosa* has small, almost round leaves. The labellum has numerous, densely packed, stout calli in four rows.

HABITAT Found mostly in moist, heavy peaty soils beside small streams and drainage areas among low shrubs and sedgy grasses, rarely on the edges of open *Eucalyptus* forest.

CONSERVATION STATUS Uncommon. Occurring within a national park. Secure.

DISCUSSION Due to its distinctive leaves, *C. turfosa* is the only *Chiloglottis* that can be positively identified out of bloom. While mostly growing in scattered groups, occasionally extensive colonies are encountered. Pollination results from pseudocopulation by male thynnid wasps. Also reproduces vegetatively from daughter tubers formed at the end of stolonoid roots. *C. turfosa* is a poorly known species and the extent of its natural distribution is yet to be determined.

Chiloglottis turfosa
Tantangara, New South Wales

26 December 1992

A plant
B flower from front
C flower from side
D flower from rear
E flower from above
F labellum flattened
G longitudinal section of labellum
H column from side
J column from front
K dorsal sepal
L lateral sepal
M petal

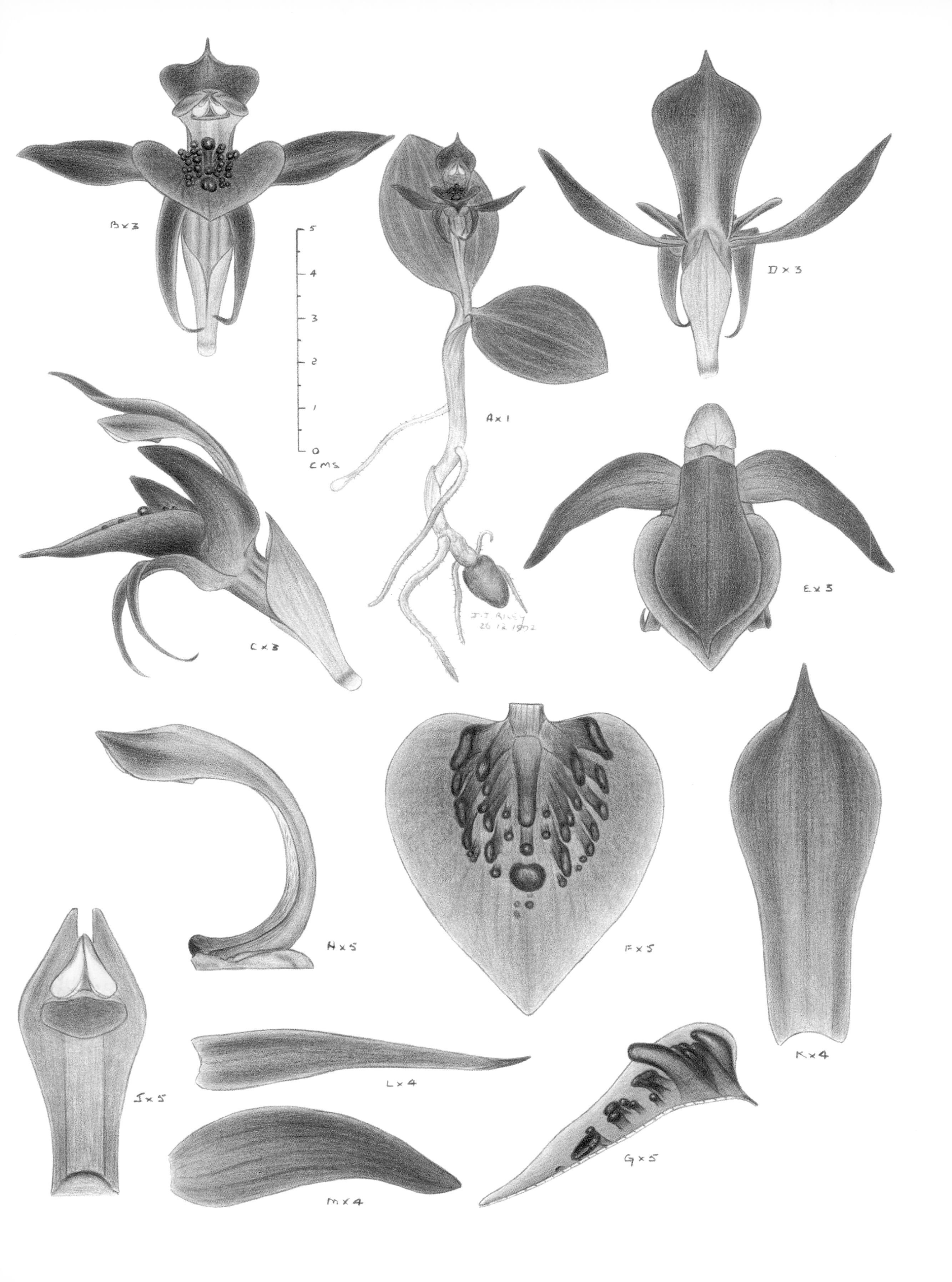
B x 3
5
4
3
2
1
0
CMS
A x 1
D x 3
C x 3
E x 3
J.J. RILEY
26 12 1992
H x 5
F x 5
K x 4
J x 5
L x 4
M x 4
G x 5

Corybas aconitiflorus Salisb.

1807 | *Paradisus Londinensis* 2:83

TYPE LOCALITY Sydney, New South Wales

ETYMOLOGY *aconitiflorus* — flowers like an *Aconitum*

FLOWERING TIME March to July

DISTRIBUTION South from Fraser Island and the adjacent mainland, Queensland, along coastal New South Wales and nearby ranges, to eastern Victoria, to the north and eastern coasts of Tasmania.

ALTITUDINAL RANGE 10 m to 800 m

DISTINGUISHING FEATURES *C. aconitiflorus* has a round, ground-hugging leaf that is dark green above and deep maroon underneath. The flower, which is pale greenish-grey, has a prominent, oval-shaped labellum. The single bloom sits on the leaf.

HABITAT This species grows in a variety of habitats, from moist heath and shrubby woodland to sheltered areas in open forest. It grows in a variety of soil types.

CONSERVATION STATUS Common. Widespread and conserved in national parks and nature reserves. Secure.

DISCUSSION Over its vast range, research may show that *C. aconitiflorus* comprises a complex of related taxa. Some areas have plants with small, narrow, reddish flowers, while in other locations the blooms are a more robust pinkish-grey. On the Cumberland Plain and Blue Mountains around Sydney, two forms occur. Both have bluish grey-green flowers. One grows in drier more exposed conditions and has larger and wider flowers. The other prefers a moister and more sheltered habitat and has smaller and narrower flowers. The illustration represents this form. *C. aconitiflorus* reproduces vegetatively from daughter tubers on the end of stolonoid roots, often forming extensive colonies. Fungus gnats pollinate the flowers. *C. aconitiflorus* is the type species for the genus *Corybas*.

Corybas aconitiflorus
Mount Keira, New South Wales

1 July 1998

A plant from side
B plant from above
C flower from front
D flower from side
E flower from rear
F longitudinal section of flower
G labellum from side
H labellum from front
J labellum from rear
K column from side
L column from front

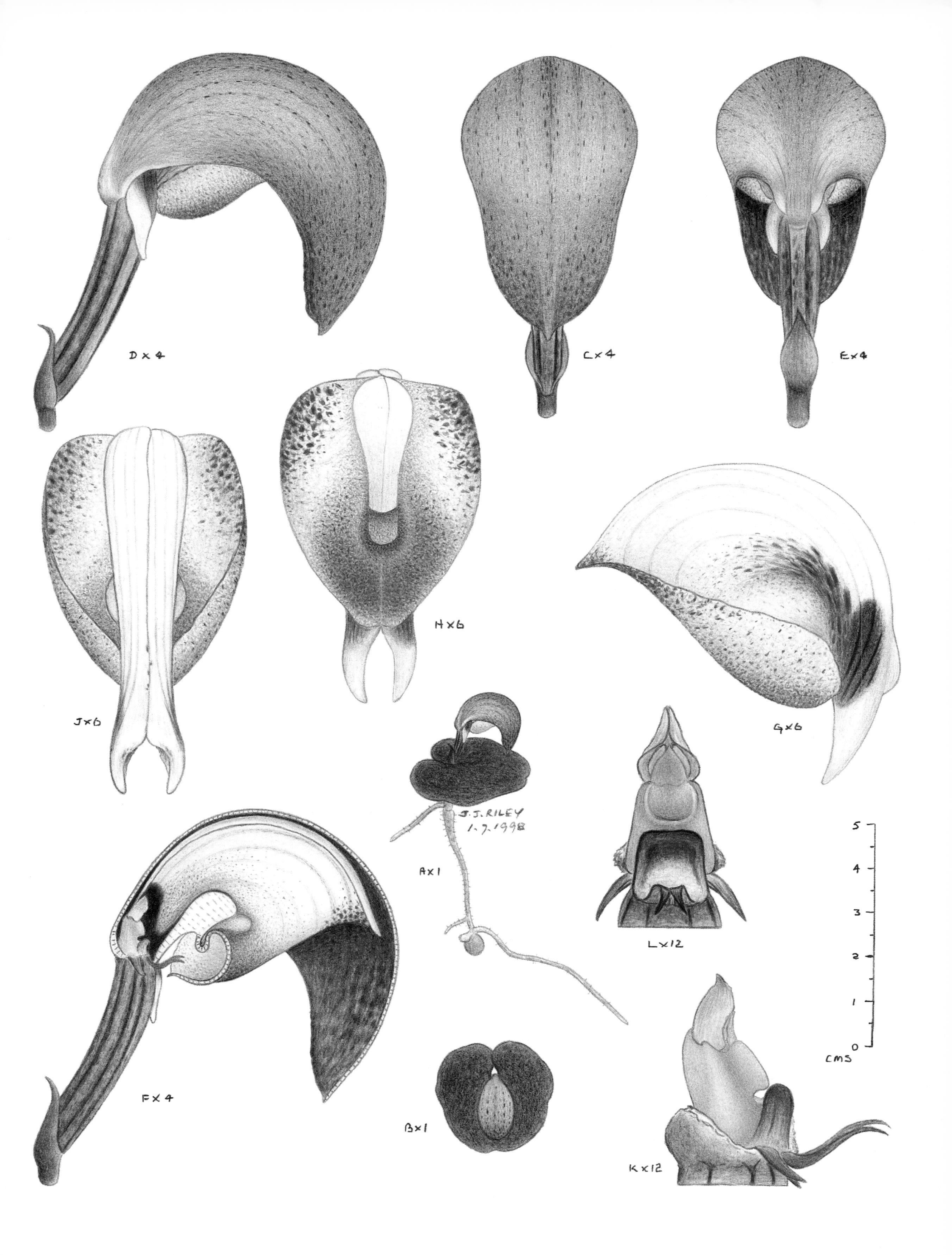
D x 4
C x 4
E x 4
H x 6
J x 6
G x 6
J. J. RILEY
1. 7. 1998
A x 1
L x 12
5
4
3
2
1
0
CMS
F x 4
B x 1
K x 12

Corybas barbarae D.L. Jones

1988 | *Austrobaileya* 2(5):548

TYPE LOCALITY Mount Tamborine, Queensland

ETYMOLOGY After Barbara Jones

FLOWERING TIME April to July

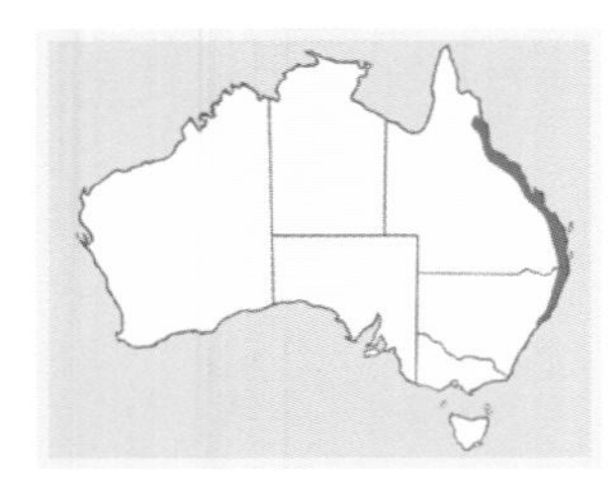

DISTRIBUTION Found from the Atherton Tableland in North Queensland, south along the coast and adjacent ranges into New South Wales as far south as the northern suburbs of Sydney.

ALTITUDINAL RANGE 10 m to 900 m

DISTINGUISHING FEATURES *C. barbarae* has a round, ground-hugging leaf that is green above and reddish-purple underneath. The flower has a prominent white labellum. The single bloom sits on the leaf.

HABITAT Grows in sheltered areas in open woodland and forest.

CONSERVATION STATUS Common. Conserved in national parks and nature reserves. Secure.

DISCUSSION *C. barbarae* is a very distinctive species with its crystalline, bulbous, white flowers. Although small, it very obviously resembles small button mushrooms scattered among the leaf litter. This orchid is not common in the southern part of its range, and is almost certainly extinct in the Sydney area. In Queensland, there is obvious morphological variation over this species' range. Further research will determine if there is more than one species involved. *C. barbarae* forms scattered colonies and reproduces vegetatively from daughter tubers on the end of stolonoid roots. Fungus gnats pollinate the flowers.

Corybas barbarae
Wardell, New South Wales

6 June 1992

A plant from side
B plant from above
C flower from front
D flower from side
E flower from rear
F longitudinal section of flower
G labellum from side
H labellum from front
J labellum from rear
K column from side
L column from front

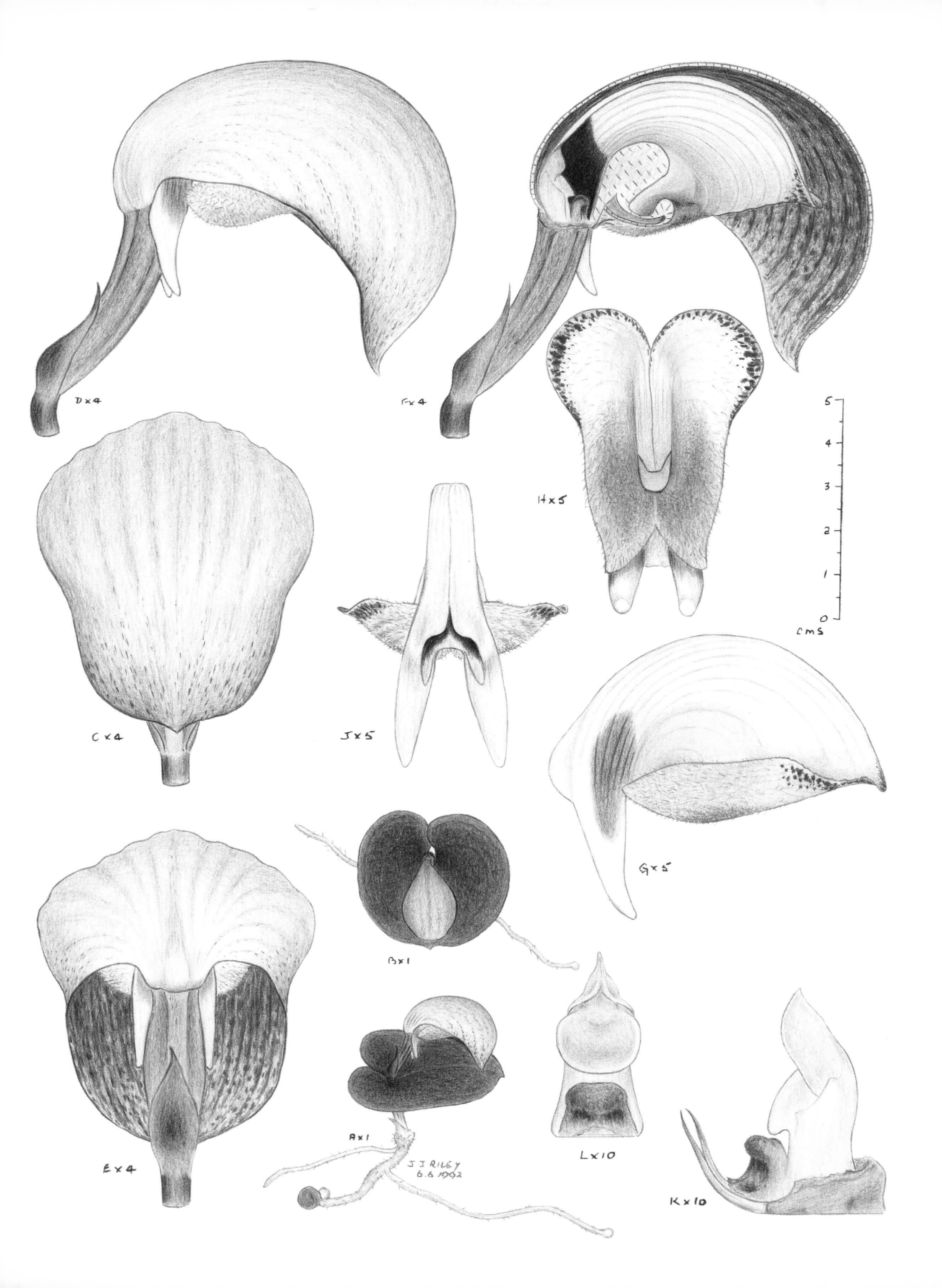

D x 4
F x 4
C x 4
J x 5
H x 5
G x 5
B x 1
A x 1
E x 4
L x 10
K x 10
5
4
3
2
1
0
CMS
J J RILEY
6.6.1992

Corybas pruinosus (Cunn.) Rchb. f.

1871 | *Beiträge zur Systematischen Pflanzenkunde*: 43

TYPE LOCALITY Parramatta, New South Wales

ETYMOLOGY *pruinosus* — frosted

FLOWERING TIME April to July

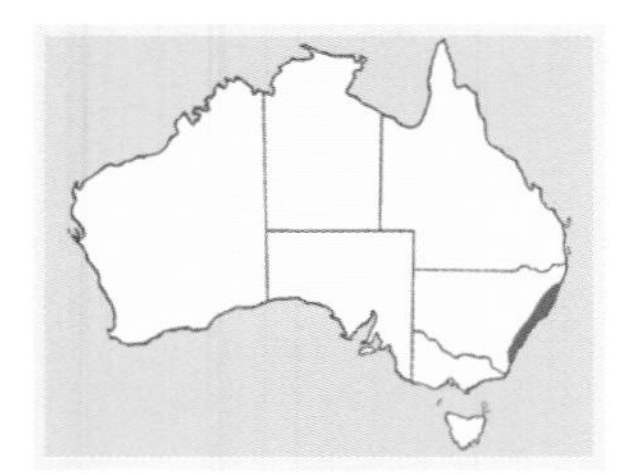

DISTRIBUTION This attractive species is endemic to New South Wales. It is found along the coast and nearby ranges from Moruya north to Coffs Harbour.

ALTITUDINAL RANGE 10 m to 700 m

DISTINGUISHING FEATURES *C. pruinosus* has a narrow, opaque, green dorsal sepal with some reddish-maroon spotting and streaking. The colours of the flowers glisten and have a frosted appearance.

HABITAT This species grows in moist woodland and forest with an understorey of ferns and shrubby plants. Soils are sand to sandy, clay loams.

CONSERVATION STATUS Common. Conserved in national parks and nature reserves. Secure.

DISCUSSION When growing in decaying leaf litter and tucked under ferns and other herbs in ideal conditions, *C. pruinosus* forms large dense colonies. With its free-flowering habit, it is almost impossible to negotiate these masses without treading on blooms. The distinctive, green dorsal sepal, contrasting with the fringed, reddish-maroon labellum, make this orchid a small, but very attractive species. It reproduces vegetatively from daughter tubers on the end of stolonoid roots. Was recently reinstated as *Corysanthes pruinosa* Cunn.

Corybas pruinosus
Nattai, New South Wales

17 May 1992

A plant
B flower from front
C flower from side
D flower from rear
E flower from above
F longitudinal section of flower
G labellum from side
H labellum from rear
J column from side
K column from front

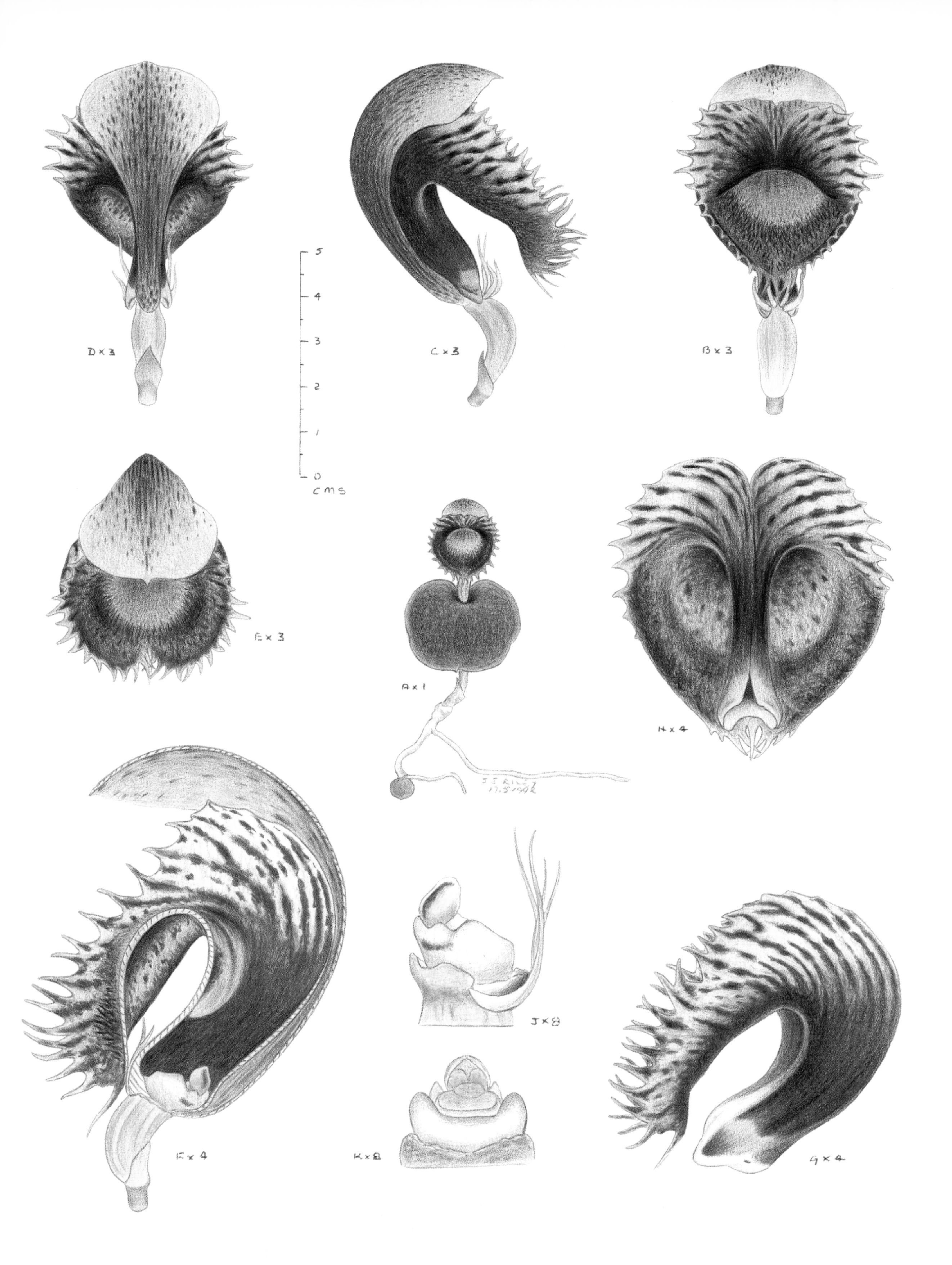
D×3
C×3
B×3
5
4
3
2
1
0
cms
E×3
A×1
H×4
F×4
J×8
K×8
G×4

Cryptostylis subulata (Labill.) Rchb. f.

1871 | *Beiträge zur Systematischen Pflanzenkunde*: 15

TYPE LOCALITY Tasmania

ETYMOLOGY *subulata* — awl-shaped

FLOWERING TIME August to April

DISTRIBUTION Found along the coast and adjacent ranges from the Paluma Range in Queensland, along the New South Wales coast to Victoria, and south-east South Australia, Kangaroo Island, the Bass Strait Islands and Tasmania.

ALTITUDINAL RANGE Up to 1000 m

DISTINGUISHING FEATURES *C. subulata* is an evergreen terrestrial orchid with long, narrow, lanceolate leaves which are green on both sides. The flowers have a large, mainly red labellum with a prominent, dark, bilobed gland.

HABITAT Grows in a wide variety of plant communities, from coastal heath, wet sclerophyll to open woodland and forest. Favours moist or seasonally wet sites, but also grows in drier areas. Often grows in poor soils with insectivorous sundews (*Drosera* spp.).

CONSERVATION STATUS This species is vulnerable in South Australia. Common elsewhere, growing in national parks and nature reserves. Secure.

DISCUSSION This tall, impressive orchid is the most common and familiar *Cryptostylis* species in eastern Australia. As the flowers open progressively, a robust plant may be in bloom for many months. It forms clumps and colonies of many individuals under favourable conditions. Male *Lissopimpla excelsa* wasps pollinate all of the five Australian species of *Cryptostylis*. In some locations, up to three or four different *Cryptostylis* species grow together and overlap with their blooming times. They must have some inhibiting mechanism preventing hybridising, despite pseudocopulation by males of the same wasp species. There have been no natural hybrids recorded, and interestingly, all man-made attempts to breed between different species have been unsuccessful.

Cryptostylis subulata
Chain Valley Bay, New South Wales

25 November 2000

A plant
B flower from side
C flower from front
D labellum flattened
E column from side
F column from front

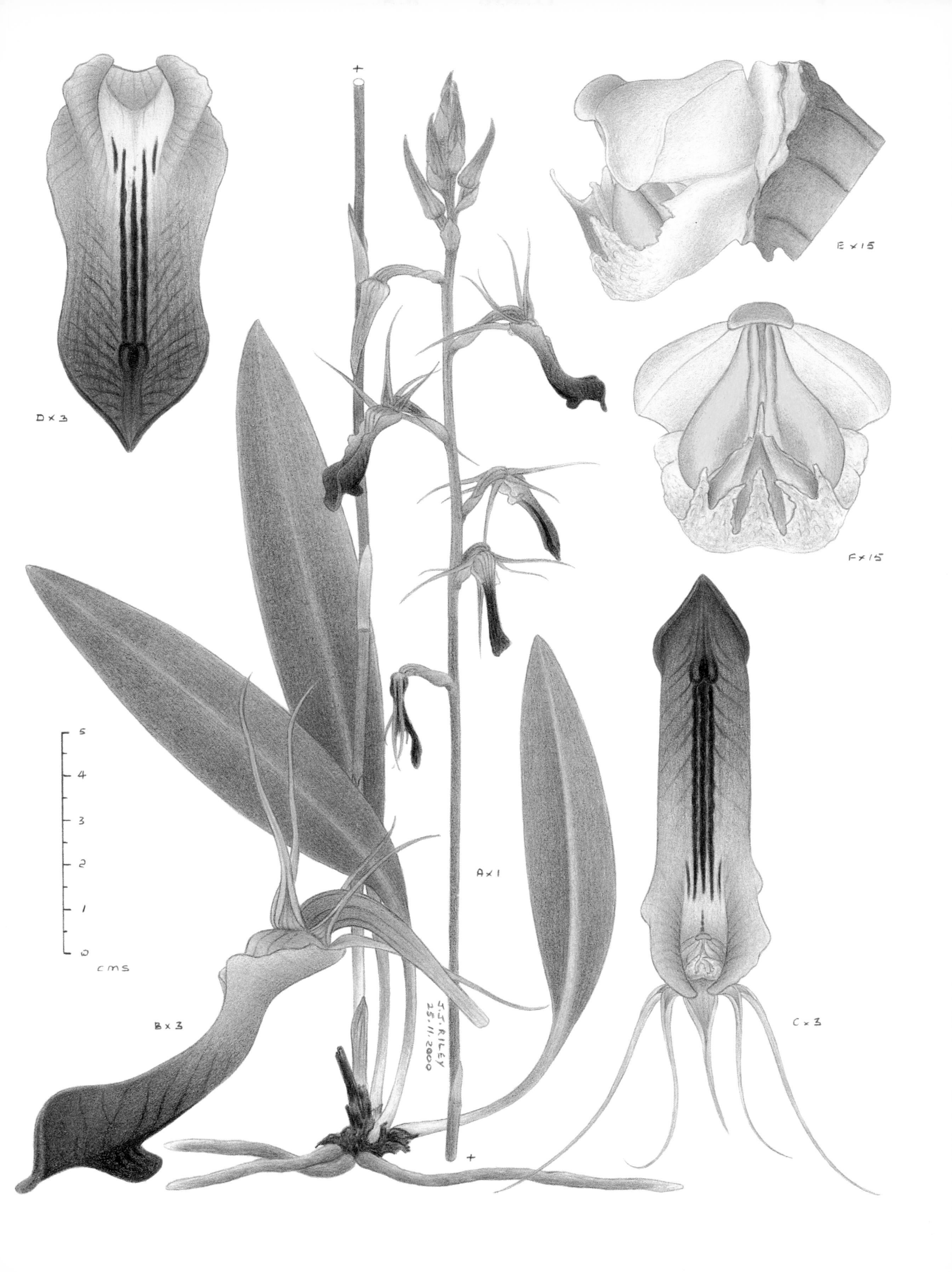
D × 3
E × 15
F × 15
A × 1
5
4
3
2
1
0
cms
B × 3
C × 3
J.J. RILEY
25.11.2000

Cryptostylis ovata R. Br.

1810 | *Prodromus Florae Novae Hollandiae*: 317

TYPE LOCALITY King George Sound, Western Australia

ETYMOLOGY *ovata* — egg-shaped

FLOWERING TIME November to April

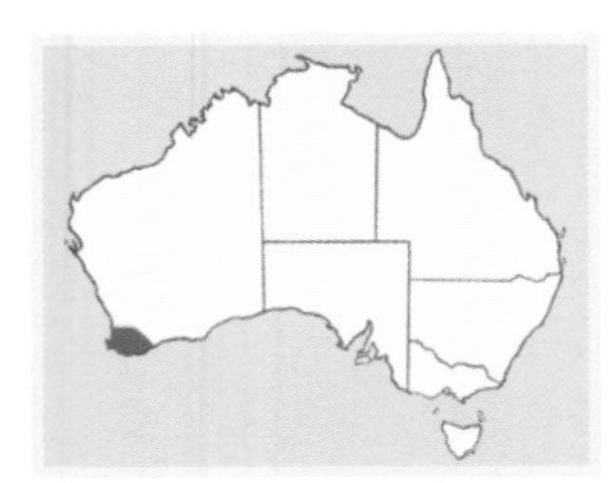

DISTRIBUTION This robust species is endemic to coastal parts of south-west Western Australia, from near Perth to Albany.

ALTITUDINAL RANGE Up to 150 m

DISTINGUISHING FEATURES *C. ovata* is the only member of this genus from Western Australia. It has a broad, green leaf with a conspicuous, pale green midrib, and reddish-maroon underneath. The labellum is mostly red with prominent darker veining.

HABITAT Found in a number of habitats including coastal heath and scrubland to woodland and moist forest. It is also found growing around swamps and seasonally moist areas, but can also grow in moderately dry sites.

CONSERVATION STATUS Locally common. Conserved in national parks and nature reserves. Secure.

DISCUSSION *C. ovata* is Western Australia's only evergreen terrestrial orchid. As the flowers open progressively, a robust plant may be in bloom for many weeks. It forms clumps and colonies of many individuals, reproducing vegetatively by producing new plants at the ends of root-like tubers. Also reproduces from seed, as this species is pollinated by male *Lissopimpla excelsa* wasps during pseudocopulation.

Cryptostylis ovata
Kwinana, Western Australia

29 November 1993

A plant
B flower from front
C flower from side
D column from side
E column from front

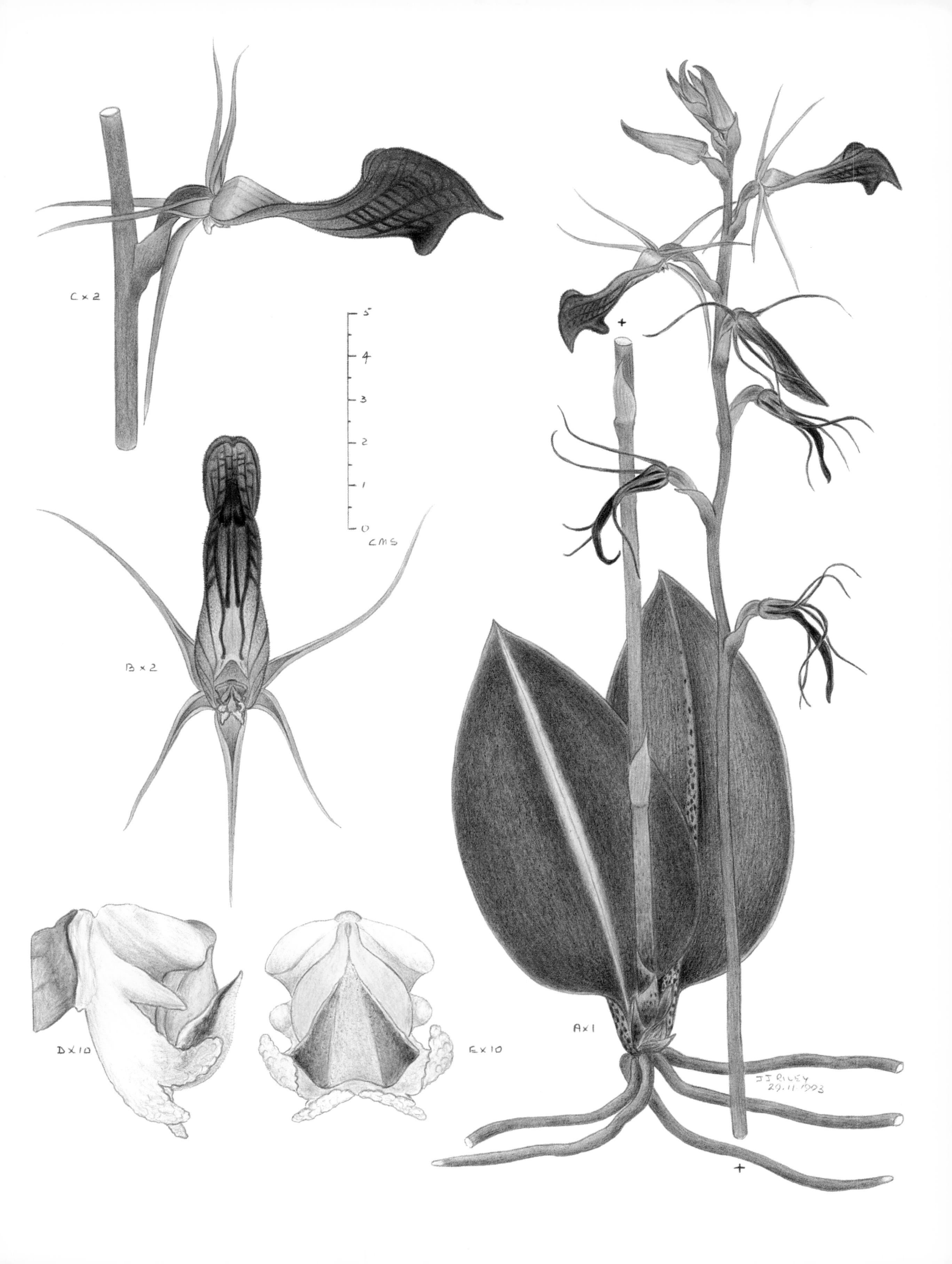
C x 2
5
4
3
2
1
0
CMS
B x 2
D x 10
E x 10
A x 1
J J RILEY
29.11.1993

Cryptostylis hunteriana Nicholls

1938 | *Victorian Naturalist* 54:182

TYPE LOCALITY Orbost, Victoria

ETYMOLOGY After original collector, W. Hunter

FLOWERING TIME August to February

DISTRIBUTION This coastal species has a sporadic range. It is found from the Tin Can Bay area, along the coast to the Glasshouse Mountains in Queensland, south along the New South Wales coast and adjacent ranges, to the east coast of Victoria.

ALTITUDINAL RANGE 50 m to 1100 m

DISTINGUISHING FEATURES *C. hunteriana* is a leafless saprophyte, the only Australian *Cryptostylis* without leaves. The labellum is held erect and is densely covered with short, red hairs.

HABITAT While it has a preference for the margins of coastal swamps, this species is also found in heathland and wallum communities to dry woodland and forest. Soils are usually deep white to grey sands to sandy loams.

CONSERVATION STATUS Rare. Considered endangered in Victoria. In New South Wales and Queensland it occurs in national parks. Vulnerable.

DISCUSSION *C. hunteriana* is a rare species which is something of an enigma. As it is leafless, it is impossible to detect unless in flower. It rarely appears in large numbers and usually emerges and blooms for one or two seasons, then disappears, to resurface after a few years in a new location. However, in some locations, such as the Nelson Bay area, *C. hunteriana* flowers regularly season after season forming small clumps and scattered colonies. It is often overlooked in the field due to this erratic flowering habit. This species flowers earlier in Queensland and later in New South Wales and Victoria. It may be more widespread, however, vast areas of its suitable habitat have been developed for housing. Male *Lissopimpla excelsa* wasps pollinate this species during pseudocopulation.

Cryptostylis hunteriana
Lemon Tree Passage, New South Wales

26 November 1995

A plant
B flower from front
C flower from side
D flower from rear
E labellum from above
F labellum from below
G column from side
H column from front
J dorsal sepal
K lateral sepal
L petal

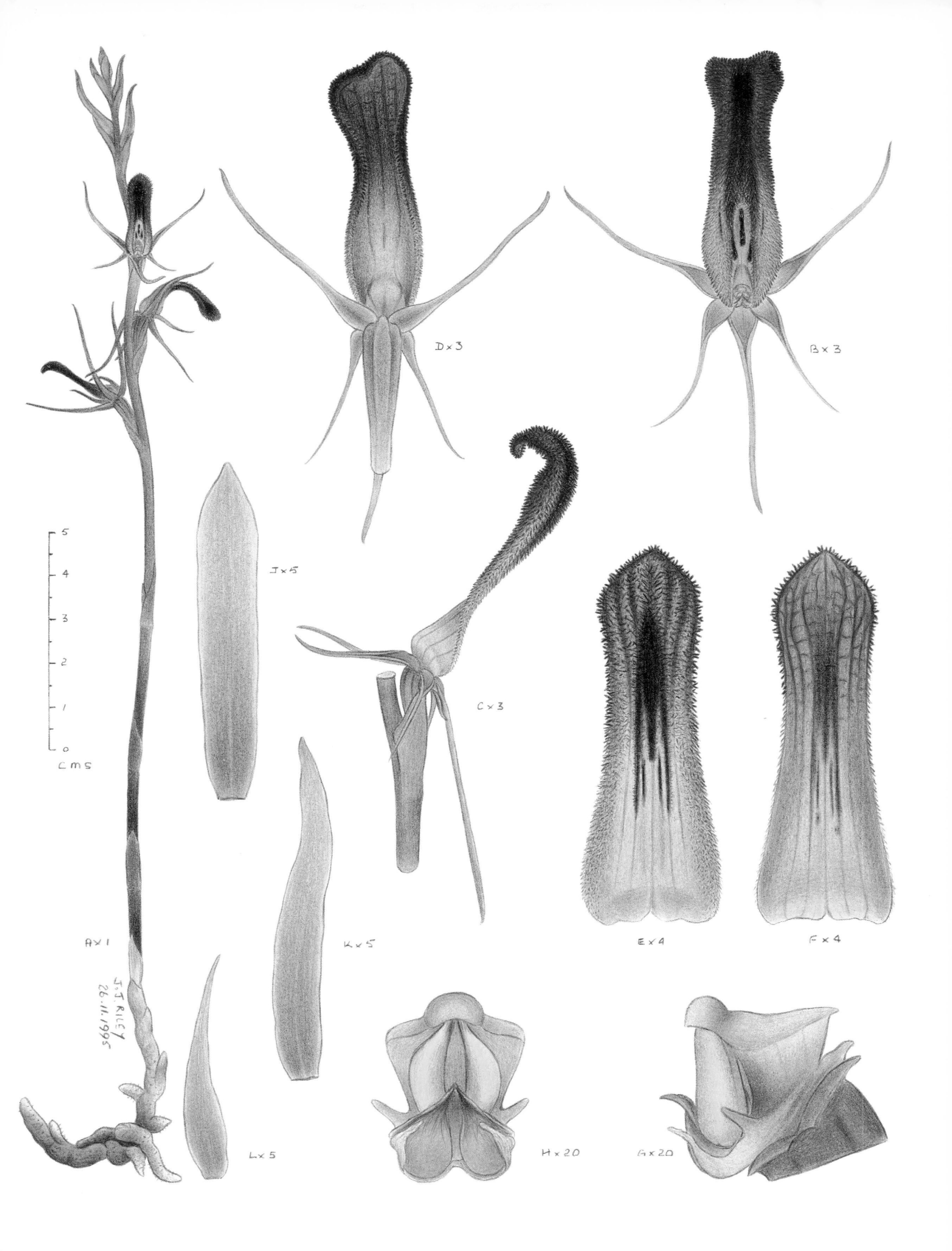

D×3
B×3
J×5
C×3
5
4
3
2
1
0
cms
E×4
F×4
A×1
K×5
J.J. RILEY
26.11.1995
L×5
H×20
G×20

Cymbidium madidum Lindl.

1839 | *Edward's Botanical Register* 26:9

TYPE LOCALITY Australia

ETYMOLOGY *madidum* — moist

FLOWERING TIME August to February

DISTRIBUTION Found from the Cape York Peninsula in North Queensland, down the coast into New South Wales, at least as far south as the Bellinger River.

ALTITUDINAL RANGE 50 m to 1100 m

DISTINGUISHING FEATURES This is an epiphyte with long flexible leaves and large pseudobulbs (not illustrated). The flowers are apple green to greenish-brown without spots. The labellum has well developed sidelobes and a yellow apex.

HABITAT *C. madidum* grows as an epiphyte in rainforest, wet sclerophyll forest and moist woodland. Often found growing from hollows in limbs or on dead trees, with its roots travelling quite a distance into the decaying heartwood. Often growing in association with epiphytic ferns from the genera *Asplenium*, *Davallia* and *Platycerium*.

CONSERVATION STATUS Common. Widespread in national parks and reserves. Secure.

DISCUSSION *C. madidum* can grow into very large plants. It generally grows in shadier and moister situations than the related *C. suave*. The arching to pendent, long inflorescences can carry up to 70 highly, and sweetly, fragrant blooms. Pollination is by native bees. Young seedlings often germinate within clumps of epiphytic ferns or on fallen timber. In the Coffs Harbour region it has been found in littoral rainforest growing as a terrestrial. Large specimens are rarely killed after bushfires and, while setback, will often re-establish from dormant pseudobulbs protected within the heart of the plant.

Cymbidium madidum
Coffs Harbour, New South Wales

7 January 1995

A flower from front
B flower from side
C flower from rear
D labellum from side
E labellum flattened
F column from side
G column from front
H dorsal sepal
J lateral sepal
K petal

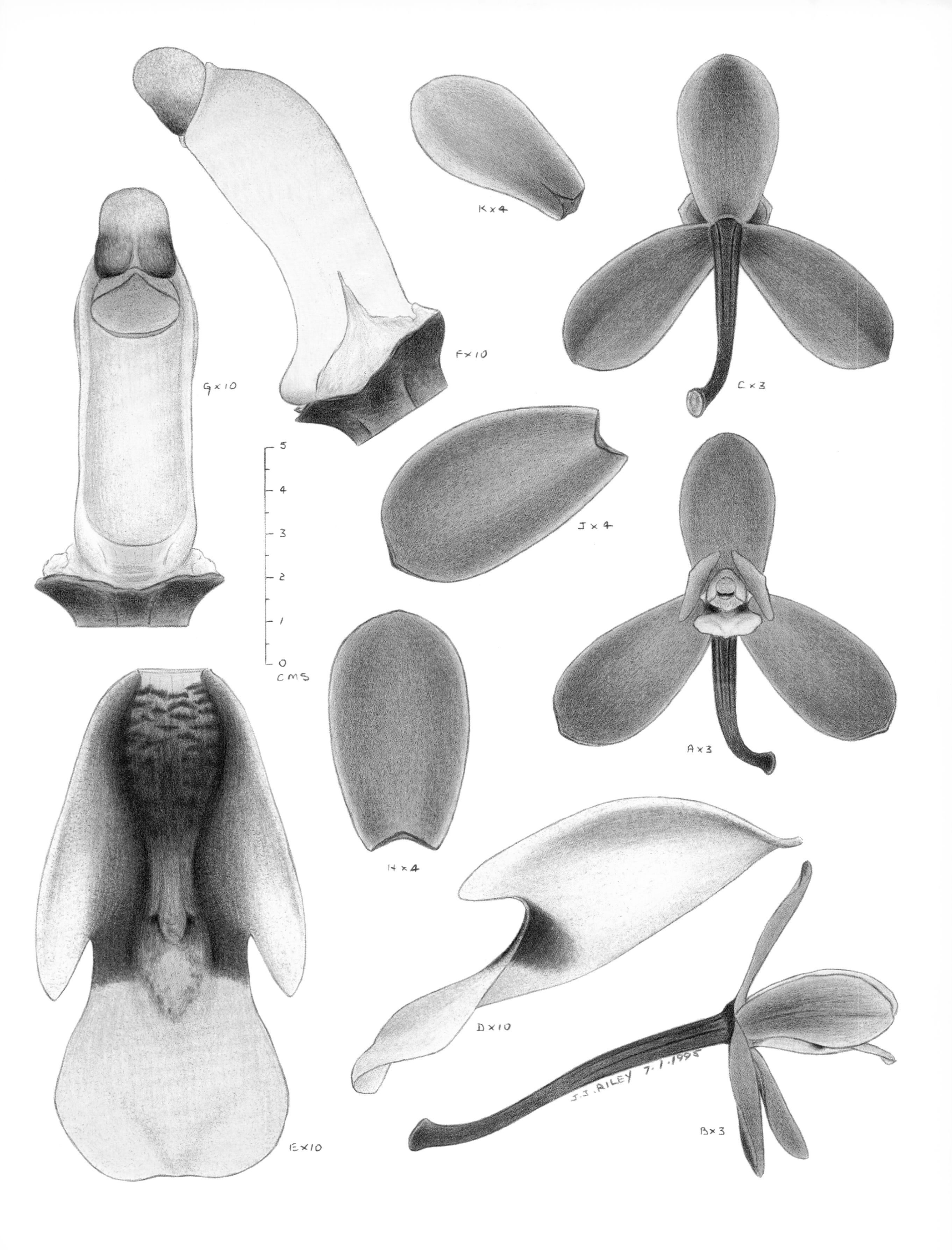
K × 4
C × 3
G × 10
F × 10
J × 4
A × 3
CMS
H × 4
D × 10
J.J. RILEY 7.1.1995
B × 3
E × 10

Cyrtostylis reniformis R. Br.

1810 | *Prodromus Florae Novae Hollandiae*: 322

TYPE LOCALITY Port Jackson, New South Wales

RECENT SYNONYMS *Acianthus reniformis* (R. Br.) Schltr.

ETYMOLOGY *reniformis* — kidney-shaped

FLOWERING TIME May to October

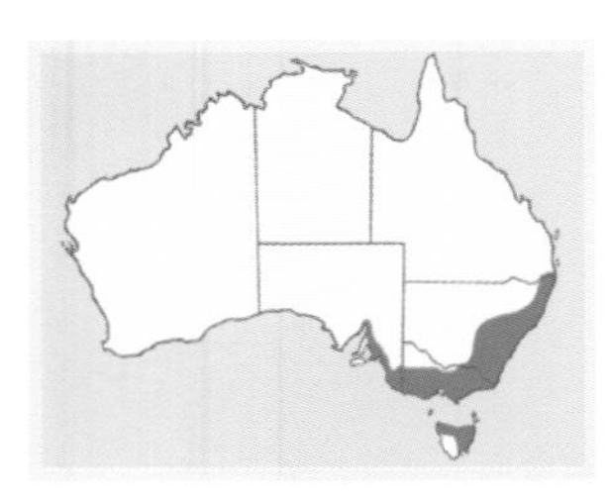

DISTRIBUTION *C. reniformis* is a widespread species, being found from south-eastern Queensland along the coast and ranges of New South Wales, to Victoria, and the south-east corner of South Australia, to the Flinders Ranges. Also occurs on the Bass Strait Islands and in Tasmania.

ALTITUDINAL RANGE 20 m to 1000 m

DISTINGUISHING FEATURES While related to *Acianthus*, the foliage and floral arrangement are quite different. *C. reniformis* has a fairly round, greyish-green leaf and the labellum has an irregular toothed and ragged apex.

HABITAT *C. reniformis* grows in moist sheltered areas in a variety of plant communities from coastal heath, open and closed woodland to montane forest. Also found among rocky outcrops in shaded situations along small creeks and drainage patterns. Soils are sand, sandy loams and clay loams.

CONSERVATION STATUS Common. Widespread and well represented in national parks and reserves. Secure.

DISCUSSION *C. reniformis* forms extensive colonies growing in moist, humus-rich sites among ferns, shrubs and grasses. When not in bloom, this species is easily recognised by its leaf. It reproduces from seed and vegetatively from daughter tubers growing on the ends of stolonoid roots. Often grows in mixed communities with other colony-forming, deciduous terrestrial orchids such as *Corybas* spp., *Acianthus* spp., and *Chiloglottis* spp.

Cyrtostylis reniformis
Fingal Bay, New South Wales

24 July 2000

A plant
B flower from side
C flower from front
D labellum flattened
E labellum and column from side
F labellum and column from front
G dorsal sepal
H lateral sepal
J petal

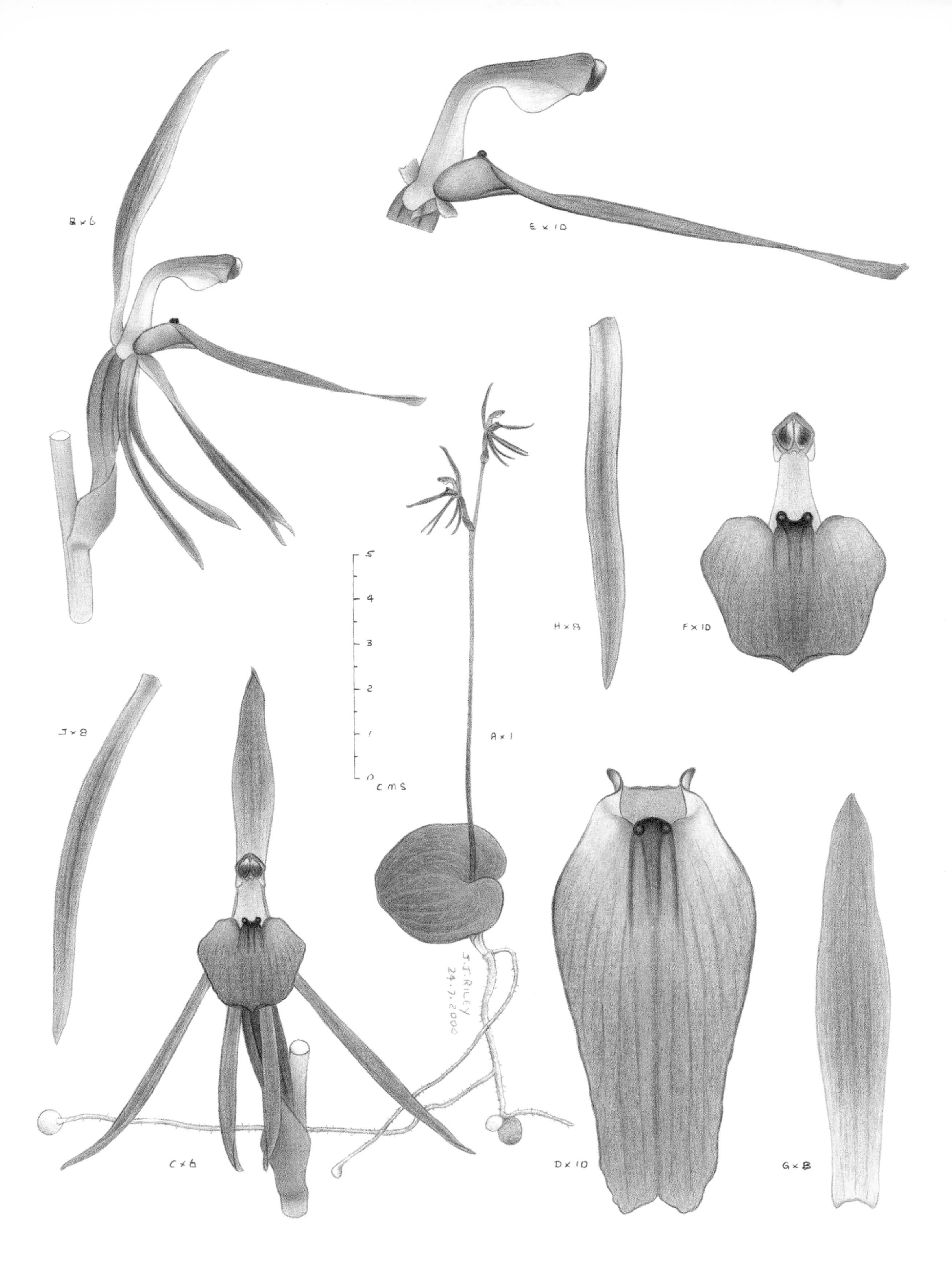

B×6
E×10
H×8
F×10
J×8
A×1
5
4
3
2
1
0
cms
J.J.RILEY
24.7.2000
C×6
D×10
G×8

Dendrobium kingianum Bidwill ex Lindl.

1844 | *Edward's Botanical Register* 30:11

TYPE LOCALITY Australia

ETYMOLOGY After Captain P.P. King

FLOWERING TIME August to November

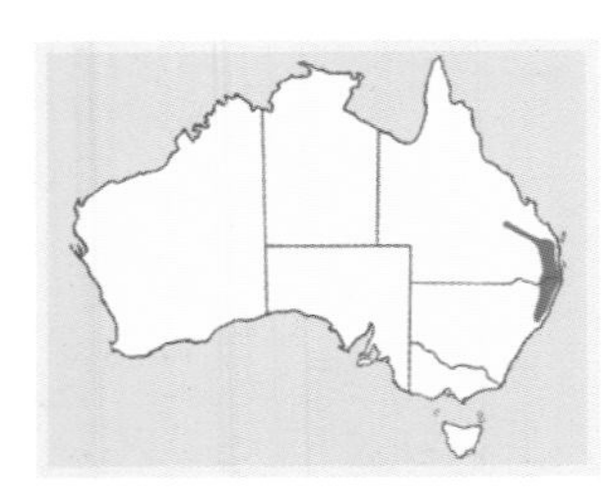

DISTRIBUTION *D. kingianum* is found from Alum Mountain at Bulahdelah, northwards along the coast and ranges to near Gympie and the outlying population at Carnarvon Gorge in central Queensland.

ALTITUDINAL RANGE 100 m to 1200 m

DISTINGUISHING FEATURES Distinctive small to medium sized lithophyte with pale pink (rarely white) to reddish-mauve, open flowers with a wide, three-lobed labellum that is heavily marked with deep purple streaks. Blooms are fragrant during the day.

HABITAT Found mostly on exposed rock outcrops and escarpments in woodland and forest as well as rock faces in wet sclerophyll forest.

CONSERVATION STATUS Common. Occurs in national parks and reserves. Secure.

DISCUSSION *D. kingianum* is one of Australia's most familiar and popular orchid species. While it is still a common orchid in the wild, wholesale collecting has decimated many former populations. This is one of our most variable orchids, with pseudobulbs varying in size from 2 cm tall at Alum Mountain, New South Wales, up to 80 cm tall in plants from the Numinbah Valley in southern Queensland. Many of these taller clones produce aerial growths. Pollination is by native bees. It is frequently cultivated by orchid enthusiasts and plant lovers and has been the subject of intensive line breeding to accentuate desirable colours and traits, in particular the dark beetroot shades. *D. kingianum* infrequently hybridises in the wild with the related species, *D. speciosum* and *D. gracilicaule*.

Dendrobium kingianum
Mullumbimby, New South Wales

1 October 1990

A plant
B flower from front
C flower from side
D flower from rear
E labellum from side
F labellum flattened
G column from side
H column from front

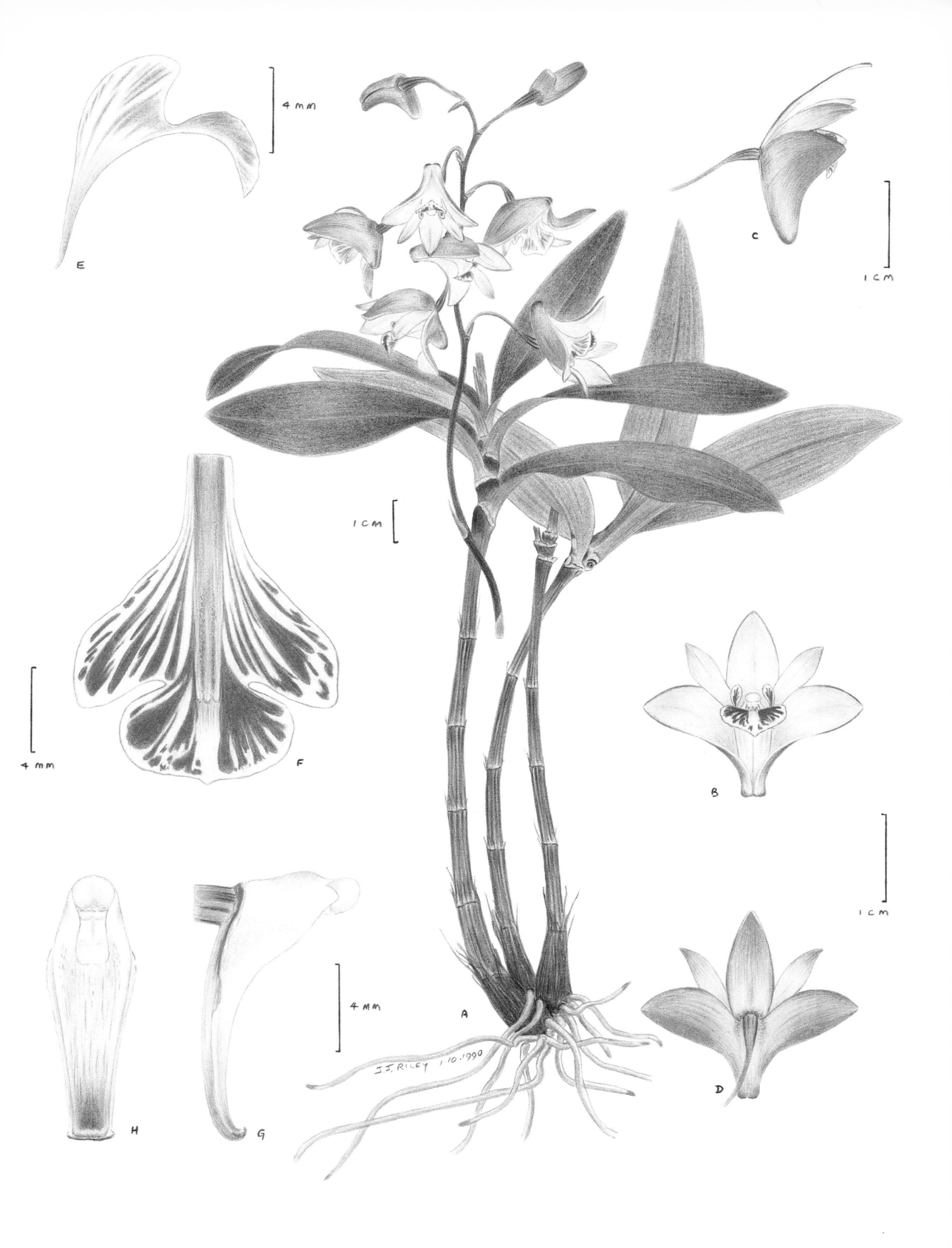

E
4 mm
C
1 cm
1 cm
F
4 mm
B
1 cm
A
4 mm
D
H
G
J.J. Riley 1.10.1990

Dendrobium moorei F. Muell.

1869 | *Fragmenta Phytographiae Australiae* 7:29

TYPE LOCALITY Lord Howe Island, off coast of New South Wales

ETYMOLOGY After Charles Moore

FLOWERING TIME April to October

DISTRIBUTION *D. moorei* is endemic to Lord Howe Island, which is about 700 km north-east of Sydney.

ALTITUDINAL RANGE 100 m to 850 m

DISTINGUISHING FEATURES The crystalline, and somewhat tubular, white flowers are slightly nodding, with the labellum similar to the other floral segments.

HABITAT Grows on rocks and trees in rainforest on the slopes of Mount Lidgbird and Mount Gower. It prefers shaded situations, but can also tolerate strong light, and plants in these positions tend to be particularly free-flowering.

CONSERVATION STATUS Uncommon. Occurs in a World Heritage Area. Secure.

DISCUSSION This unique species occurs in two different growth forms, yet the flowers on these are identical. One has short, squat pseudobulbs while the other has long, slender pseudobulbs and also freely produces aerial growths. The main blooming season is from autumn to spring, however, larger plants can produce flowers at other times. It is unusual among dendrobiums in having a labellum that is only slightly modified and is the same colour as the other segments. It is frequently found growing with *Dendrobium comptonii*.

Dendrobium moorei
Lord Howe Island

3 April 1992

A plant
B flower from front
C flower from side
D flower from above
E labellum flattened
F labellum and column from side
G column from front

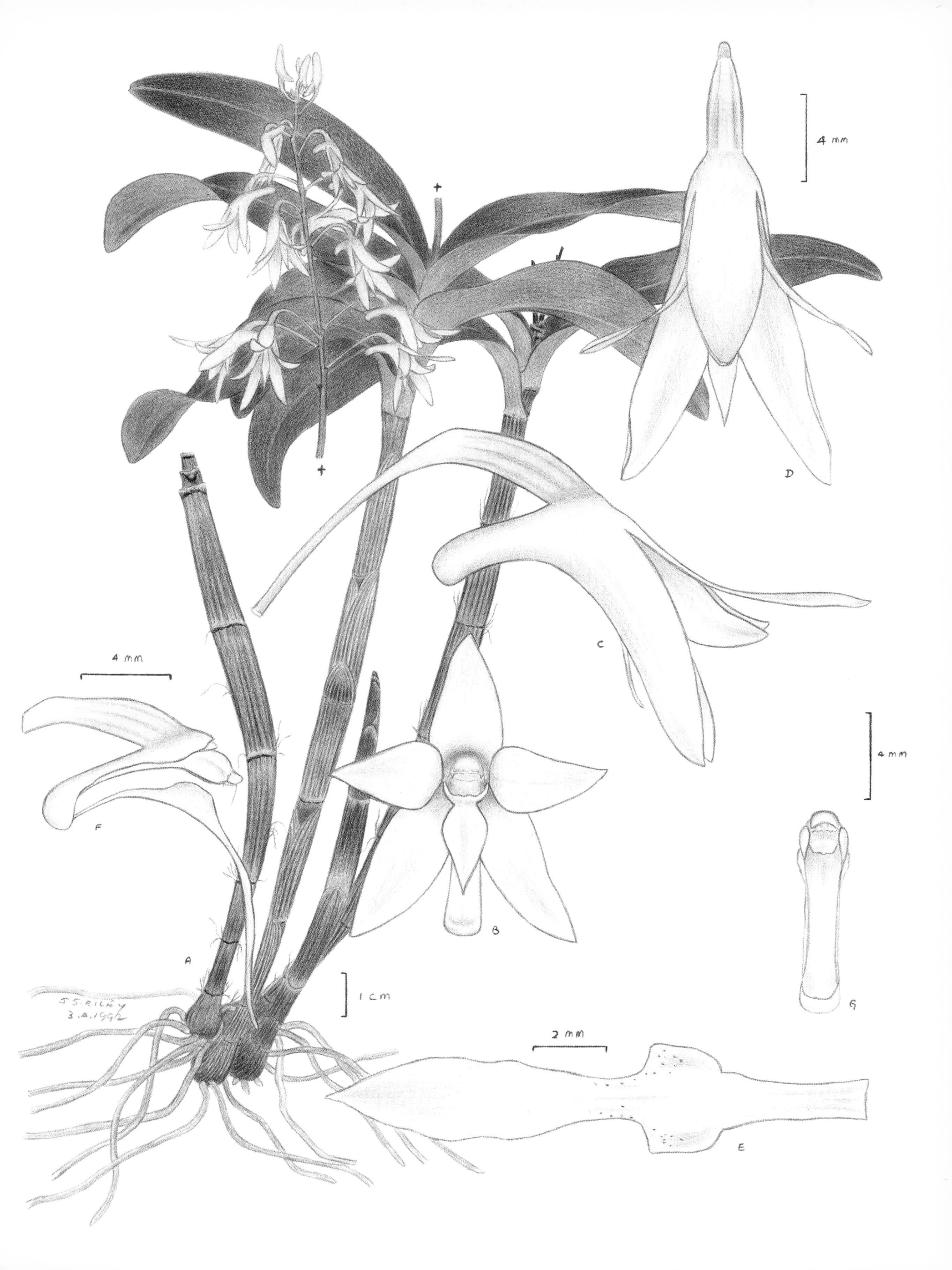
4 mm
D
C
4 mm
F
4 mm
B
A
1 cm
J.S. RILEY
3.4.1992
G
2 mm
E

Dendrobium fellowsii F. Muell.

1870 | *Fragmenta Phytographiae Australiae* 7:63

TYPE LOCALITY Rockingham Bay, Queensland

RECENT SYNONYMS *Dendrobium bairdianum* Bailey

ETYMOLOGY After original collector, Mr Fellows

FLOWERING TIME November to January

DISTRIBUTION *D. fellowsii* is endemic to North Queensland where it is found on the ranges between Townsville and the Bloomfield River.

ALTITUDINAL RANGE 450 m to 600 m

DISTINGUISHING FEATURES A distinctive epiphytic species with green flowers and a labellum that is green with prominent calli ridges, marked heavily with purple. Unlikely to be confused with any other Australian *Dendrobium*.

HABITAT This species grows mainly on Casuarina and Eucalypt spp. in dry sclerophyll forest. Rarely found growing on rocks in open forest and woodland.

CONSERVATION STATUS Uncommon. Occurs in national parks and state forests. Vulnerable.

DISCUSSION *D. fellowsii* is the only member of section *Eleutheroglossum* within the genus *Dendrobium* found in Australia. The other three related species occur in New Caledonia. This has never been a common species and is seldom encountered by orchid enthusiasts. It is probably more widespread, but its remoteness and the limited access to its preferred habitat have hindered detailed searches to assess its true conservation status.

Dendrobium fellowsii
Atherton Tableland, Queensland

24 December 1997

A plant
B flower from front
C flower from side
D labellum from side
E labellum flattened
F column from side
G column from front
H dorsal sepal
J lateral sepal
K petal

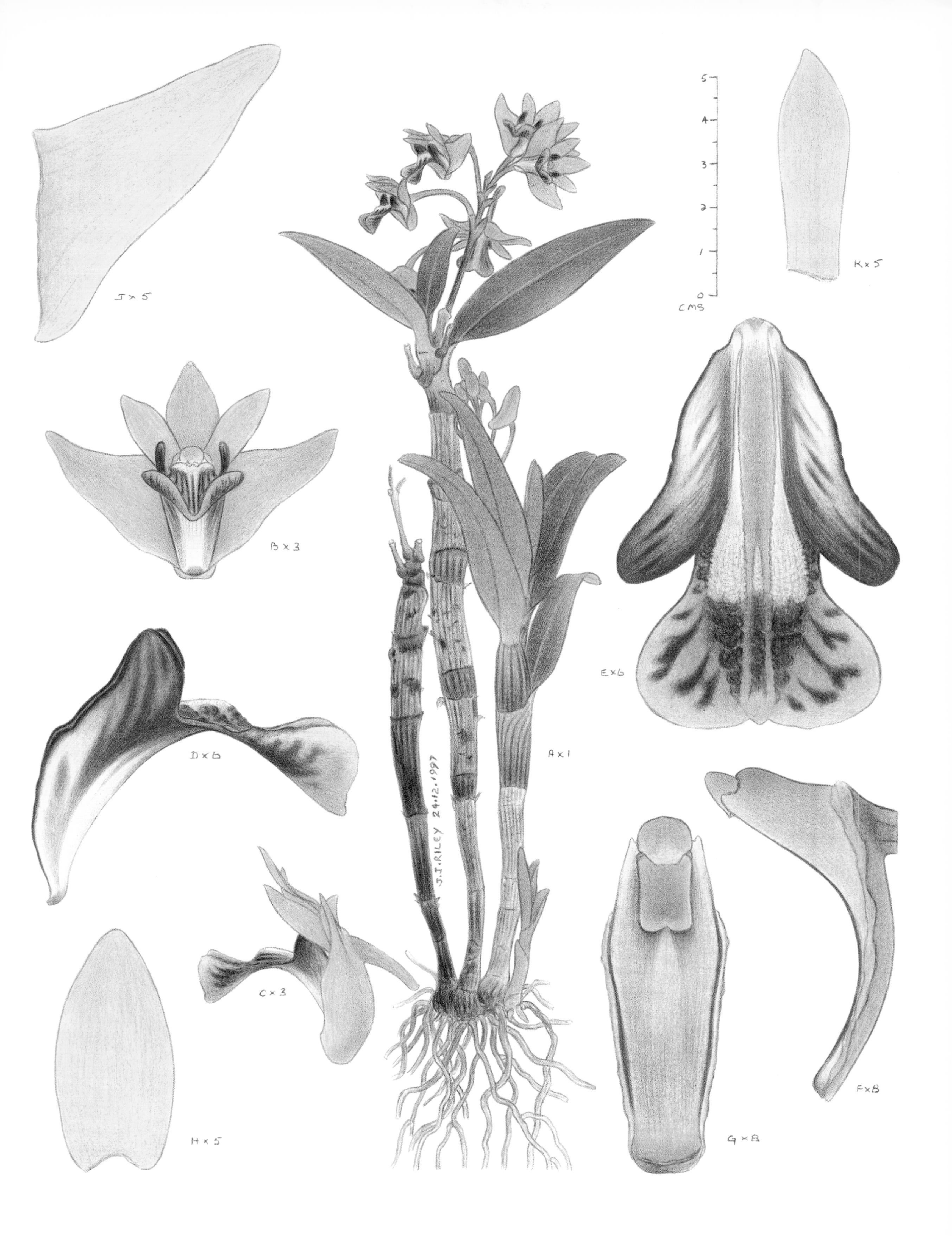
J × 5
5
4
3
2
1
0
CMS
K × 5
B × 3
E × 6
D × 6
A × 1
J.J. RILEY 24.12.1997
C × 3
H × 5
G × 8
F × 8

Dendrobium toressae (Bailey) Dockr.

1964 | *Orchadian* 1:64

TYPE LOCALITY Mount Bellenden Ker, Queensland

ETYMOLOGY After Toressa Meston

FLOWERING TIME Mostly September to January but can be sporadic throughout the year.

DISTRIBUTION *D. toressae* is endemic to North Queensland. It is found on the ranges between Innisfail and Cooktown.

ALTITUDINAL RANGE 50 m to 1000 m

DISTINGUISHING FEATURES This is a creeping plant without pseudobulbs. The grain-sized leaves are slightly wrinkly, and are tightly packed together on the plant. The individual blooms are often larger than the leaf. *D. toressae* is unlikely to be confused with any other orchid.

HABITAT *D. toressae* grows mostly as a lithophyte in wet sclerophyll forest from the coast to nearby ranges, often in exposed positions. It is less common on trees in rainforest or their fringes.

CONSERVATION STATUS Locally common. Conserved in national parks and state forests. Secure.

DISCUSSION This species grows in one of the wettest parts of Australia, with constant high humidity. The Bellenden Ker Ranges have the highest annual rainfall in Australia. Wet sclerophyll forest and lush rainforest surround numerous rocky outcrops and ridges, where *D. toressae* often forms large mats. It grows with a host of ferns and other epiphytic orchids, particularly from the genera *Bulbophyllum* and *Dendrobium*. *D. toressae* is a distinctive species with an unusual growth and flowering habit, and has no close relatives.

Dendrobium toressae
South Johnstone River, Queensland

1 February 1995

A plant
B leaf
C flower from front
D flower and leaf from side
E flower from above
F labellum from side
G labellum flattened
H column from side
J column from front
K dorsal sepal
L petal

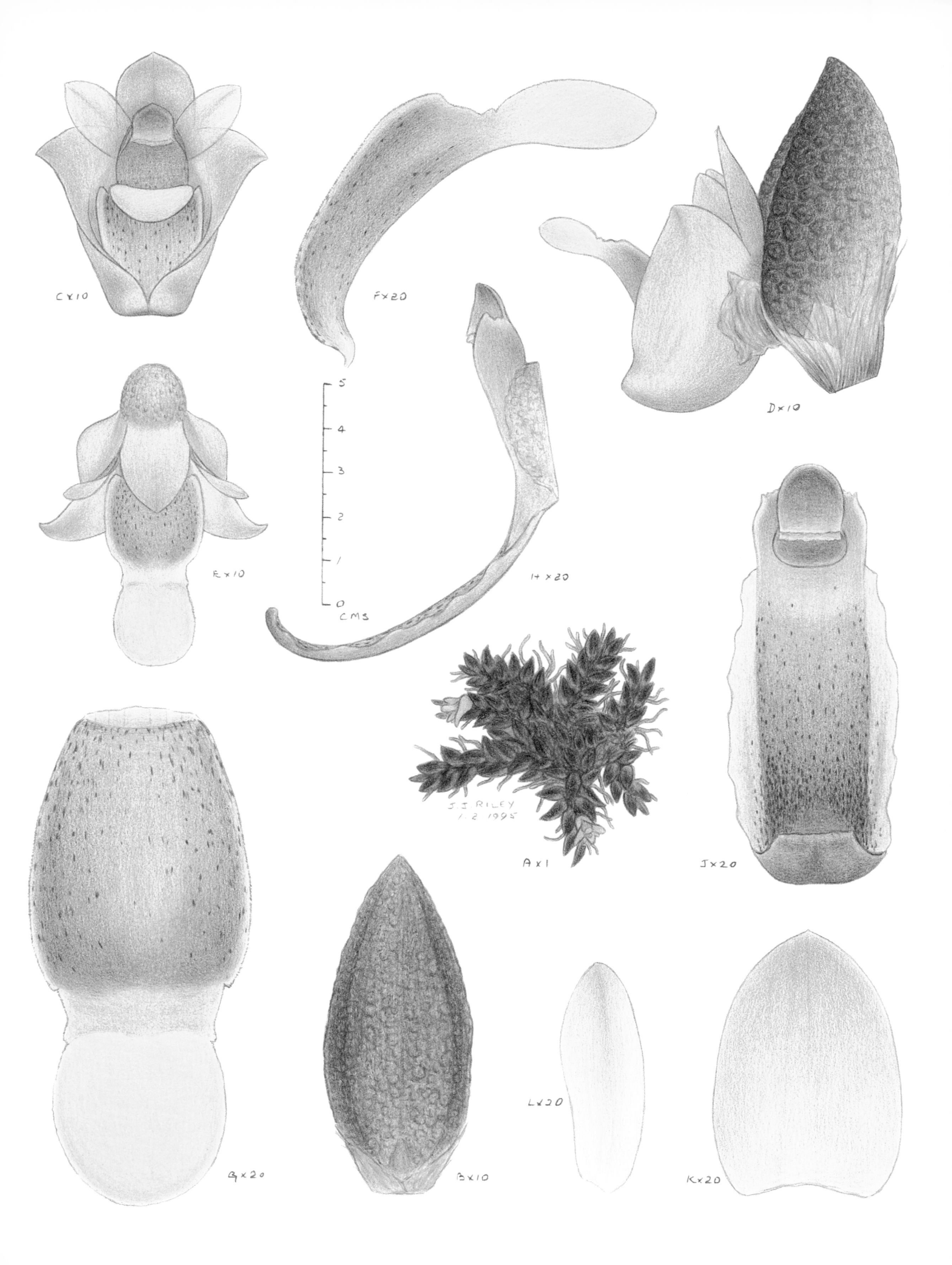

C×10
F×20
D×10
E×10
5
4
3
2
1
0
CMS
H×20
J×20
J.J. RILEY
1.2 1995
A×1
G×20
B×10
L×20
K×20

Dipodium hamiltonianum Bailey

1881 | *Proceedings of the Linnean Society of New South Wales* 6:140

TYPE LOCALITY Stradbroke Island, Queensland

ETYMOLOGY After James Hamilton

FLOWERING TIME October to February

DISTRIBUTION This species has a sporadic distribution. It is found south from Hervey Bay, along the Queensland coast, including Fraser Island and Stradbroke Island, through eastern New South Wales to north-eastern Victoria.

ALTITUDINAL RANGE 10 m to 1000 m

DISTINGUISHING FEATURES *D. hamiltonianum* is a leafless saprophyte with yellow to greenish-yellow flowers that are overlaid with maroon spots.

HABITAT Grows in a wide variety of plant communities, including coastal heath and among grasses on the banks of streams. On the ranges it occurs in open *Eucalyptus* woodland and forest, while on the western slopes it occurs in dry *Eucalyptus* and *Callitris* woodland. Occurs in soils ranging from sand to gravelly loams.

CONSERVATION STATUS Sporadic and uncommon throughout its range, being considered rare in Victoria. Known to occur in national parks and reserves. Vulnerable.

DISCUSSION *D. hamiltonianum* is a spectacular species that is rarely encountered in the wild. It is uncommon with plants occurring as isolated individuals. It is more frequently encountered on the western portion of the New England Tableland, the North-Western Slopes and the Pilliga Scrub. With this species flowering in mid-summer, in remote areas that are hot and dry, few people venture in search of it.

Dipodium hamiltonianum
Coonabarabran, New South Wales

23 December 1988

A plant

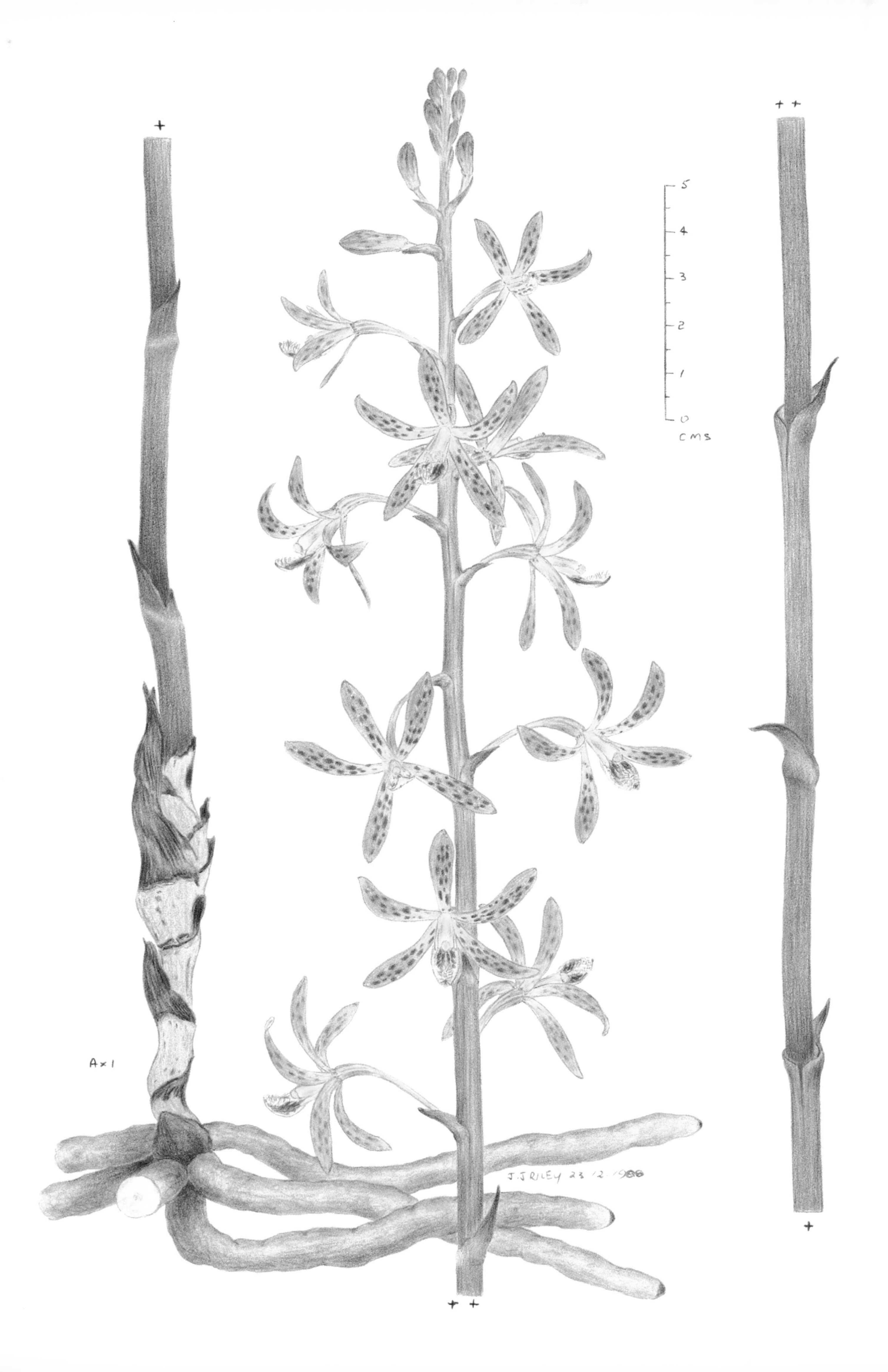

5
4
3
2
1
0
CMS
A x 1
J.J RILEY 23/12/1988

Dipodium punctatum (Sm.) R. Br.

1810 | *Prodromus Florae Novae Hollandiae*: 331

TYPE LOCALITY Port Jackson, New South Wales

ETYMOLOGY *punctatum* — spotted

FLOWERING TIME November to February

DISTRIBUTION This species is found from Tin Can Bay in Queensland, south through New South Wales, disjunct in eastern and western Victoria and eastern South Australia.

ALTITUDINAL RANGE 10 m to 1100 m

DISTINGUISHING FEATURES *D. punctatum* is a saprophytic species producing reddish-pink flowers with conspicuous, darker spots. The petals and sepals are not reflexed. The labellum has spots with hairs on the long and narrow midlobe.

HABITAT A common species mostly found growing in coastal regions in heathland and heathy woodland. Less frequently seen on the ranges where it favours open woodland and forest. It can be often found growing along tracks, the sides of roads in open forest and other disturbed sites.

CONSERVATION STATUS Common. Conserved in national parks. Secure.

DISCUSSION *D. punctatum* is an attractive, tall, brightly coloured orchid that can be locally common. It, and related species, are often referred to as hyacinth orchids. It occurs as individuals or small groups growing at the base of, or close to, large *Eucalyptus* trees. A number of taxa, previously included within *D. punctatum* have been given specific status. This includes *D. variegatum*, *D. roseum* and *D. campanulatum*.

Dipodium punctatum
Seven Mile Beach, New South Wales

14 January 1990

A plant
B flower from front
C flower from side
D flower from rear
E labellum flattened
F column from side
G column from front
H column from rear
J apex of column

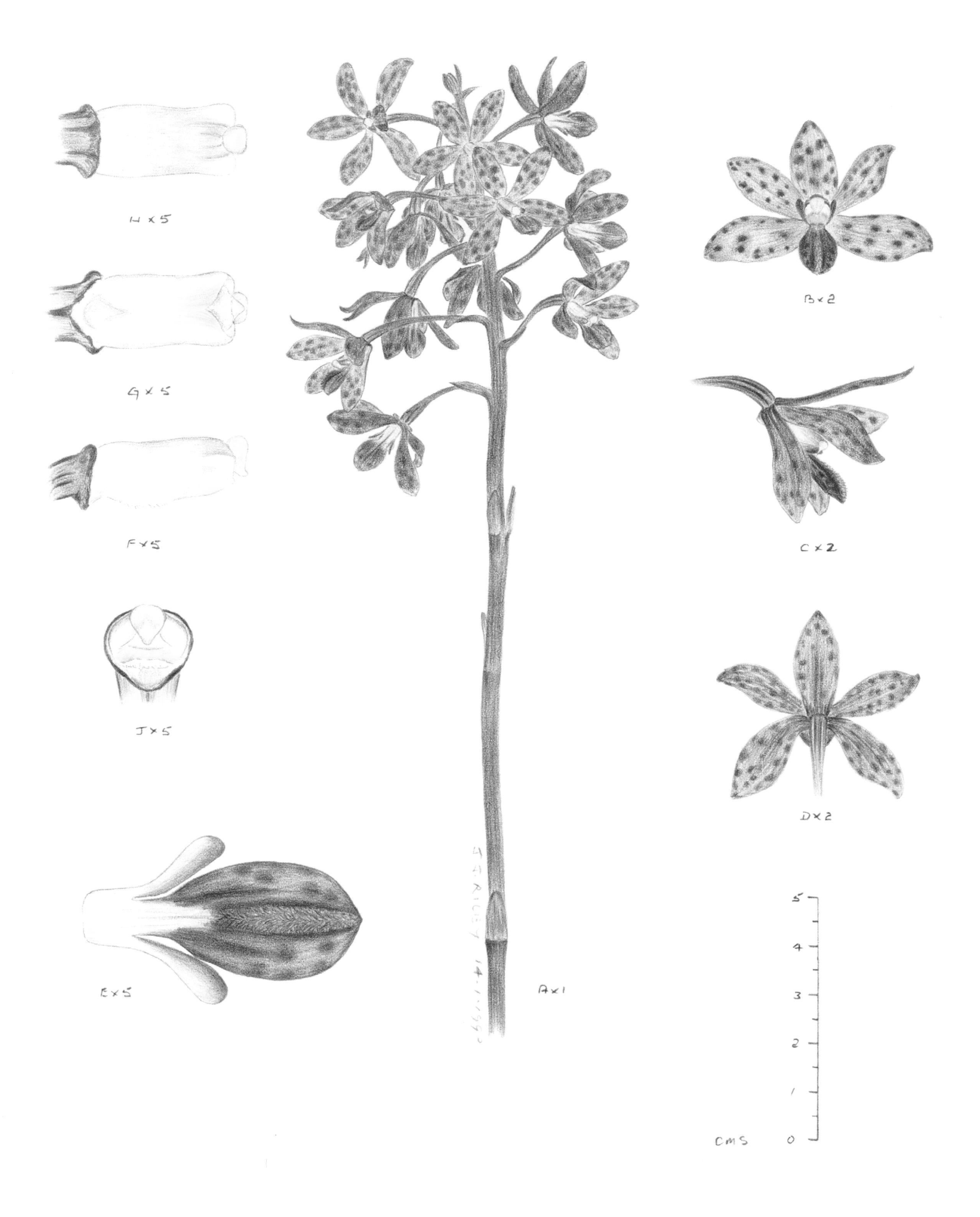
H x 5
G x 5
F x 5
J x 5
E x 5
A x 1
B x 2
C x 2
D x 2
5
4
3
2
1
0
CMS

Dipodium roseum D.L. Jones & M.A. Clem.

1991 | *Australian Orchid Research* 2:51

TYPE LOCALITY Montrose, Victoria

ETYMOLOGY *roseum* — rose pink

FLOWERING TIME November to March

DISTRIBUTION Found from south-eastern Queensland along the coast and ranges of New South Wales, to the southern half of Victoria, eastern South Australia, across to the north and eastern coasts of Tasmania.

ALTITUDINAL RANGE 10 m to 1400 m

DISTINGUISHING FEATURES *D. roseum* is a saprophytic species producing rose pink flowers with small, indistinct spots. The labellum has dark pink lines and ridges of hairs on the midlobe, that is wide and prominent towards the apex.

HABITAT *D. roseum* grows in a range of habitats, from coastal heathy woodland to moist forest and dry open woodland. In the northern part of its range, it is rare on the coast, becoming increasingly plentiful in montane forest at higher altitude.

CONSERVATION STATUS Widespread. Conserved in national parks and reserves. Secure.

DISCUSSION *D. roseum* is a most attractive, summer-flowering terrestrial orchid. It often grows in tight groups with a number of inflorescences blooming at the same time. Like many of the saprophytic dipodiums, it is found growing at the base of, or in the immediate proximity of, mature *Eucalyptus* trees. *D. roseum* is still commonly mistaken for *D. punctatum*.

Dipodium roseum
Boyd Plateau, New South Wales

1 February 1990

A plant
B flower from front
C flower from side
D labellum from above
E column from side
F column from front
G column from rear
H apex of column

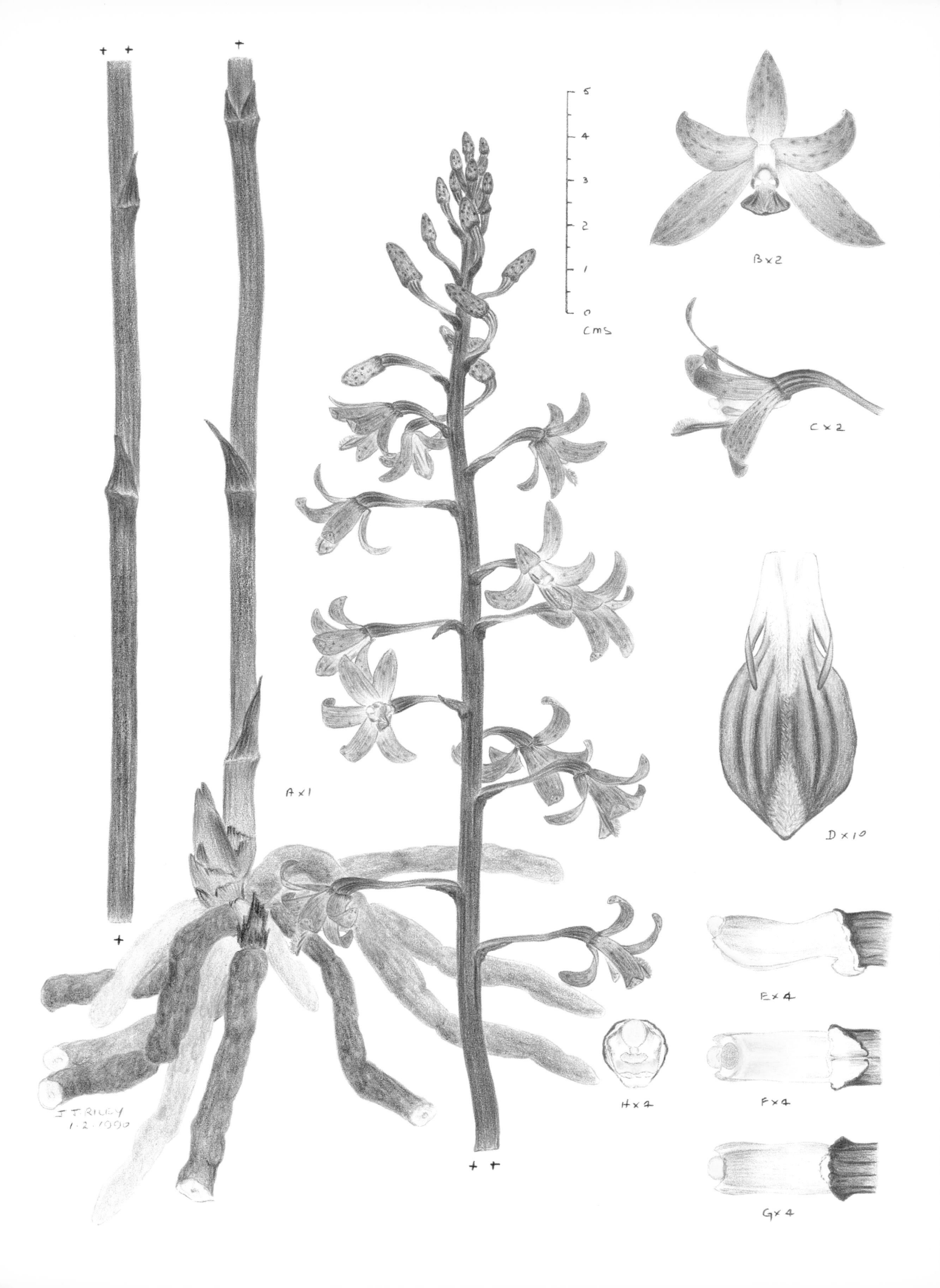
5
4
3
2
1
0
cms
B × 2
C × 2
D × 10
A × 1
E × 4
H × 4
F × 4
G × 4
J J RILEY
1·2·1990

Dipodium variegatum M.A. Clem. & D.L. Jones

1987 | *Proceedings of the Royal Society of Queensland* 98:128

TYPE LOCALITY Beenleigh, Queensland

ETYMOLOGY *variegatum* — variegated, of mixed colours

FLOWERING TIME October to February

DISTRIBUTION Found from the Windsor Tableland in North Queensland along the coast and ranges of New South Wales to far eastern Victoria. It is usually found on the ranges in the northern parts of its distribution, gradually becoming more coastal to the south.

ALTITUDINAL RANGE Up to 800 m

DISTINGUISHING FEATURES *D. variegatum* is a saprophytic species producing very pale pink flowers distinctly spotted with dark purple. The flowers are crowded at the top of the inflorescence. The labellum has indistinct spotting and a ridge of prominent hairs widely scattered towards the apex.

HABITAT Found in coastal heath, heathy woodland and moist to dry woodland with a grass and shrub understorey.

CONSERVATION STATUS Common. Conserved in national parks. Secure.

DISCUSSION *D. variegatum* is a distinctive species which is readily identified. It is one of the most common species of hyacinth orchid. *D. variegatum* and *D. punctatum* grow together in some localities and natural hybrids have been recorded. The saprophytic dipodiums are pollinated by small bees.

Dipodium variegatum
Oakdale, New South Wales

23 January 1990

A plant
B flower from front
C flower from side
D flower from rear
E labellum from above
F column from side
G column from front
H column from rear

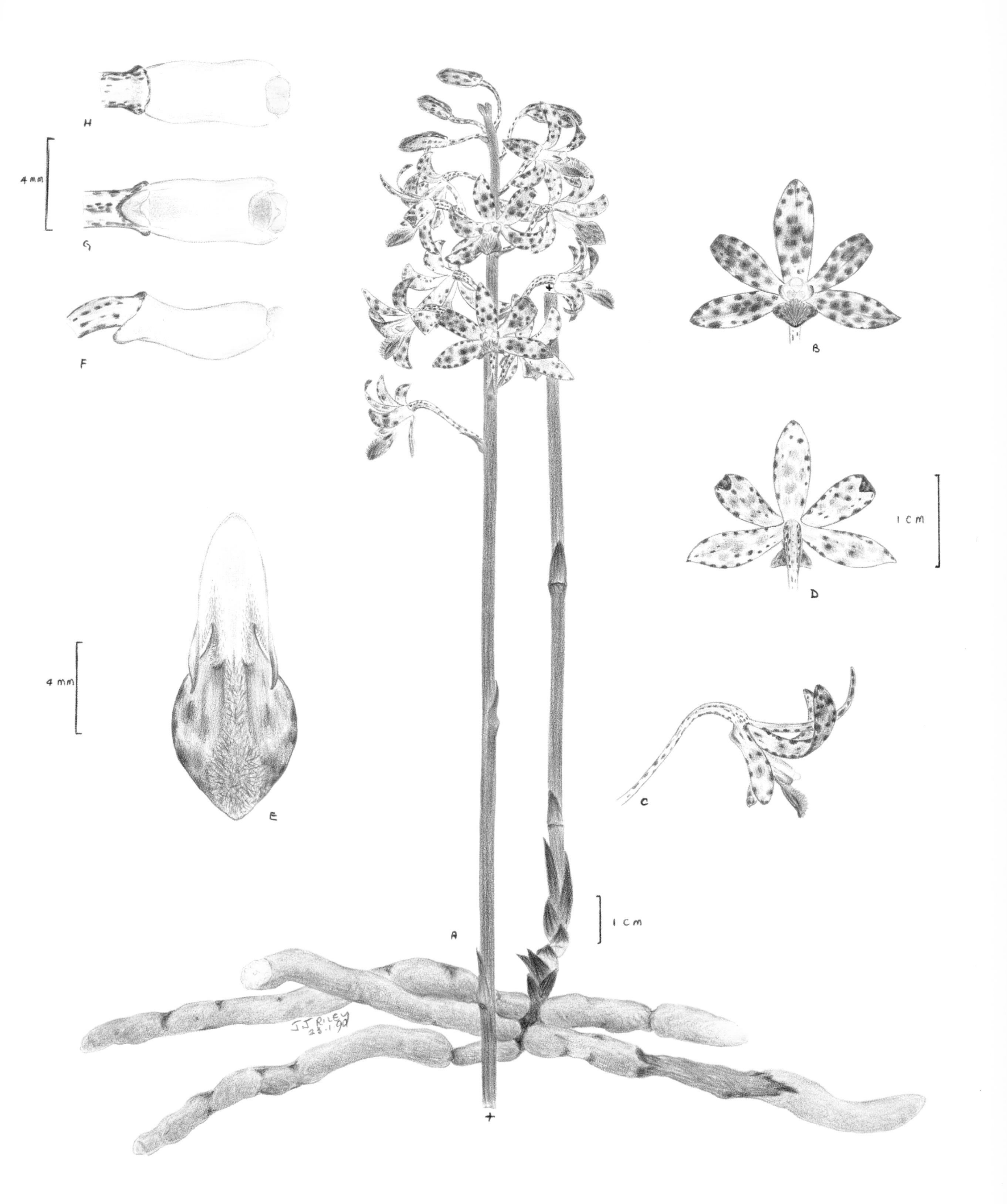
H
4 mm
G
F
B
D
1 cm
4 mm
E
C
A
1 cm
J J RILEY
23.1.90

Dipodium ensifolium F. Muell.

1865 | *Fragmenta Phytographiae Australiae* 5:42

TYPE LOCALITY Rockingham Bay, Queensland

ETYMOLOGY *ensifolium* — sword-shaped leaves

FLOWERING TIME October to January

DISTRIBUTION *D. ensifolium* is endemic to North Queensland, being found from Ingham to Cooktown. It occurs on the coast and the adjacent ranges.

ALTITUDINAL RANGE 40 m to 900 m

DISTINGUISHING FEATURES *D. ensifolium* is an evergreen terrestrial with distinctive foliage. It has creamy pink blooms with prominent, rich purple spotting.

HABITAT Occurs in coastal, lowland sclerophyll forest with a grassy understorey on sandy loams. To the north it grows on the slopes and ridges of hills on the ranges supporting open sclerophyll forest with a grass understorey in stony clay loams. The region has a monsoonal climate, with dry winters and wet summers.

CONSERVATION STATUS Conserved in national parks. Secure.

DISCUSSION This is one of two species of evergreen, non-saprophytic dipodiums in Australia. The other species is *D. pictum*. *D. ensifolium* is common in open forest and forms loose colonies and groups, occasionally reproducing vegetatively from the base of the plant and off stolon-like roots. This process may be a response to the frequent fires that occur in the spear grass dominated understorey, which burn all the above ground parts. The plants surprisingly recover very quickly after fires.

Dipodium ensifolium
Atherton Tableland, Queensland

31 December 1997

A plant
B flower from side
C flower from front
D labellum flattened from above
E labellum flattened from rear
F labellum from side
G column from front
H column from rear
J column from side
K dorsal sepal
L lateral sepal
M petal

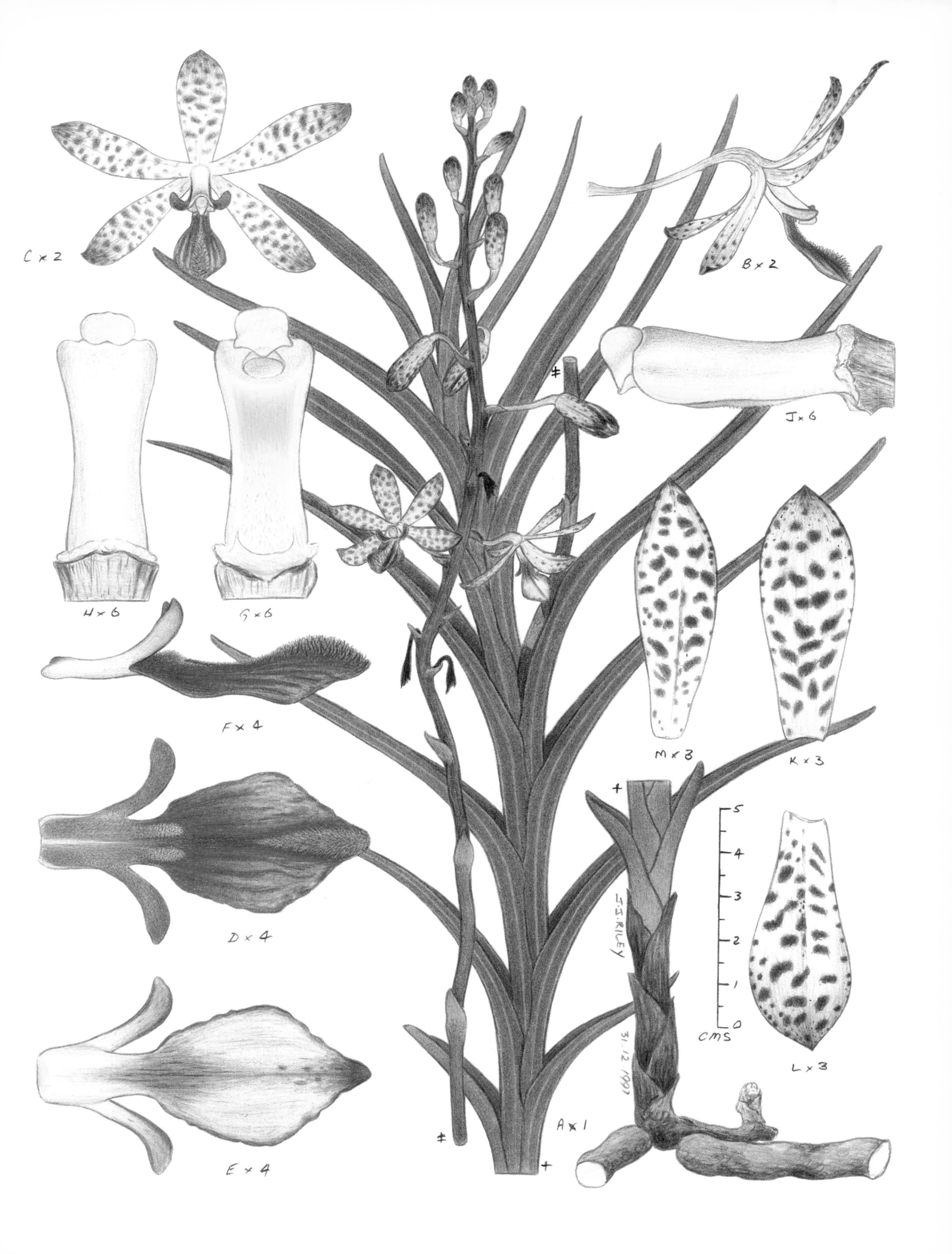
C × 2
B × 2
J × 6
H × 6
G × 6
F × 4
M × 3
K × 3
D × 4
5
4
3
2
1
0
CMS
J.J. RILEY
31.12.1997
L × 3
A × 1
E × 4

Diuris maculata Sm.

1804 | *Exotic Botany* 1:57

TYPE LOCALITY Port Jackson, New South Wales

ETYMOLOGY *maculata* — spotted

FLOWERING TIME July to September

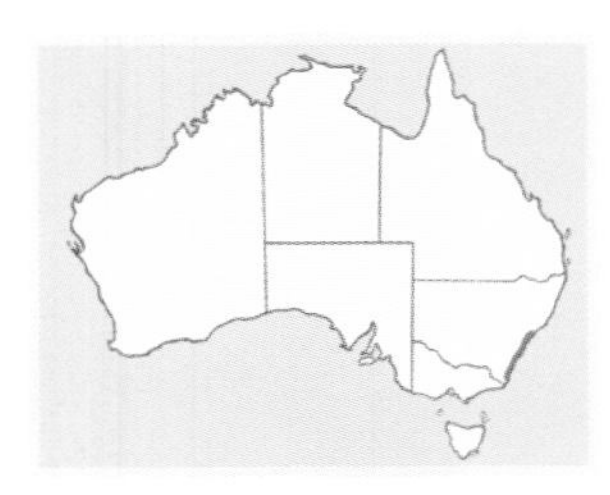

DISTRIBUTION Endemic to New South Wales, being found south from the Hawkesbury River, to Bega on the South Coast. It is mostly coastal but does extend to the lower Blue Mountains.

ALTITUDINAL RANGE 50 m to 550 m

DISTINGUISHING FEATURES *D. maculata* has small yellow flowers, that are heavily blotched with brown, primarily on the labellum and back of the petals. The labellum has a folded midlobe that has a pronounced humped outline when viewed from the side. The labellum, when flattened, is wider than long with prominent sidelobes and a wide, blunt midlobe.

HABITAT Found close to the coast in open woodland with a grassy, heathy understorey on sandstone. Soils are shallow sandy loams. Closer to the ranges, it occurs in open forest growing in clay loams.

CONSERVATION STATUS Common. Conserved in national parks, reserves and other protected land. Secure.

DISCUSSION Before settlement, *D. maculata* would have had a much wider distribution. The difficult broken terrain of the Sydney Sandstone was unsuited to agriculture in the early days of settlement and remained uncleared. Fortunately large portions of this undamaged land were set aside for national parks, military reserves and water catchments, saving this, and many other restricted plant species, from encroaching suburbia. This was not the case on the Cumberland Plain to the west of Sydney, where orchids, including *D. maculata,* have become quite isolated and rare on the clay soils. *D. maculata* reproduces from seed and, like most of the donkey orchids, is pollinated by small native bees.

Diuris maculata
Heathcote, New South Wales

16 August 1998

A plant
B flower from front
C flower from side
D labellum from side
E labellum flattened
F column from side
G column from front
H column from rear
J dorsal sepal from rear
K dorsal sepal from front
L lateral sepal
M petal from front
N petal from rear

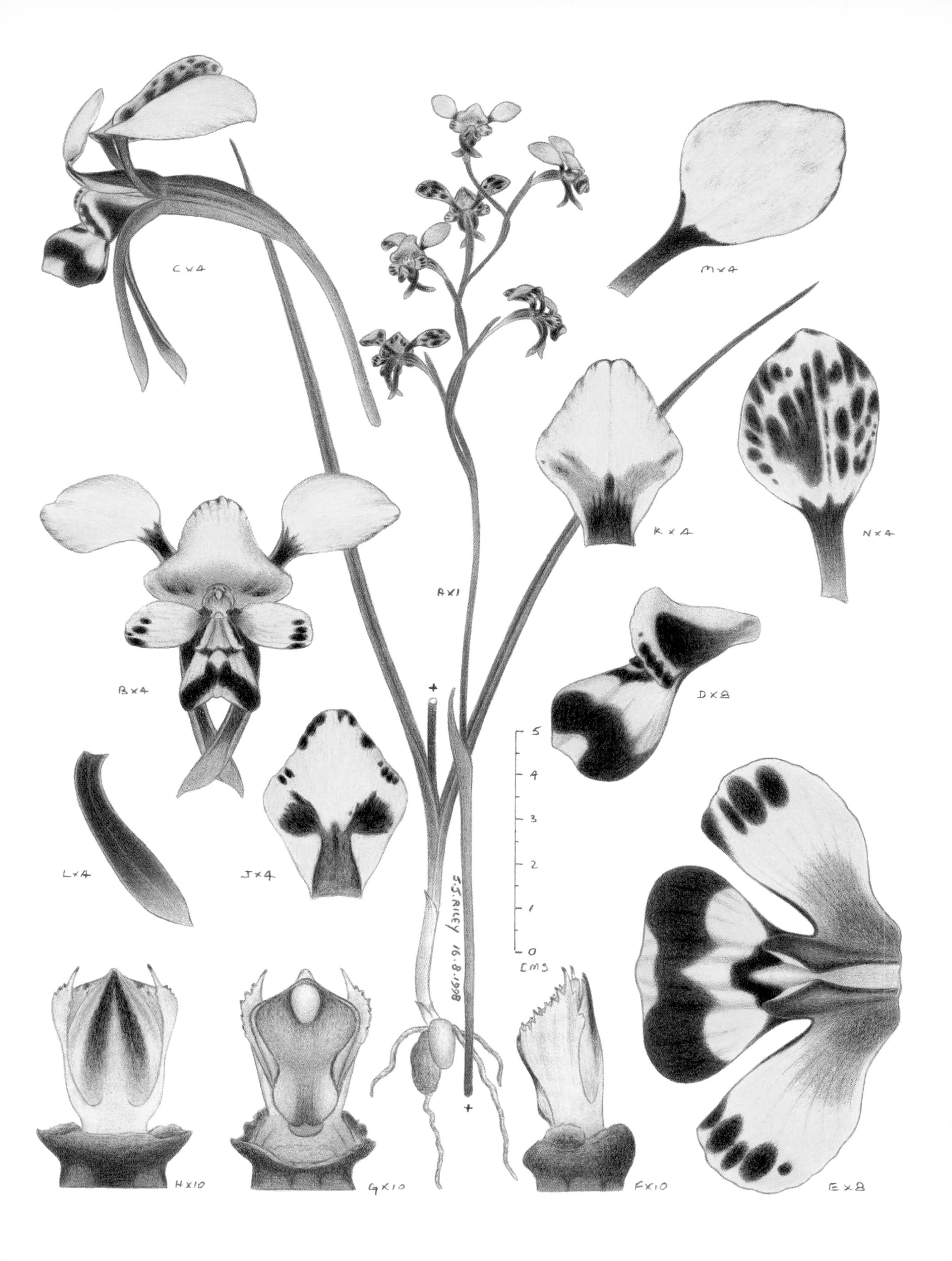
C×4
M×4
A×1
K×4
N×4
B×4
D×8
5
4
3
2
1
0
CMS
L×4
J×4
J.J. RILEY 16.8.1998
H×10
G×10
F×10
E×8

Diuris aequalis F. Muell. ex Fitzg.

1876 | *Australian Orchids* 1(2)

TYPE LOCALITY Liverpool, New South Wales

ETYMOLOGY *aequalis* — of similar size

FLOWERING TIME November to December

DISTRIBUTION *D. aequalis* is endemic to New South Wales, being found near Jenolan Caves and in the Braidwood area.

ALTITUDINAL RANGE 700 m to 1400 m

DISTINGUISHING FEATURES *D. aequalis* has bright, golden yellow flowers, with rounded petals and green lateral sepals. The labellum has narrow and very prominent sidelobes.

HABITAT Found in open montane woodland and forest with a grassy and sparse shrubby understorey. Soils are gravelly loams.

CONSERVATION STATUS Uncommon. Represented in national parks and reserves. Secure.

DISCUSSION *D. aequalis* was named from plants collected at low altitude near Liverpool, New South Wales in the 1870s. Thorough searches have failed to relocate it in this district. All known existing populations are restricted to montane and sub-alpine areas. Flowering is best in years with reliable rainfall; in dry years few plants bloom. This species reproduces from seed and, like many *Diuris* species, mimics flowers of the pea family, Fabaceae. Pollination is by small native bees. Children often remark how the floral shape of this species resembles the head of the cartoon character Mickey Mouse.

Diuris aequalis
Boyd Plateau, New South Wales

17 November 1991

A plant
B flower from front
C flower from side
D flower from rear
E flower from above
F labellum flattened
G column from side
H column from front
J column from rear
K dorsal sepal
L lateral sepal
M petal

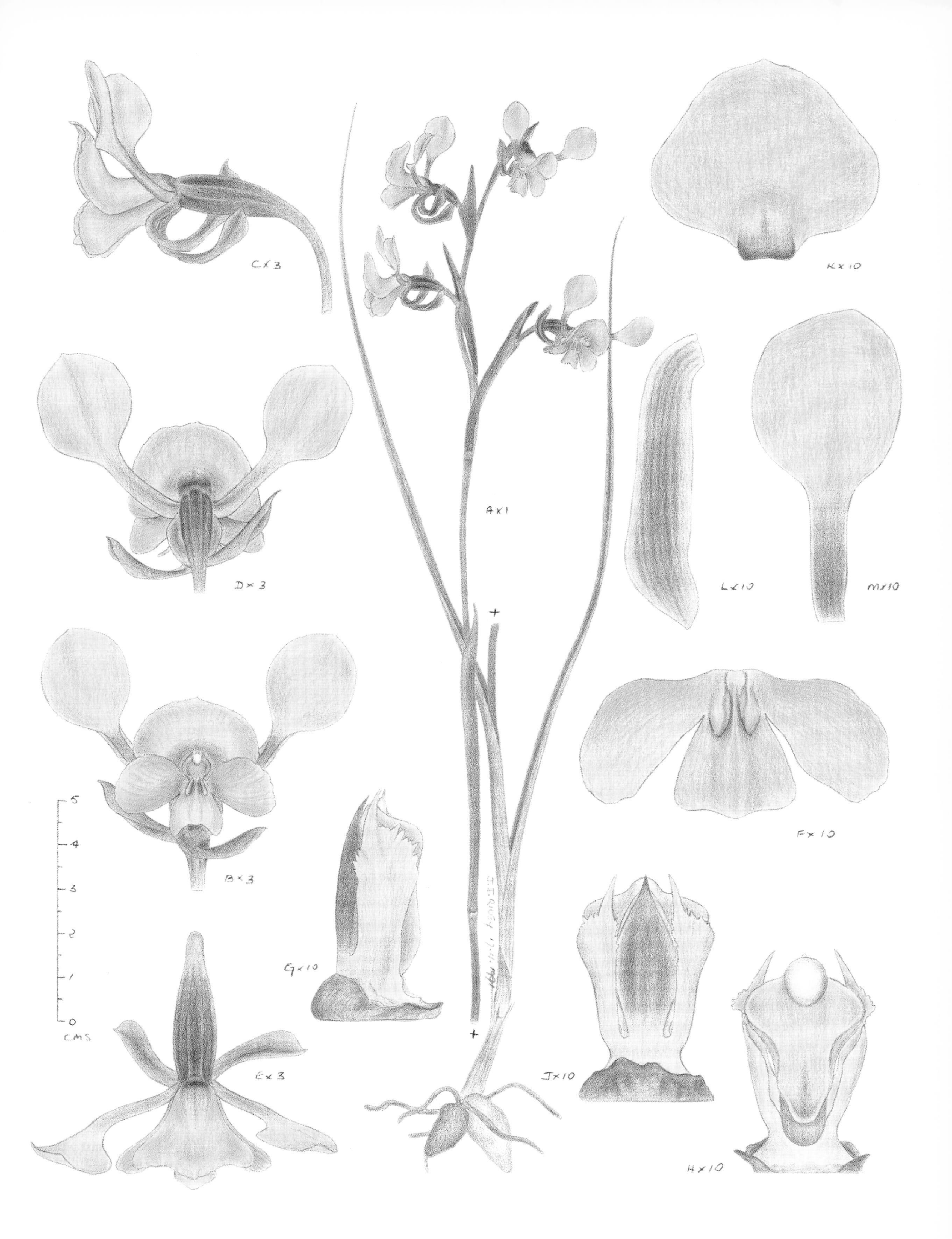

C×3
K×10
A×1
D×3
L×10
M×10
F×10
B×3
5
4
3
2
1
0
CMS
G×10
E×3
J×10
H×10

Diuris oporina D.L. Jones

1991 | *Australian Orchid Research* 2:59

TYPE LOCALITY Watsonville, Queensland

ETYMOLOGY *oporina* — autumnal

FLOWERING TIME March to July

DISTRIBUTION *D. oporina* is endemic to North Queensland. It is found on the Herberton Range and drier areas on the western side of the Atherton Tableland.

ALTITUDINAL RANGE 900 m to 1100 m

DISTINGUISHING FEATURES *D. oporina* has a single leaf which is a distinctive blue-green with a reddish-maroon base. It has small, white flowers with lilac markings and olive green lateral sepals.

HABITAT This species grows on steep slopes and ridges in very open, dry sclerophyll woodland with a grassy understorey. The region has dry winters with cold nights, and wet summers.

CONSERVATION STATUS Uncommon. Not known to occur in a national park, but does grow in state forests.

DISCUSSION *D. oporina* is part of the *D. punctata* complex and closely resembles *D. alba*, a coastal species. While never occurring in large numbers, it is found as individual plants or small groups in patches of tall grasses. It can be hard to detect in bloom. This is the most northern member of the *D. punctata* complex. This species reproduces from seed and is pollinated by small native bees.

Diuris oporina
Atherton Tableland, Queensland

19 May 1996

A plant
B flower from front
C flower from side
D labellum from side
E labellum flattened
F column from side
G column from front
H column from rear
J dorsal sepal
K petal

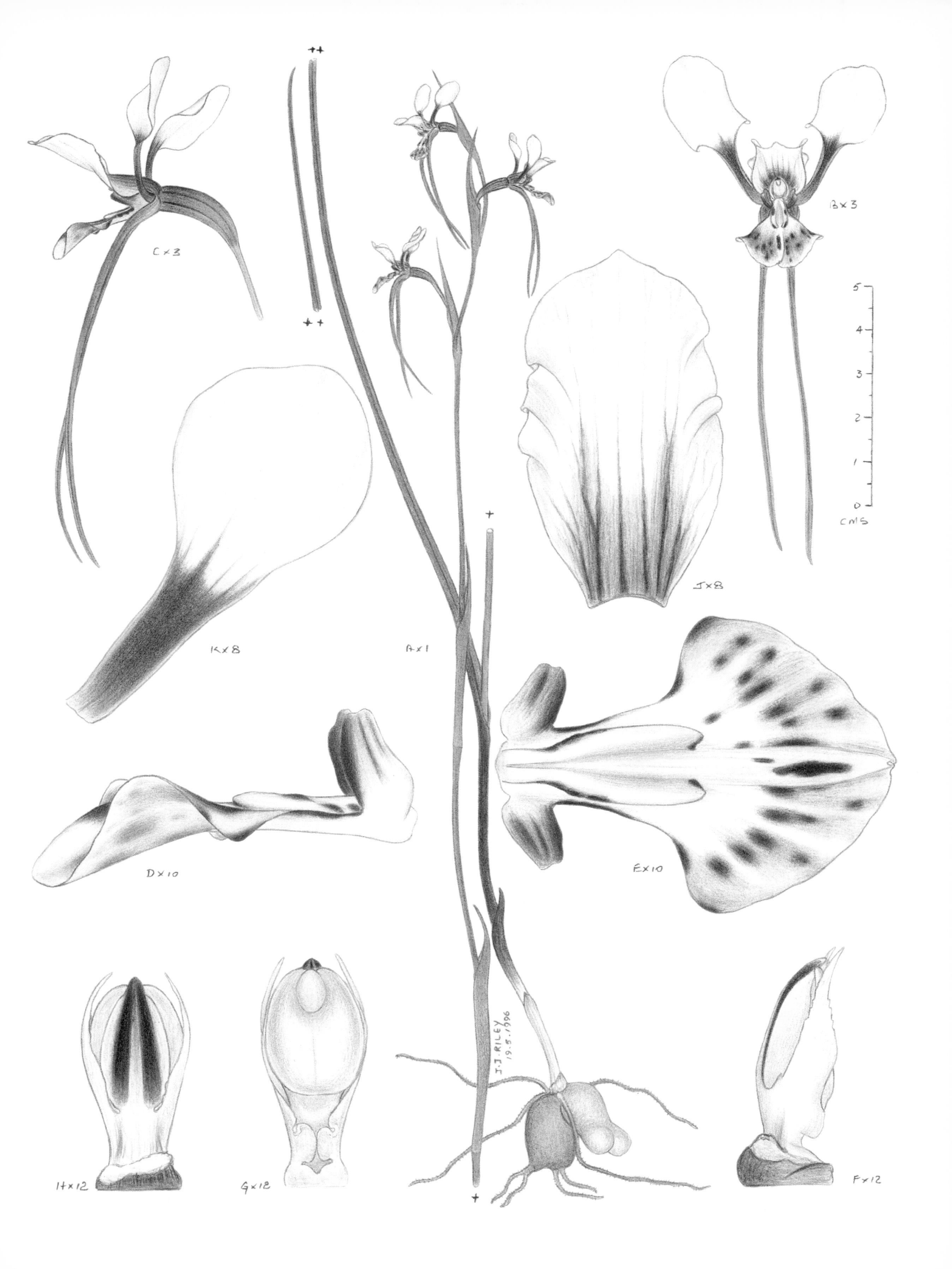
C×3
A×1
B×3
5
4
3
2
1
0
CMS
K×8
J×8
D×10
E×10
J.J. RILEY
19.5.1996
H×12
G×12
F×12

Diuris praecox D.L. Jones

1991 | *Australian Orchid Research* 2:60

TYPE LOCALITY Merewether Heights, New South Wales

ETYMOLOGY *praecox* — early

FLOWERING TIME July to September

DISTRIBUTION *D. praecox* is endemic to New South Wales, being found from near Wyong to Port Stephens.

ALTITUDINAL RANGE 30 m to 300 m

DISTINGUISHING FEATURES *D. praecox* is an early-flowering species that has small, lemon yellow flowers with a rounded labellum midlobe.

HABITAT This coastal species is most frequently seen growing in heathland among bracken fern and grasses on deep grey white sand and shallow soils derived from laterite and conglomerate. May also be found in scrubby woodland on the tops and slopes of rocky hills.

CONSERVATION STATUS Rare. Occurs in national parks and reserves. Threatened.

DISCUSSION *D. praecox* is a rare species that is always found close to the coast line. In the past *D. praecox* has been confused with, and identified as, *D. abbreviata*, an inland species that flowers much later and is found on the ranges. Generally found as individual specimens and sometimes in loose groups of three or four plants. This species reproduces from seed and is pollinated by small native bees.

Diuris praecox
Merewether Heights, New South Wales

11 August 1999

A plant
B flower from side
C flower from front
D labellum from side
E labellum flattened
F column from rear
G column from front
H column from side
J dorsal sepal
K lateral sepal
L petal

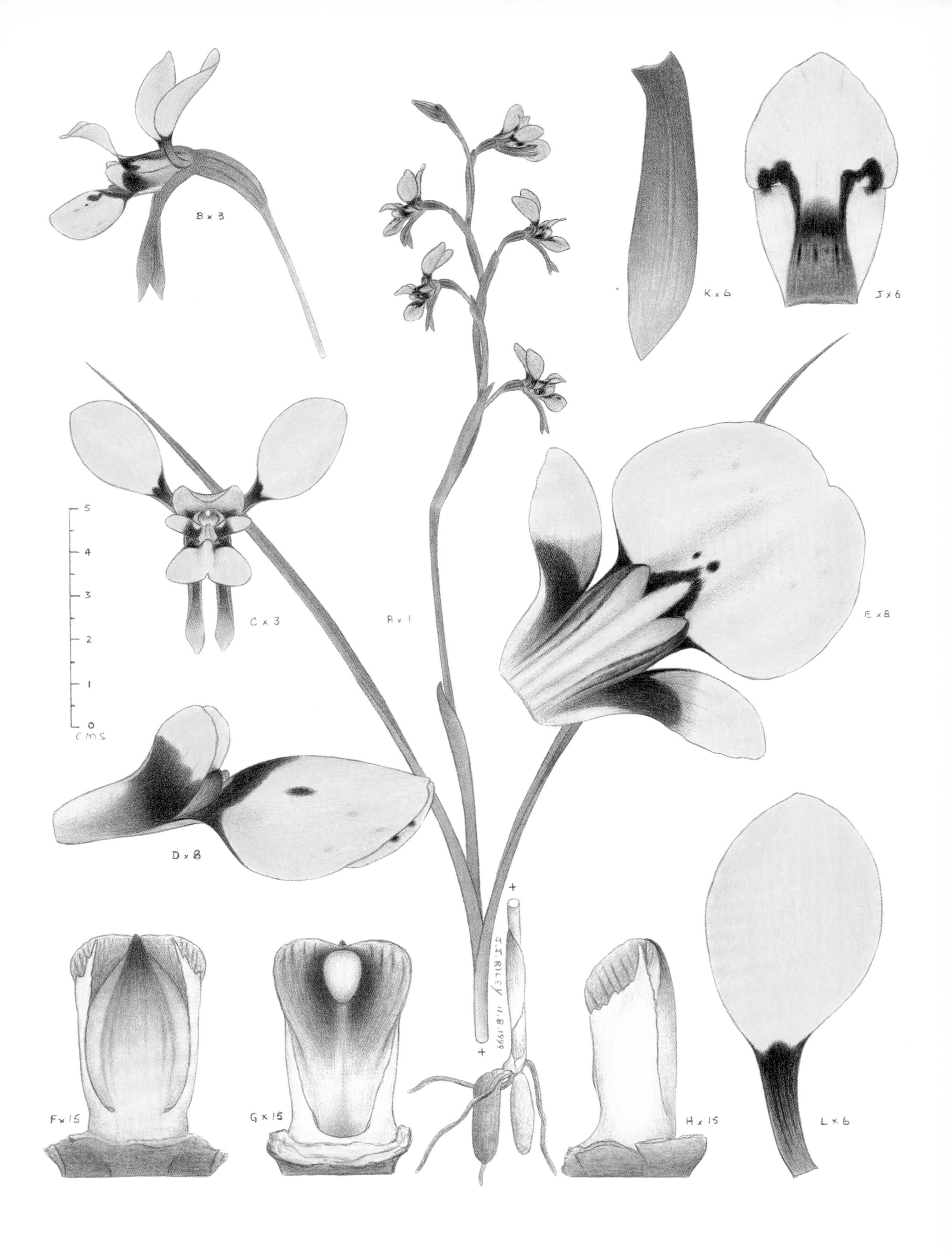
B x 3
K x 6
J x 6
5
4
3
2
1
0
CMS
C x 3
A x 1
E x 8
D x 8
F x 15
G x 15
H x 15
L x 6
J. I. RILEY 11.8.1999

Diuris luteola D.L. Jones

1991 | *Australian Orchid Research* 2:57

TYPE LOCALITY Herberton Range, Queensland

ETYMOLOGY *luteola* — pale yellow

FLOWERING TIME August to September

DISTRIBUTION *D. luteola* is endemic to North Queensland. It is found from the Blackdown Tableland to near the Windsor Tableland.

ALTITUDINAL RANGE 900 m to 1150 m

DISTINGUISHING FEATURES *D. luteola* has pale lemon yellow flowers with a slender labellum and a narrow midlobe.

HABITAT This species grows on slopes and stony ridges in dry sclerophyll woodland, with a grassy understorey. Soils range from sand to gravelly clay loams. The region experiences dry winters with cold nights, and wet summers.

CONSERVATION STATUS Widespread. Conserved in national parks. Secure.

DISCUSSION *D. luteola* has the most northerly occurrence of the *D. aurea* complex. This species, which can be locally common, is not well known as it grows in remote and sparsely populated areas. Throughout its range, there are populations of *Diuris* that don't morphologically fit comfortably under *D. luteola*. Research is required to establish the status of these populations. This species reproduces from seed and is pollinated by small native bees.

Diuris luteola
Ravenshoe, Queensland

16 August 1996

A plant
B flower from front
C flower from side
D labellum from side
E labellum flattened
F column from side
G column from front
H column from rear
J dorsal sepal
K petal

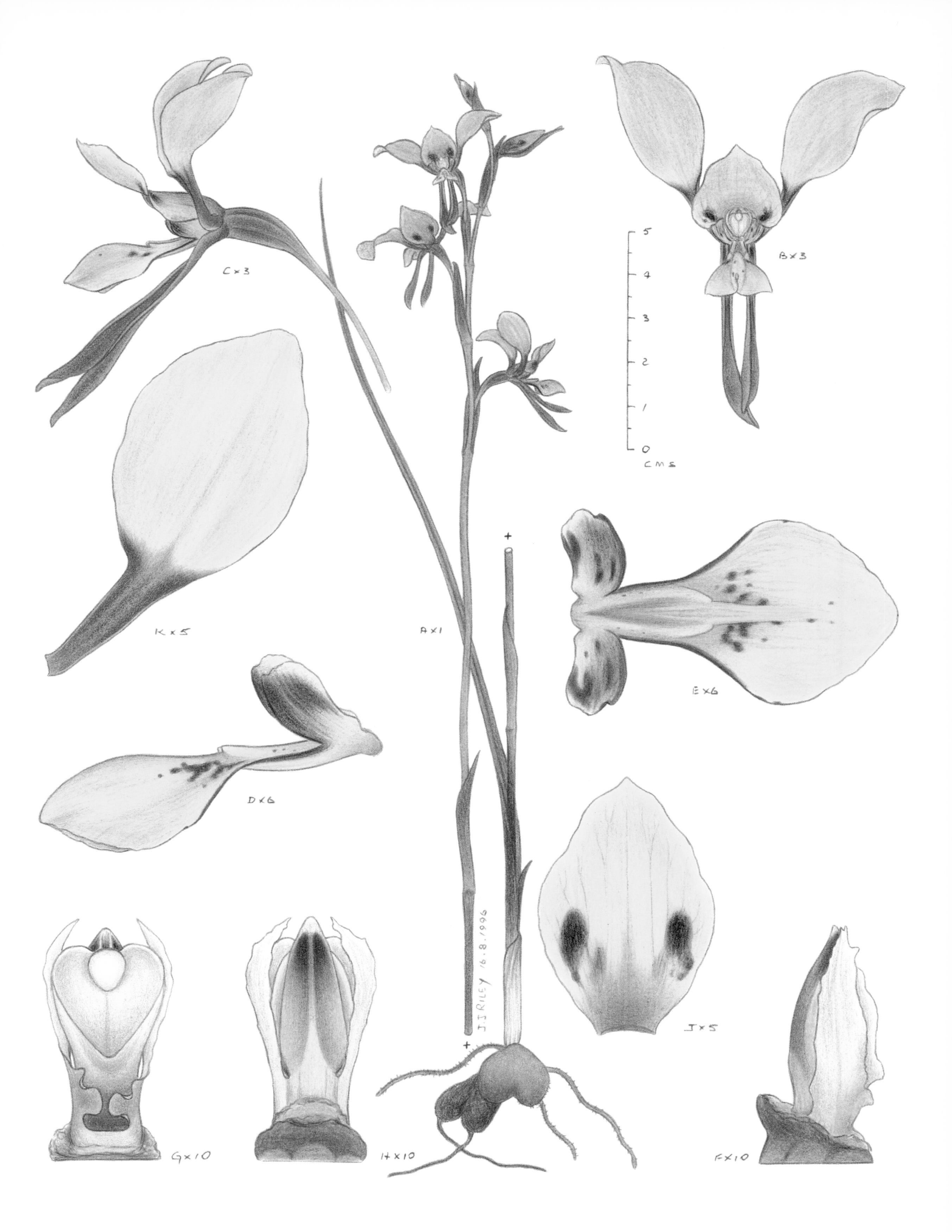
C×3
B×3
5
4
3
2
1
0
CMS
K×5
A×1
E×6
D×6
J×5
G×10
H×10
F×10
J.J. RILEY 16.8.1996

Diuris conspicillata D.L. Jones

1991 | *Australian Orchid Research* 2:54

TYPE LOCALITY Esperance, Western Australia

ETYMOLOGY *conspicillata* — spectacled

FLOWERING TIME August to September

DISTRIBUTION *D. conspicillata* is endemic to Western Australia. It is found from the Fitzgerald River to near Israelite Bay. The largest populations are near Esperance, where it is locally common.

ALTITUDINAL RANGE 20 m to 250 m

DISTINGUISHING FEATURES *D. conspicillata* has bright, conspicuous flowers and well developed sidelobes on the labellum, with distinct, dark blotches. It has a narrow, wedge-shaped labellum midlobe.

HABITAT This species can be found among scrubby heath and between boulders on granite outcrops. Also in coastal heath and heathy woodland on limestone sandy soils. In this habitat winters are wet and summers dry.

CONSERVATION STATUS Known to occur in national parks and nature reserves. Secure.

DISCUSSION *D. conspicillata* is part of the *D. corymbosa* complex, with most species endemic to Western Australia. While the flowers are normally bright yellow with brownish-purple markings, the illustrated specimen has more brown than usual. This species reproduces vegetatively from plants formed on the ends of the long root-like tubers. Colonies can be extensive with a high percentage of mature plants blooming, making an attractive display. It also reproduces from seed and is pollinated by small native bees. The large labellum sidelobes and midlobe, with their prominent brown markings, give the impression of the orchid wearing glasses.

Diuris conspicillata
Esperance, Western Australia

18 August 1999

A plant
B flower from side
C flower from front
D labellum flattened
E labellum from side
F dorsal sepal from front
G dorsal sepal from rear
H petal from front
J petal from rear
K lateral sepal
L column from front
M column from rear
N column from side

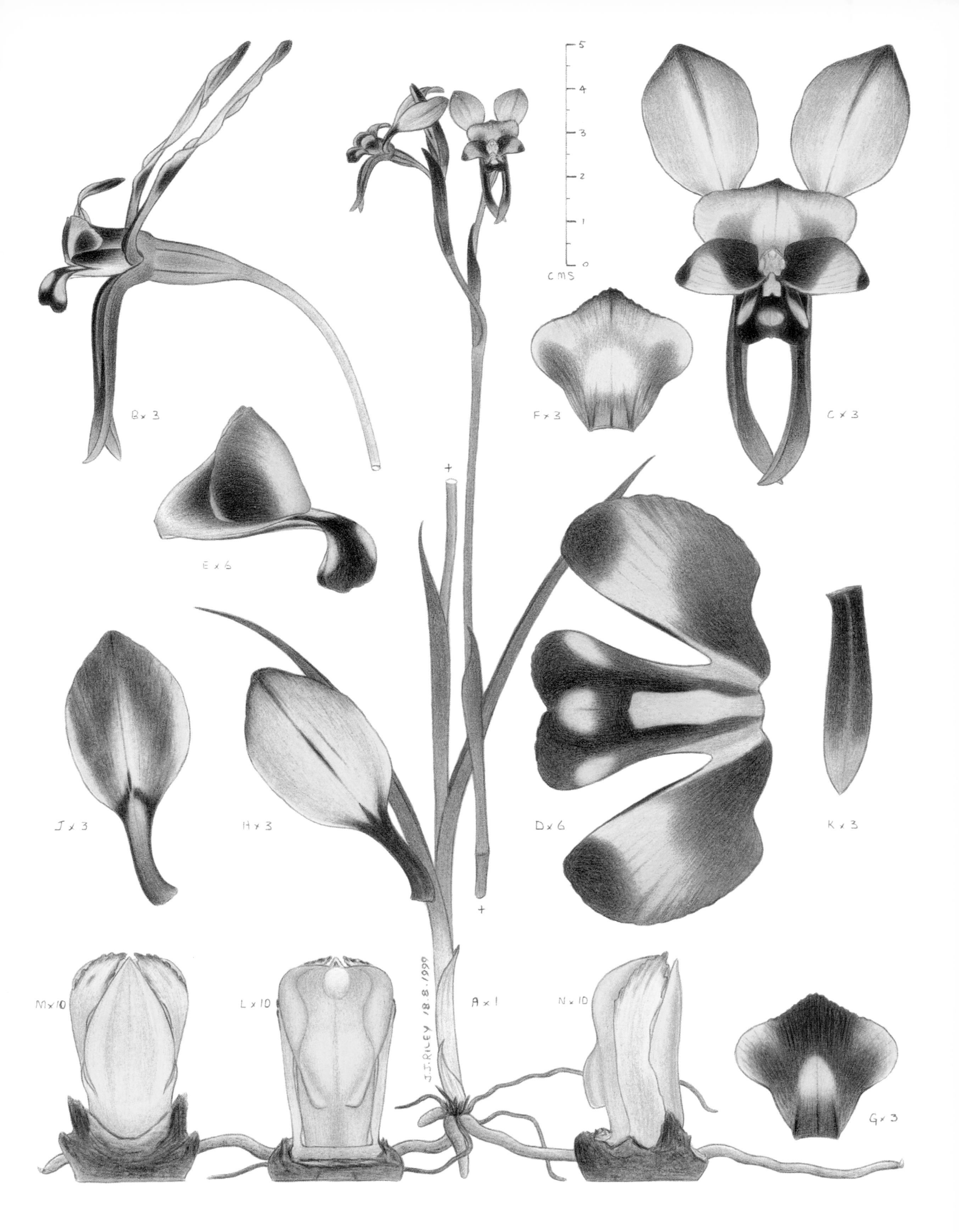
B x 3
C x 3
F x 3
E x 6
J x 3
H x 3
D x 6
K x 3
M x 10
L x 10
A x 1
N x 10
G x 3
CMS
0
1
2
3
4
5
J.J. RILEY 18.8.1999

Dockrillia cucumerina (MacLeay ex Lindl.) Brieger

1981 | *Schltr., Die Orchideen* 3(1):745

TYPE LOCALITY New South Wales

RECENT SYNONYMS *Dendrobium cucumerinum* MacLeay ex Lindl.

ETYMOLOGY *cucumerina* — cucumber-like

FLOWERING TIME November to March

DISTRIBUTION *D. cucumerina* can be found from near Camden and the upper Blue Mountains of New South Wales, northwards along the ranges to the Bunya Mountains in southern Queensland. It is of sporadic occurrence, with many, seemingly suitable habitats for this species devoid of plants.

ALTITUDINAL RANGE 100 m to 600 m

DISTINGUISHING FEATURES Easily recognised by its gherkin-like, succulent leaves.

HABITAT Found almost exclusively on mature river oaks (*Casuarina cunninghamiana*), that line rivers and streams in the valleys draining to the east of the Great Dividing Range.

CONSERVATION STATUS Sporadic. Conserved in national parks and reserves. Secure.

DISCUSSION The type plants of *D. cucumerina* were probably collected from the Mount Hunter area near Camden in New South Wales. W. MacLeay, who named this unusual species as *Dendrobium cucumerinum* in 1842, was Colonial Secretary for New South Wales and owned a large property at Mount Hunter. This orchid still grows in a couple of gullies in the area, but is now quite scarce. Its favourite hosts are large trees of *Casuarina cunninghamiana*, but it can occasionally be found on other tree species, including the introduced weeping willow (*Salix babylonica*). With its creeping rhizomes, *D. cucumerina* can form large mats, usually on the underside of major horizontal branches and on the sheltered sides of trunks of mature trees.

Dockrillia cucumerina
Mount Hunter, New South Wales

22 April 1991

A plant
B flower from front
C flower from rear
D flower from side
E labellum flattened from above
F labellum from side
G column from side
H column from front

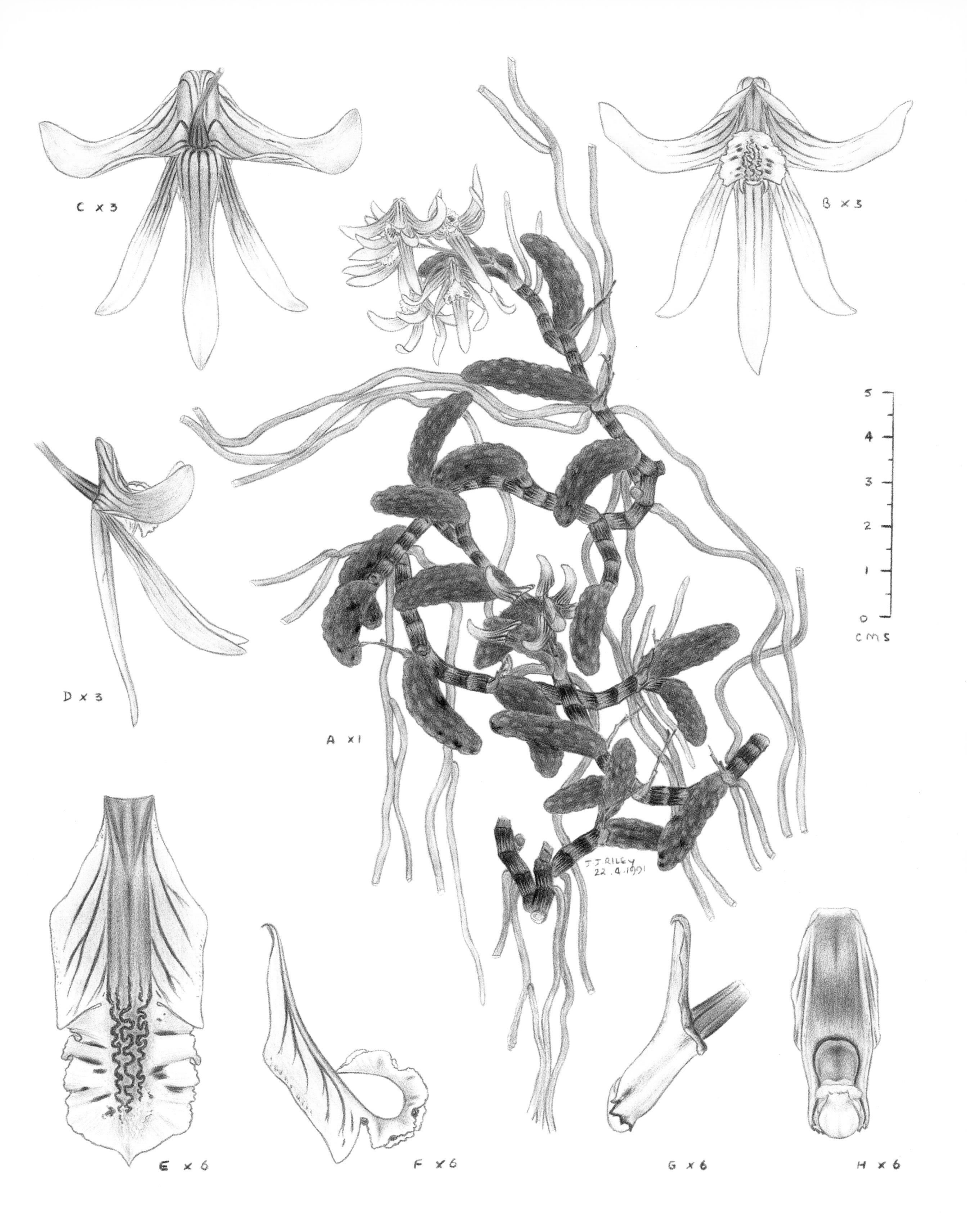
C x 3
B x 3
D x 3
A x1
5
4
3
2
1
0
CMS
T.J. RILEY
22.4.1991
E x 6
F x 6
G x 6
H x 6

Dockrillia striolata (Rchb. f.) Rauschert

1983 | *Feddes Repert* 94(7–8):447

TYPE LOCALITY No location cited

RECENT SYNONYMS *Dendrobium striolatum* Rchb. f.

ETYMOLOGY *striolata* — striped

FLOWERING TIME September to November

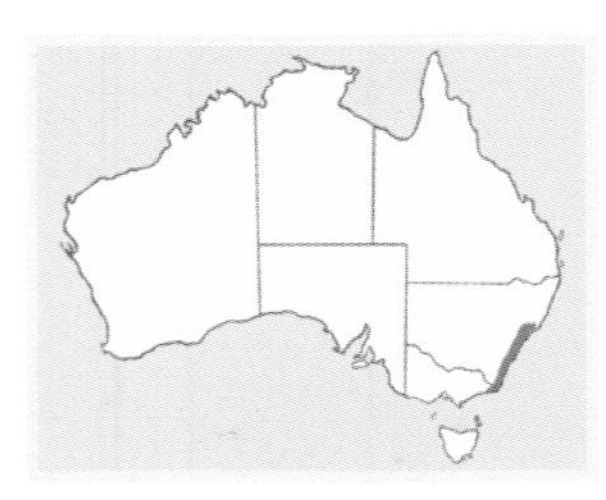

DISTRIBUTION *D. striolata* subsp. *striolata* is found from Cape Barren Island and Flinders Island, Tasmania, across Bass Strait to the Gippsland area of Victoria, and along the coast and ranges at least as far north as the Watagan Mountains, New South Wales. There are unconfirmed reports of this species from the Barrington Tops region of New South Wales.

ALTITUDINAL RANGE 50 m to 1100 m

DISTINGUISHING FEATURES *D. striolata* is a lithophytic species that has short, terete, pendent, and often slightly curved, leaves. The flowers are produced singly or in pairs and are markedly streaked on the back of the sepals. The pure white labellum, devoid of markings, has a distinctly frilly edge. The closely related *D. striolata* subsp. *chrysantha* is endemic to eastern Tasmania and has bright yellow blooms with only faint stripes on the back of the flower.

HABITAT Mostly found on rocky outcrops, along sandstone cliff faces and in protected areas on large boulders near creeks in scrubby woodland to open forest. It is sometimes found on the coast, but is more common on the adjacent hills and higher ranges. In some New South Wales locations, such as Mount York and Jenolan Caves, plants may be dusted with snow in winter. This is one of Australia's few frost hardy epiphytes.

CONSERVATION STATUS Common. Conserved in national parks and reserves. Secure.

DISCUSSION No location was given for the type specimen of *D. striolata*. In favourable conditions this orchid can grow into large, thick, compact clumps. Across its presently recognised range there are at least two distinct variations of plants. The form from the Watagan Mountains differs by having larger, mushroom pink-grey blooms with a huge labellum and a distinctly pendent growth habit with longer leaves and aerial roots along the rhizome.

Dockrillia striolata
Nattai, New South Wales

15 October 1990

A plant
B flower from front
C flower from side
D flower from rear
E labellum from side
F labellum flattened
G column from side
H column from front

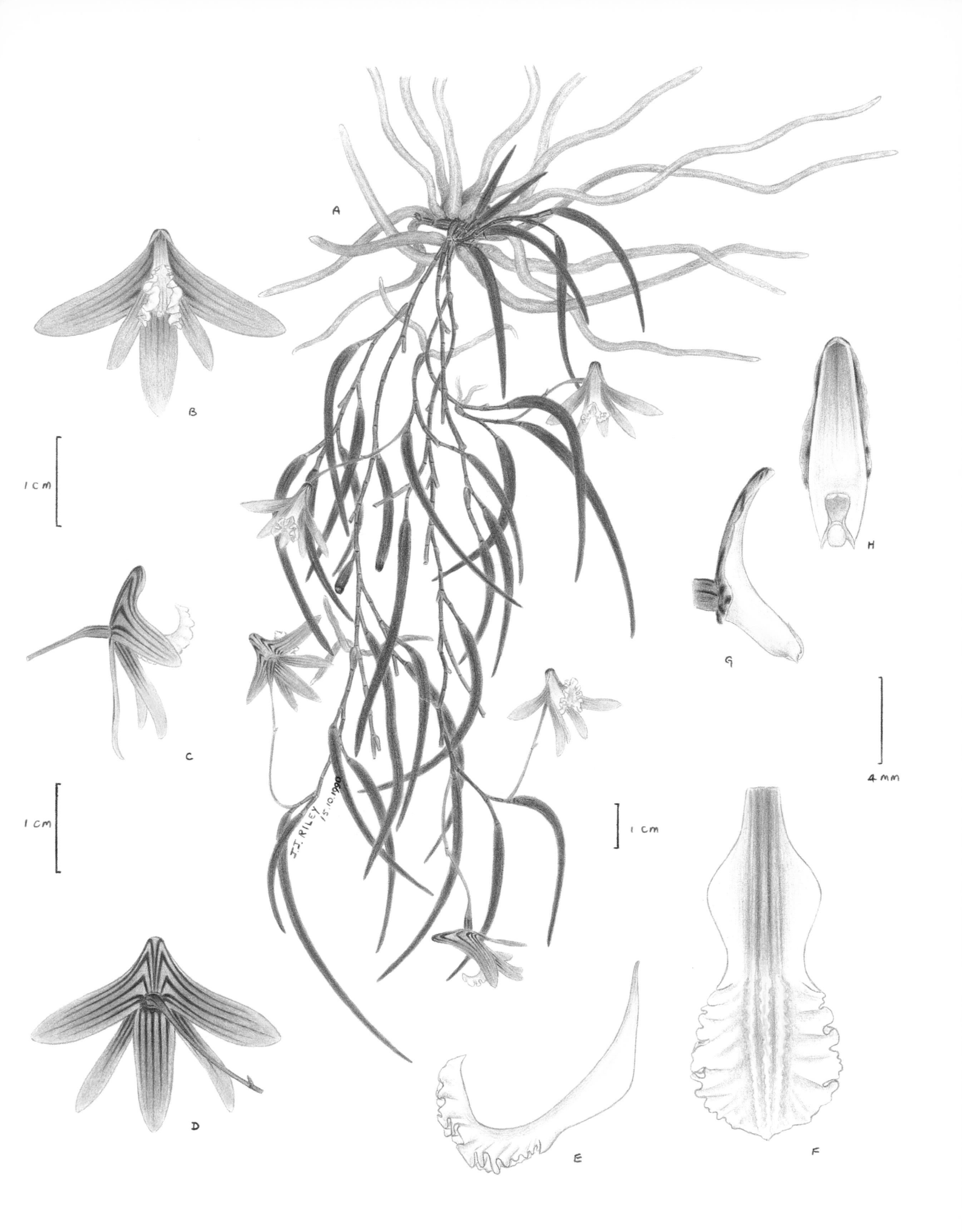
A
B
1 cm
C
1 cm
D
J.J. RILEY 15.10.1990
1 cm
E
F
G
H
4 mm

Dockrillia bowmanii (Benth.) M.A. Clem. & D.L. Jones

1996 | *Lasianthera* 1:17

TYPE LOCALITY Berserker Range, Queensland

RECENT SYNONYMS *Dendrobium bowmanii* Benth.

ETYMOLOGY After E. Bowman

FLOWERING TIME Mostly February to June but can be sporadic

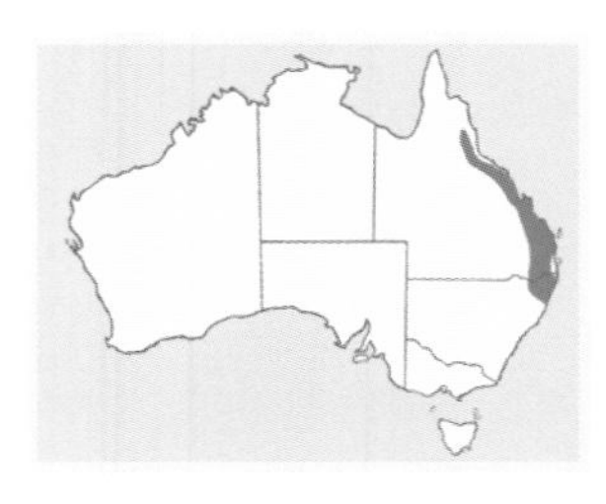

DISTRIBUTION Found from Mount Garnet in North Queensland, south along the coast and ranges of Queensland to the catchment area of the Clarence River in northern New South Wales.

ALTITUDINAL RANGE 50 m to 750 m

DISTINGUISHING FEATURES *D. bowmanii* is a terete-leafed species with an untidy growth habit. Produces pairs of apple green to yellowish-green blooms with a white labellum.

HABITAT This species grows as an epiphyte and less frequently on rocks in dry, scrubby woodland, vine thickets and in dry coastal rainforest. It can also be seen in patches of dry scrub on the western slopes of the Great Dividing Range, well inland from the coast.

CONSERVATION STATUS Common. Conserved in national parks and reserves. Secure.

DISCUSSION *D. bowmanii* is mostly an autumn and winter bloomer, but can be sporadic with a couple of flushes of blooms throughout the year, sometimes only a few weeks apart. It has a straggly growth habit with the rhizome and leaves growing in a number of different directions. This is a very hardy species that can grow in exposed positions and take long periods of hot and dry weather, with very low humidity. *D. bowmanii* was also previously well known as *Dendrobium mortii*, which to confuse issues, actually relates to the species formerly known as *Dendrobium tenuissimum*, now *Dockrillia mortii*.

Dockrillia bowmanii
Brisbane Forest Park, Queensland

20 May 1990

A plant
B flower from front
C flower from side
D flower from rear
E labellum flattened
F column from side
G column from front

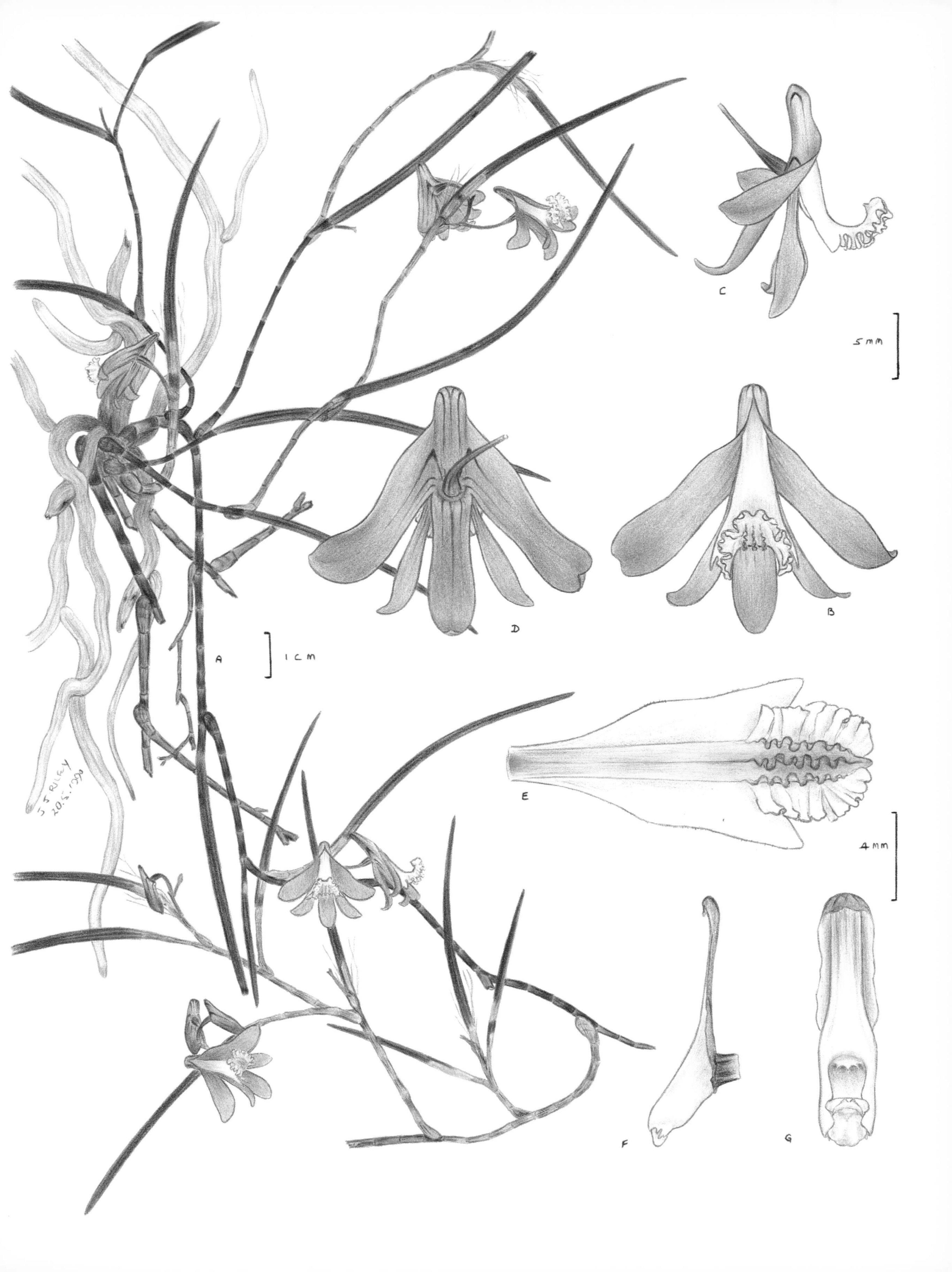
C
5 MM
D
B
A
1 CM
E
4 MM
F
G
J J RILEY
20.5.1990

Epipogium roseum (D. Don) Lindl.

1857 | *Journal of the Proceedings of the Linnean Society, Botany* 1:177

TYPE LOCALITY Nepal

ETYMOLOGY *roseum* — rose pink

FLOWERING TIME December to March

DISTRIBUTION Found from Cape York Peninsula in North Queensland, south along the coast and ranges of Queensland, to around Wauchope, on the mid North Coast of New South Wales.

ALTITUDINAL RANGE 50 m to 700 m

DISTINGUISHING FEATURES *E. roseum* is a leafless saprophyte with pale, insignificant blooms that don't open fully and are somewhat nodding on the inflorescence.

HABITAT This species grows in high rainfall areas along the coast and ranges. It is mostly found in rainforest in heavy shade growing in a moist humus-rich soil. Can sometimes be found in moist grassland.

CONSERVATION STATUS Sporadic. Recorded from national parks and reserves. Should be secure.

DISCUSSION *E. roseum* is a short-lived, saprophytic orchid, growing in leaf litter and decaying organic material. When it first emerges, the flowering stem is bent like a candy cane but straightens as the blooms open. The plant is pale and generally delicate and brittle, with the stem and tuber more or less hollow. While preferring native forest, *E. roseum* has been recorded growing under the introduced *Pinus radiata*. The illustrated specimen was found growing near rainforest on a slashed road verge in front of houses. It emerges after heavy, soaking summer rains and develops rapidly, flowers, sets and disperses seed, and dies in a very short period, often only a matter of weeks. This unpredictability, coupled with its short life cycle, has ensured this poorly understood species is often overlooked in the field. The type specimen of *E. roseum* is from Nepal. Further research is required as it is unlikely that this would represent the same taxon as the Australian species.

Epipogium roseum
Mount Tamborine, Queensland

30 December 1999

A plant
B flower from side
C flower from above
D labellum from above
E longitudinal section of labellum
F column from front
G column from side
H dorsal sepal
J lateral sepal
K petal

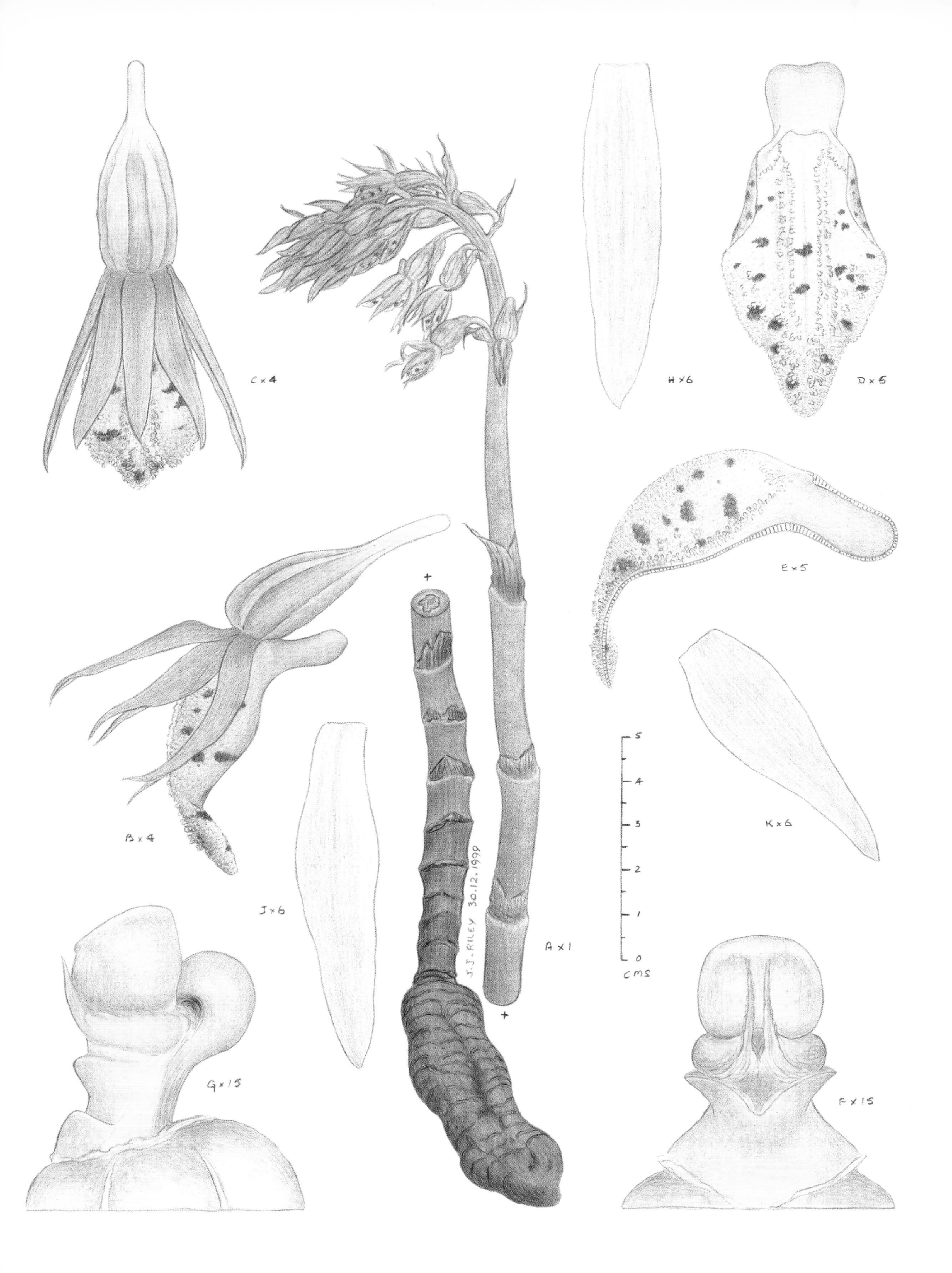

C×4
H×6
D×5
E×5
B×4
K×6
J×6
A×1
J.J. RILEY 30.12.1999
CMS
G×15
F×15

Eriochilus cucullatus (Labill.) Rchb. f.

1871 | *Beiträge zur Systematischen Pflanzenkunde*: 27

TYPE LOCALITY Tasmania

ETYMOLOGY *cucullatus* — hooded

FLOWERING TIME January to May

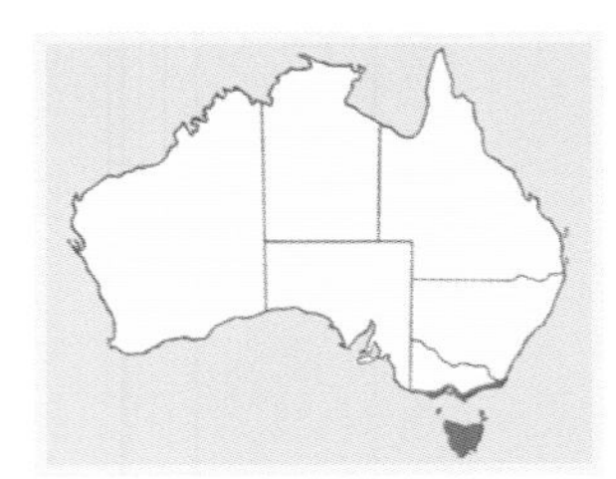

DISTRIBUTION Found from Tasmania and the Bass Strait islands, across to southern Victoria and the south-eastern corner of New South Wales.

ALTITUDINAL RANGE 30 m to 1000 m

DISTINGUISHING FEATURES *E. cucullatus* has small, white, pinkish-white to musk pink blooms with large, well-developed, lateral sepals. The leaf, which is smooth and narrow ovate-lanceolate, is absent or just emerging at flowering.

HABITAT Grows in a number of different habitats including swamps and seasonally wet drainage areas, heath, grassland, open forest, woodland and in shallow sandy soil among rocks.

CONSERVATION STATUS Widespread. Common in national parks and reserves. Secure.

DISCUSSION Many texts quote a wider distribution for this species. *E. cucullatus* has also been recorded from southern Queensland, throughout New South Wales, Victoria and South Australia. We feel these records may represent a complex of related species. Although looking similar at first glance, there are morphological differences with the labellum, column and major leaf variation, which is usually not present when blooming. Research is needed to determine the status of these plants. The illustrated specimen is of a Tasmanian plant and the distribution notes refer to this form. This species is pollinated by small native bees and reproduces from seed.

Eriochilus cucullatus
Blackmans Bay, Tasmania

20 March 1994

A plant
B leaf
C flower from front
D flower from side
E flower from rear
F labellum from side
G labellum from above
H labellum flattened
J column from side
K column from front
L dorsal sepal
M lateral sepal
N petal

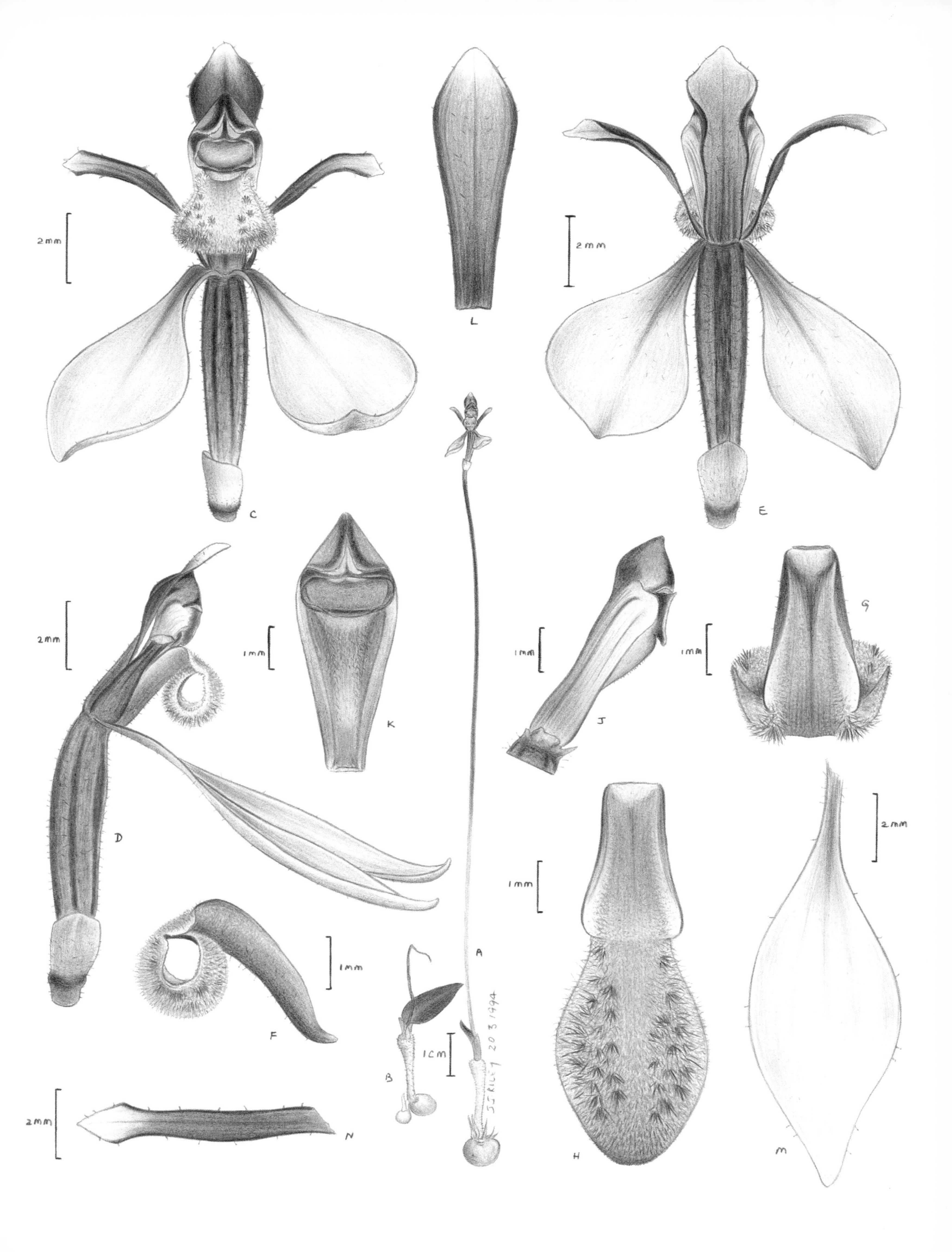
2mm
L
2mm
C
E
2mm
1mm
1mm
1mm
G
K
J
D
2mm
1mm
1mm
A
F
1cm
B
SJ Riley 20 3 1994
2mm
N
H
M

Eriochilus multiflorus Lindl.

1840 | *Edward's Botanical Register*, Appendix to Vols 1–23

TYPE LOCALITY Swan River, Western Australia

RECENT SYNONYMS *Eriochilus dilatatus* Lindl. subsp. *multiflorus* Lindl. (Hopper & A.P. Brown)

ETYMOLOGY *multiflorus* — many flowers

FLOWERING TIME March to June

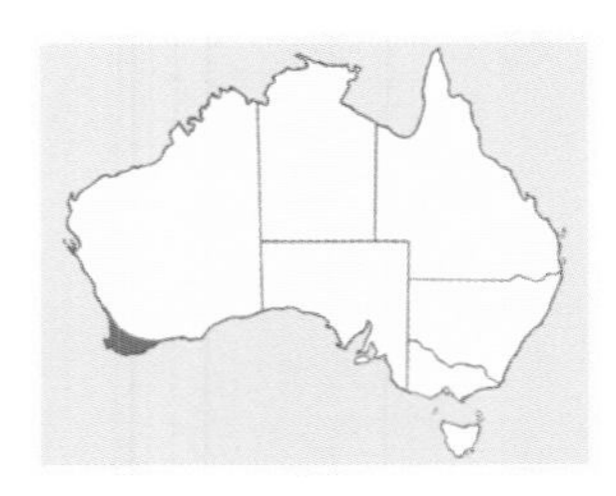

DISTRIBUTION Endemic to the south-western corner of Western Australia, from Perth to Albany.

ALTITUDINAL RANGE 50 m to 200 m

DISTINGUISHING FEATURES Flowering plants have a very small leaf and up to sixteen white flowers with prominent lateral sepals and a labellum covered in fine hairs. Non-flowering plants have significantly larger leaves.

HABITAT Grows in coastal *Banksia* woodland and tall *Eucalyptus* forest, experiencing wet winters and dry summers.

CONSERVATION STATUS Common. Conserved in national parks and reserves. Secure.

DISCUSSION *E. multiflorus* is an early-flowering species that is well known and often locally common. While not needing the stimulus of fire, flowering is enhanced the year after a burn. Interestingly, immature or non-blooming plants have a significantly larger leaf than flowering plants. This species is pollinated by small native bees and reproduces from seed.

Eriochilus multiflorus
Joondalup, Western Australia

6 May 1993

A plant
B non-flowering plant
C flower from front
D flower from side
E flower from rear
F labellum from side
G labellum from above
H labellum flattened
J column from side
K column from front
L dorsal sepal
M lateral sepal
N petal

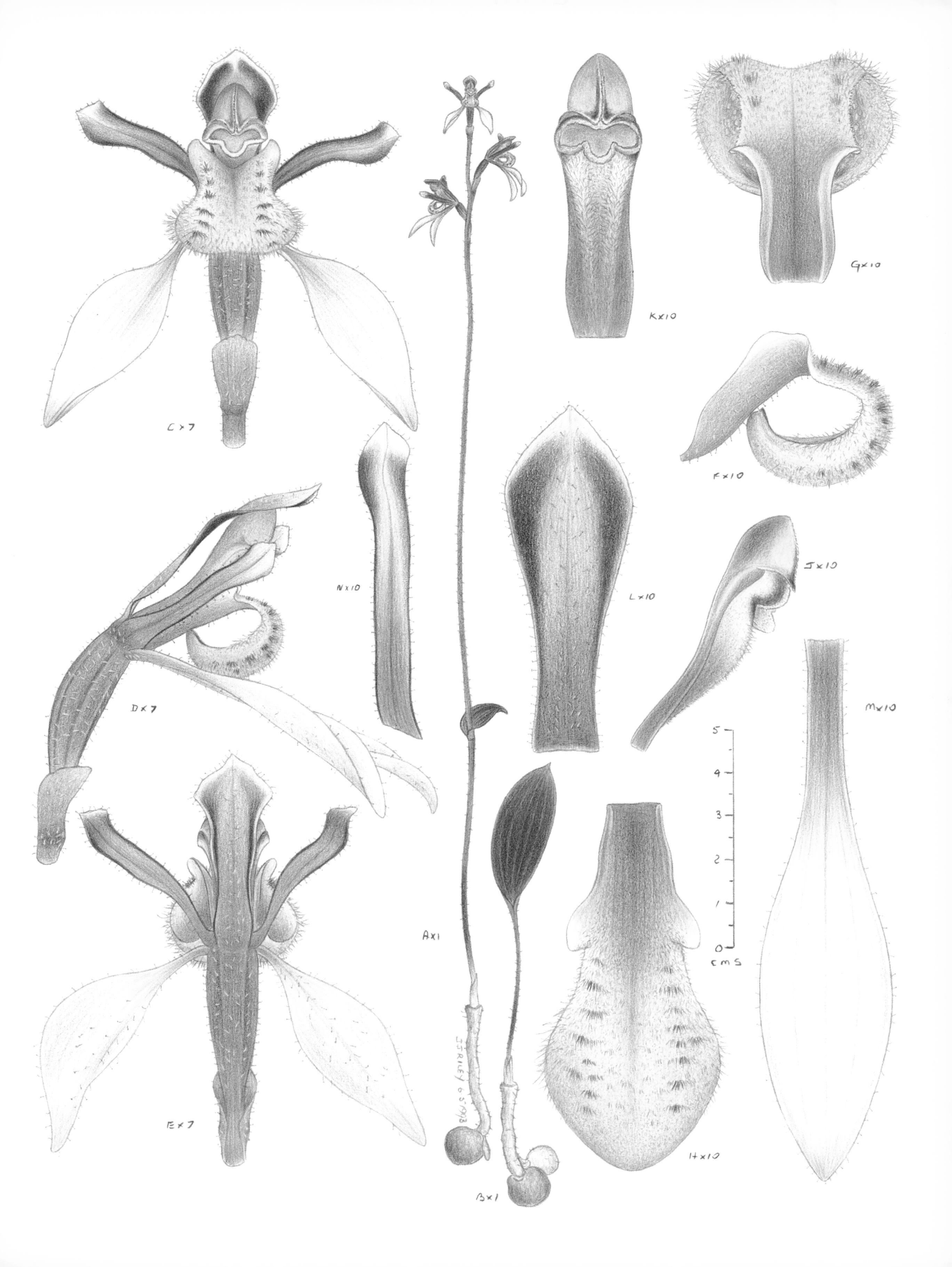

C×7
D×7
E×7
N×10
K×10
G×10
F×10
J×10
L×10
M×10
H×10
A×1
B×1
5
4
3
2
1
0
cms

Gastrodia sesamoides R. Br.

1810 | *Prodromus Florae Novae Hollandiae*: 319

TYPE LOCALITY Port Jackson, New South Wales

ETYMOLOGY *sesamoides* — resembling sesame

FLOWERING TIME October to February

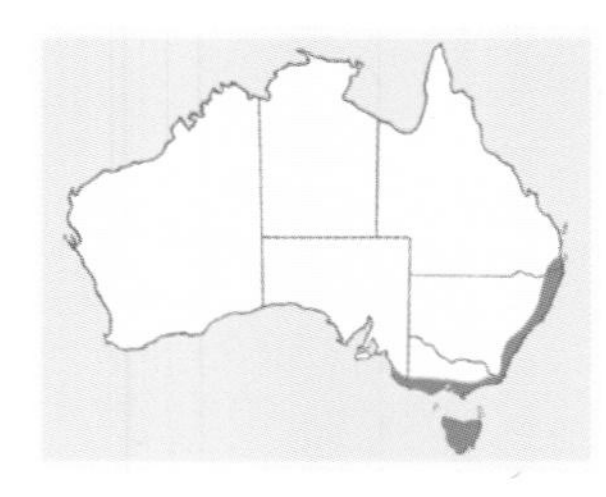

DISTRIBUTION Found from extreme southern Queensland along the coast and ranges of New South Wales, to the southern half of Victoria, to the south-east of South Australia and across to Tasmania.

ALTITUDINAL RANGE 30 m to 1100 m

DISTINGUISHING FEATURES *G. sesamoides* is a leafless saprophyte with a slender growth habit. It has a few, uncrowded tubular flowers on a flower stem that starts off bent like a candy cane, straightening as the blooms open.

HABITAT This species grows in a wide range of plant communities, including heath, shrubby and grassy woodland and forest. It is found from the coast to the ranges and nearby inland slopes. Soils are generally rich in humus, with a covering of leaf litter.

CONSERVATION STATUS Common. Conserved in national parks and reserves. Secure.

DISCUSSION This saprophytic terrestrial is usually of sporadic occurrence. At times it can be locally common with small groups of plants growing close to large *Eucalyptus* trees or decaying timber. *G. sesamoides* has blooms with a pleasant cinnamon fragrance during warm weather. Flowering is greatly increased by fires the previous year. *G. sesamoides* appears to be self-pollinating but may be pollinated by small native bees. It reproduces mainly from seed.

Gastrodia sesamoides
Sunny Corner, New South Wales

30 November 2000

A plant
B flower from front
C flower from side
D flower from rear
E labellum and column from side
F labellum flattened
G column from front

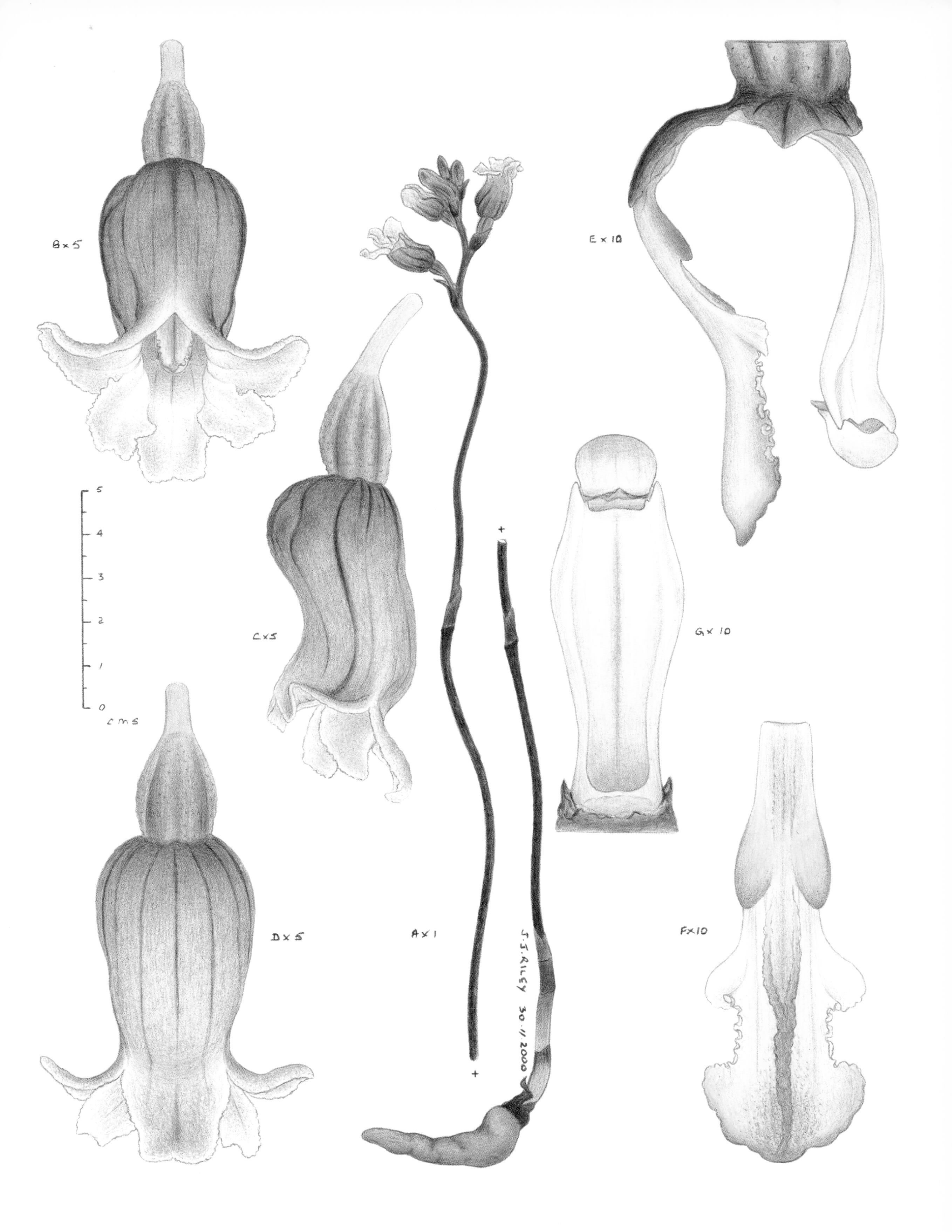

B×5
E×10
C×5
G×10
cms
D×5
A×1
F×10
J.J. RILEY 30.11.2000

Genoplesium baueri R. Br.

1810 | *Prodromus Florae Novae Hollandiae*: 310

TYPE LOCALITY Port Jackson, New South Wales

ETYMOLOGY After Ferdinand Bauer

FLOWERING TIME December to May

DISTRIBUTION Endemic to New South Wales, found from Port Stephens south to Jervis Bay.

ALTITUDINAL RANGE 30 m to 400 m

DISTINGUISHING FEATURES *G. baueri* is a short, leafless saprophyte, with up to five, relatively large, fleshy flowers with prominent, yellow-green, upswept sepals. It also has an unusual and distinctive tuber.

HABITAT This species usually grows in sand or sandy loams in heathland to shrubby woodland. Sometimes found in moss gardens, growing in shallow soils over sandstone.

CONSERVATION STATUS Uncommon. Known to occur in national parks and reserves.

DISCUSSION *G. baueri* is a poorly known species that is very sporadic in occurrence. It was probably more common along the coast of New South Wales before housing and development reduced its preferred habitat. It is most frequently seen after fires the previous season, with the individual plants scattered among the charred ground. The number of blooming plants drops quickly as the surrounding vegetation recovers after the fire. After a couple of years the sighting of a single flowering specimen may be very rare, until the next burn. This is the type species of the genus *Genoplesium*, which may prove to be monotypic. It is possibly pollinated by tiny flies and reproduces from seed.

Genoplesium baueri
Waterfall, New South Wales

18 March 1998

A plant
B flower from front
C flower from side
D flower from above
E labellum from side
F labellum from front
G labellum from rear
H column from side
J column from front
K column from rear
L dorsal sepal from front
M dorsal sepal from side
N lateral sepal
O petal
P floral bract

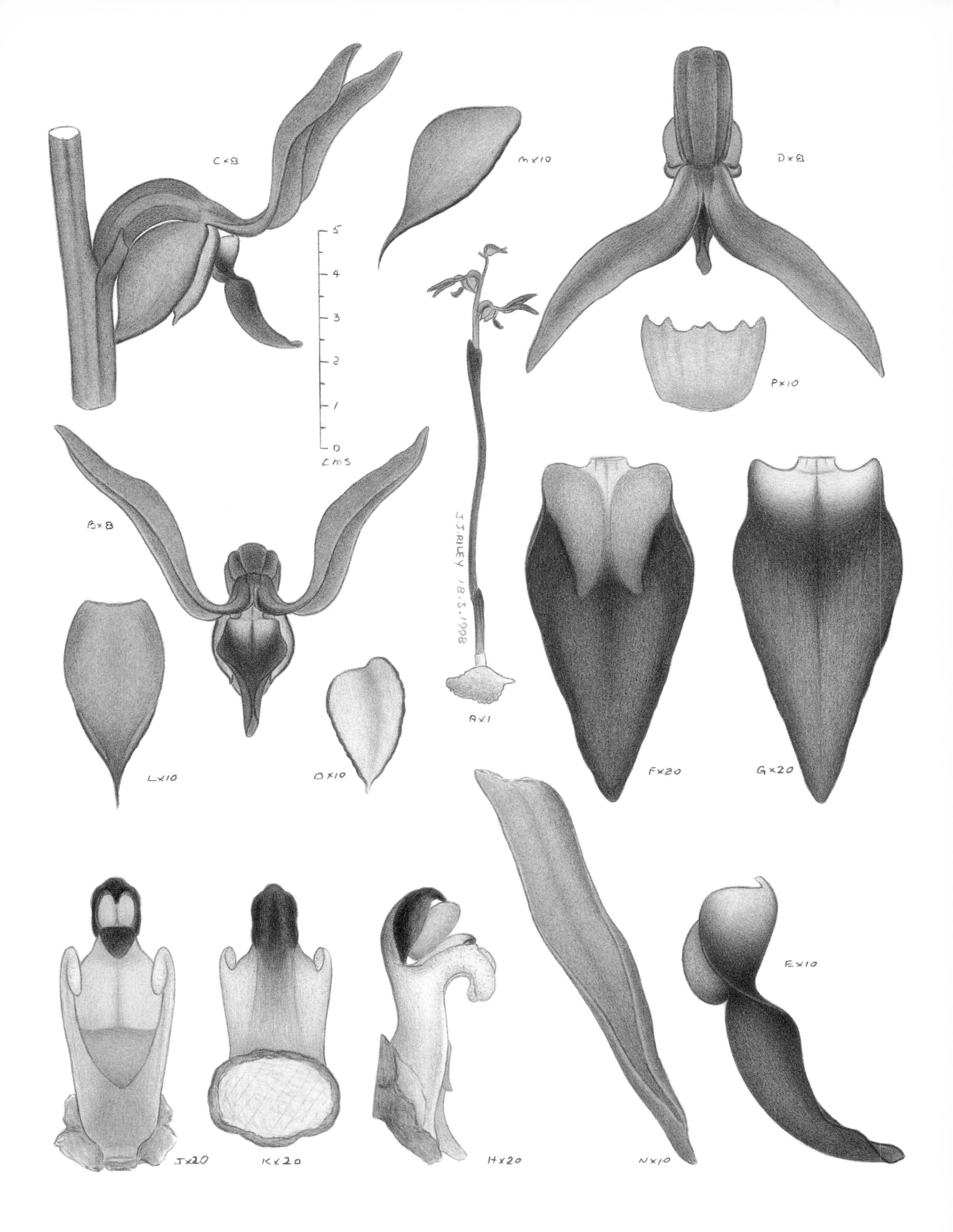

C×8
M×10
D×8
5
4
3
2
1
0
CMS
P×10
B×8
J.J.RILEY 18.3.1998
L×10
O×10
A×1
F×20
G×20
E×10
J×20
K×20
H×20
N×10

Genoplesium nigricans (R. Br.) D.L. Jones & M.A. Clem.

1989 | *Lindleyana* 4(3):143

TYPE LOCALITY Port Lincoln, South Australia

RECENT SYNONYMS *Prasophyllum nigricans* R. Br.

ETYMOLOGY *nigricans* — black

FLOWERING TIME February to May

DISTRIBUTION This species is endemic to South Australia, being found on the lower Eyre Peninsula and Kangaroo Island.

ALTITUDINAL RANGE 30 m to 150 m

DISTINGUISHING FEATURES *G. nigricans* is a short plant with dark flowers. It has a greenish-brown, ovate labellum and brown petals with parallel sides and a rounded apex tipped with a gland.

HABITAT Found in heath, heathy to shrubby mallee and heathy woodland, often on limestone. Soils are sand or sandy loams. Can also be seen growing in the shallow soils on limestone outcrops.

CONSERVATION STATUS Conserved in national parks and conservation reserves. Secure.

DISCUSSION In the past, there has been some misunderstanding regarding the true identity of *G. nigricans*, with many, already named taxa being mistaken for this species. This has led to some texts quoting a wider distribution. *G. nigricans* is usually seen growing as scattered individuals or in loose groups at the base of small shrubs or among mosses and coral lichens in clearings among the heathy plants. It is pollinated by tiny flies and reproduces from seed. Was recently reclassified as *Corunastylis nigricans* (R. Br.) D.L. Jones & M.A. Clem.

Genoplesium nigricans
Kangaroo Island, South Australia

28 March 1998

A plant
B flower from front
C flower from side
D flower from above
E labellum from side
F labellum from front
G labellum from rear
H column from side
J column from front
K column from rear
L dorsal sepal from front
M dorsal sepal from side
N lateral sepal
O petal
P floral bract

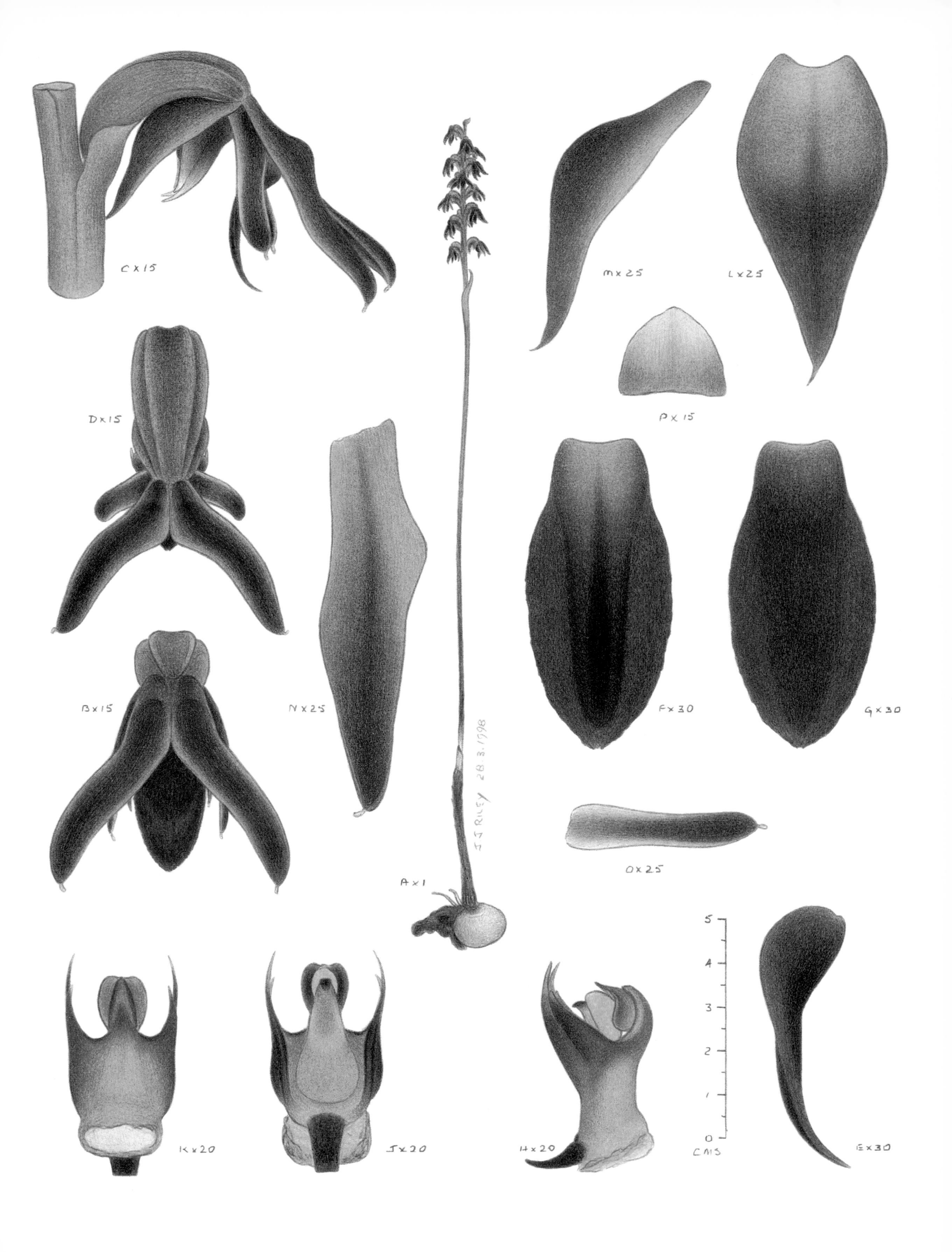

C x 15
M x 25
L x 25
D x 15
P x 15
B x 15
N x 25
J J RILEY 28.3.1998
F x 30
G x 30
O x 25
A x 1
5
4
3
2
1
0
CMS
K x 20
J x 20
H x 20
E x 30

Genoplesium citriodorum D.L. Jones

1991 | *Australian Orchid Research* 2:67

TYPE LOCALITY Woodford, New South Wales

ETYMOLOGY *citriodorum* — lemon-scented

FLOWERING TIME December to April

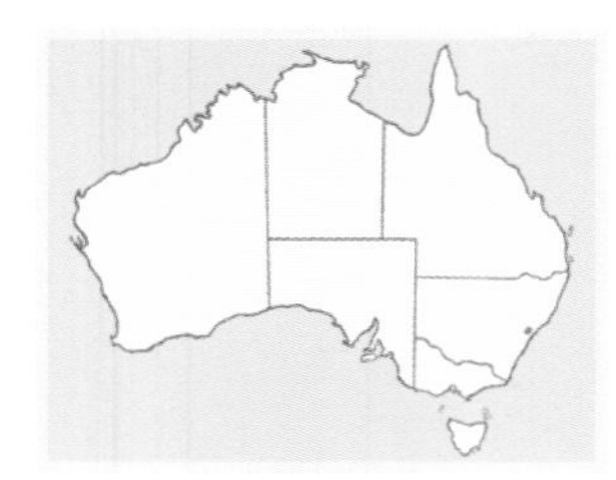

DISTRIBUTION This species is endemic to the upper Blue Mountains, New South Wales.

ALTITUDINAL RANGE 700 m to 1000 m

DISTINGUISHING FEATURES *G. citriodorum* is a tall *Genoplesium* with up to forty blooms crowded at the top of the upright flowering stem. The labellum is short and wide with a fringe of long, purple hairs. The flowers have a very noticeable, and strong, lemon fragrance.

HABITAT Mostly shrubby woodland growing in moss gardens on horizontally bedded, flat sandstone ridges in shallow soils. It can be locally common with a preference for disturbed sites, sometimes colonising the edges of walking trails and the edges of dirt roads.

CONSERVATION STATUS Uncommon and sporadic. Is found in a national park.

DISCUSSION *G. citriodorum* favours slightly open and exposed sites, with little competition from grasses and thick low scrub. It is mostly seen among coral lichens and mosses in open spaces between scraggly shrubs. *G. citriodorum* has a delightful lemon fragrance that is particularly strong and noticeable on warm days. It is pollinated by tiny flies and reproduces from seed. Was recently reclassified as *Corunastylis citriodora* (D.L. Jones & M.A. Clem.) D.L. Jones & M.A. Clem.

Genoplesium citriodorum
Mount Tomah, New South Wales

13 February 1990

A plant
B flower from front
C flower from side
D flower from above
E labellum from front
F labellum from rear
G column from side
H column from front
J dorsal sepal from front
K dorsal sepal from side
L petal

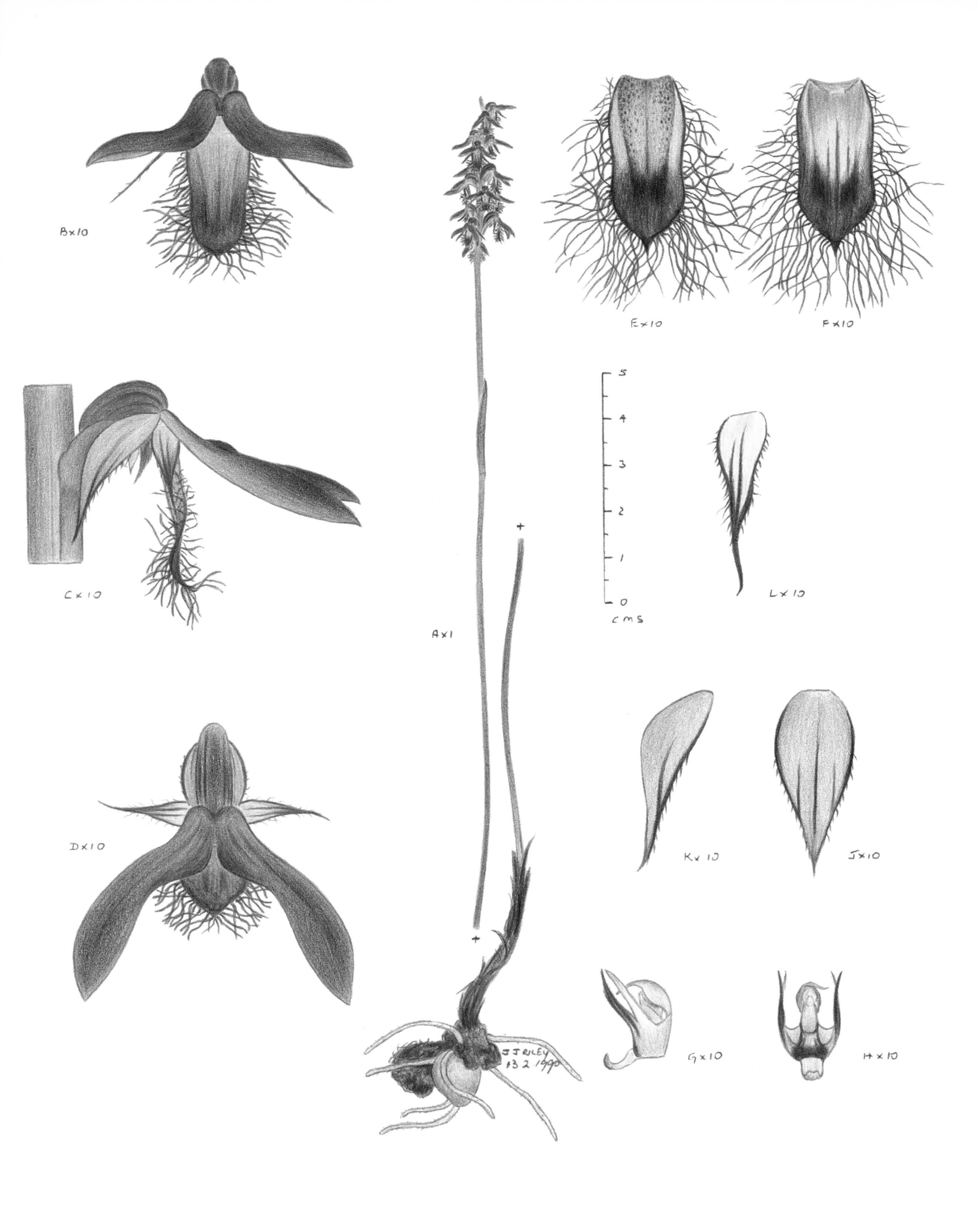

B×10
E×10
F×10
C×10
A×1
L×10
cms
D×10
K×10
J×10
G×10
H×10
J J RILEY
13 2 1990

Genoplesium plumosum (Rupp) D.L. Jones & M.A. Clem.

1989 | *Lindleyana* 4(3):144

TYPE LOCALITY Botany Bay, New South Wales

ETYMOLOGY *plumosum* — feathery

FLOWERING TIME February to April

DISTRIBUTION This species is endemic to New South Wales, and has only been recorded from Kurnell and Tallong.

ALTITUDINAL RANGE 10 m to 700 m

DISTINGUISHING FEATURES *G. plumosum* is short in stature and only produces a few flowers. The floral segments are long and narrow with the lateral sepals swept upwards. There are numerous, fine ciliate hairs on the front lobe of the labellum.

HABITAT The habitat of the Kurnell population is unknown. At Tallong, it occurs in low scrubby heath that is dominated by *Kunzea capitata*. It grows in shallow soils on flat to slightly sloping, conglomerate outcrops.

CONSERVATION STATUS Very rare. Not conserved in national parks or reserves. Vulnerable.

DISCUSSION *G. plumosum* was initially named as *Prasophyllum plumosum* by Rupp in 1947 from material collected in 1928 at Kurnell, on the southern shores of Botany Bay. Very little is known of its habitat and abundance at that location. Repeated searches over many years have been unsuccessful. This fact, coupled with the amount of development for housing and industry, means this species is almost certainly extinct at the type-site. Fortunately another population was located in the late 1980s. It is rare at Tallong and only occurs as scattered individuals or small groups. It is mostly found tucked under or among heathy shrubs on shallow coral lichen covered soils. It is pollinated by tiny flies and reproduces from seed. Was recently reclassified as *Corunastylis plumosa* (Rupp) D.L. Jones & M.A. Clem.

Genoplesium plumosum
Tallong, New South Wales

7 March 1996

A plant
B flower from front
C flower from side
D flower from above
E labellum from side
F labellum from front
G column from front
H column from rear
J column from side
K dorsal sepal from front
L dorsal sepal from side
M lateral sepal
N petal
O floral bract

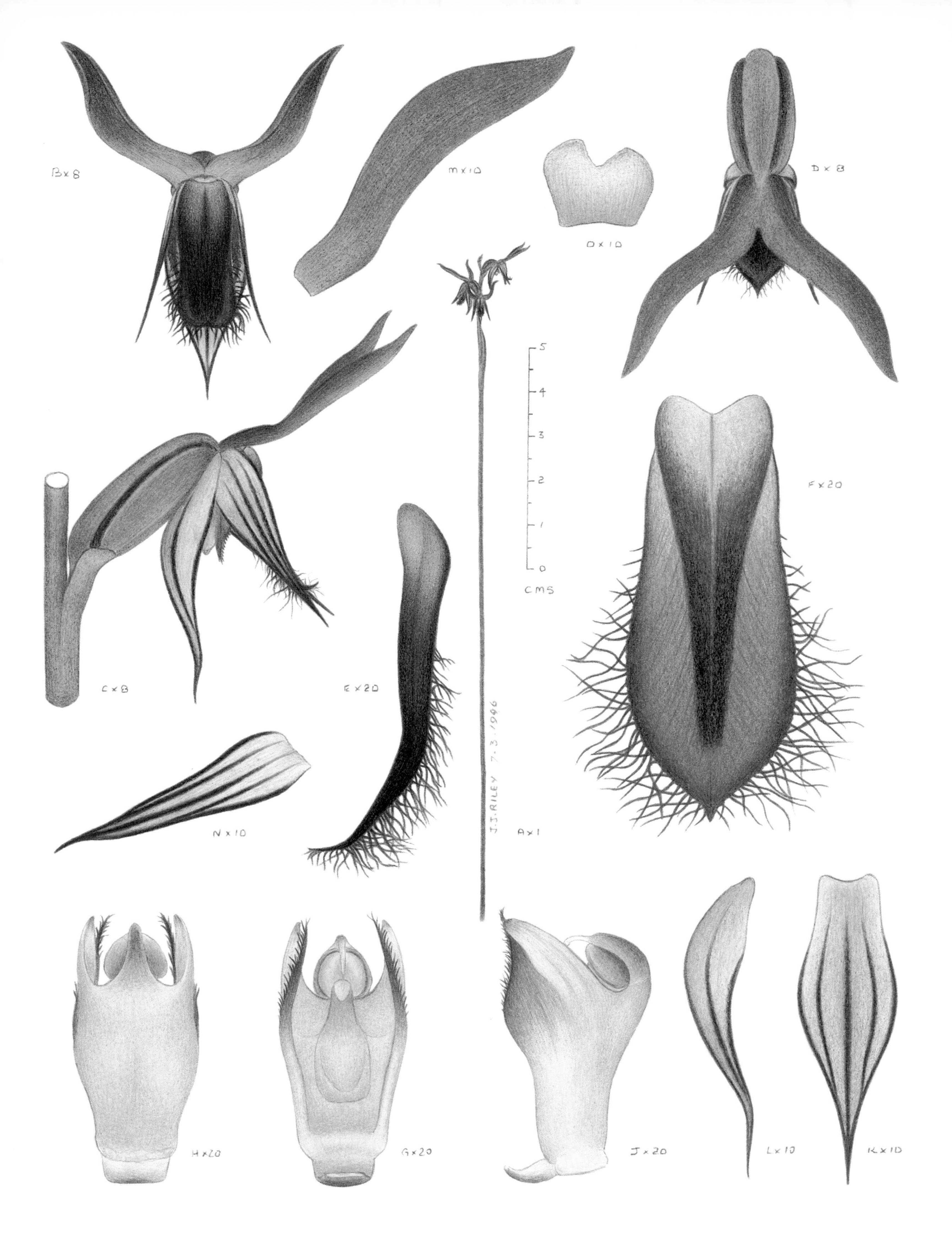
B x 8
M x 10
O x 10
D x 8
C x 8
E x 20
F x 20
5
4
3
2
1
0
CMS
N x 10
J.J. RILEY 7.3.1996
A x 1
H x 20
G x 20
J x 20
L x 10
K x 10

Genoplesium alticolum D.L. Jones & B. Gray

1991 | *Australian Orchid Research* 2:66

TYPE LOCALITY Cook District, Queensland

ETYMOLOGY *alticolum* — growing at high elevations

FLOWERING TIME December to February

DISTRIBUTION This species is only known from North Queensland, and is presently known only from the Herberton Range and Walshs Pyramid.

ALTITUDINAL RANGE 600 m to 850 m

DISTINGUISHING FEATURES *G. alticolum* has a large, oval labellum that is densely fringed with short fine cilia.

HABITAT This species grows among clumps of grass and small shrubs, usually on the tops of ridges and on slopes among rock outcrops in sparse open forest. Soils are shallow sandy or gravelly clay loams. Sometimes plants can be observed growing on banks and cuttings along roads and tracks. Winters are dry, with most of the rain falling in the summer.

CONSERVATION STATUS Uncommon. Known to occur in a national park. Secure.

DISCUSSION *G. alticolum* is a poorly known species that may be more common and widespread. This small orchid is difficult to detect in the field among the grasses and low herbs. The summer temperatures are very high which would discourage searches for this orchid. Occurs as scattered individuals or small groups. It is pollinated by tiny flies and reproduces from seed. Was recently reclassified as *Corunastylis alticola* (D.L. Jones & B. Gray) D.L. Jones & M.A. Clem.

Genoplesium alticolum
Atherton Tableland, Queensland

20 February 1998

A plant
B flower from side
C flower from above
D flower from front
E labellum from front
F labellum from rear
G labellum from side
H column from front
J column from side
K column from rear
L dorsal sepal from front
M dorsal sepal from side
N lateral sepal
O petal
P floral bract

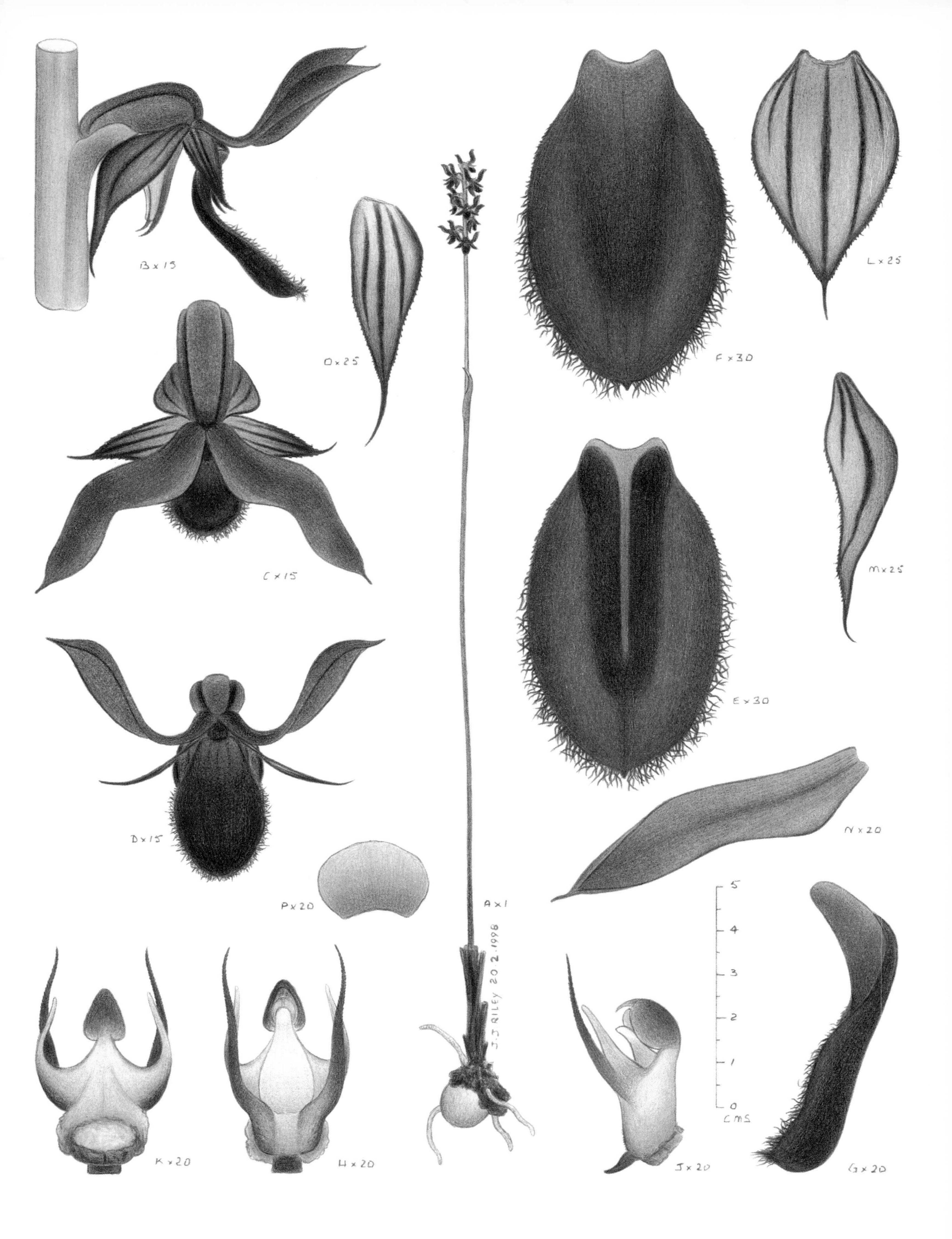
B x 15
O x 25
F x 30
L x 25
C x 15
M x 25
E x 30
D x 15
N x 20
P x 20
A x 1
J.J. RILEY 20.2.1998
5
4
3
2
1
0
CMS
K x 20
H x 20
J x 20
G x 20

Leporella fimbriata (Lindl.) A.S. George

1971 | *Nuytsia* 1(2):183

TYPE LOCALITY Swan River, Western Australia

RECENT SYNONYMS *Leptoceras fimbriata* Lindl.

ETYMOLOGY *fimbriata* — fringed

FLOWERING TIME March to August

DISTRIBUTION This widespread species is found in Victoria from near Melbourne along the coast and inland to the Grampians, the south-east of South Australia including the Mount Lofty Ranges, lower Flinders Ranges, Eyre Peninsula and Kangaroo Island, then across to south-west Western Australia.

ALTITUDINAL RANGE 50 m to 300 m

DISTINGUISHING FEATURES *L. fimbriata* is a distinctive species with a large, wide, fringed labellum and prominent petals that are held erect.

HABITAT Grows in a variety of plant communities, including coastal *Banksia* heath, scrubby woodland and damp heathy areas in open forest.

CONSERVATION STATUS Widespread and common. Conserved in national parks and reserves. Secure.

DISCUSSION *L. fimbriata* has a somewhat disjunct distribution. It has one or two bluish-green leaves (not illustrated) that are broadly lanceolate and edged with red maroon. This species forms extensive colonies vegetatively but can be a shy-bloomer. Flowering is greatly enhanced the year after a bushfire. *L. fimbriata* also reproduces from seed and is pollinated by male flying ants that attempt to copulate with the labellum of the flower.

Leporella fimbriata
Bibra Lake, Western Australia

3 May 1995

A plant
B flower from front
D flower from side
E labellum from side
F labellum flattened
G column from side
H column from front
J dorsal sepal
K lateral sepal
L petal

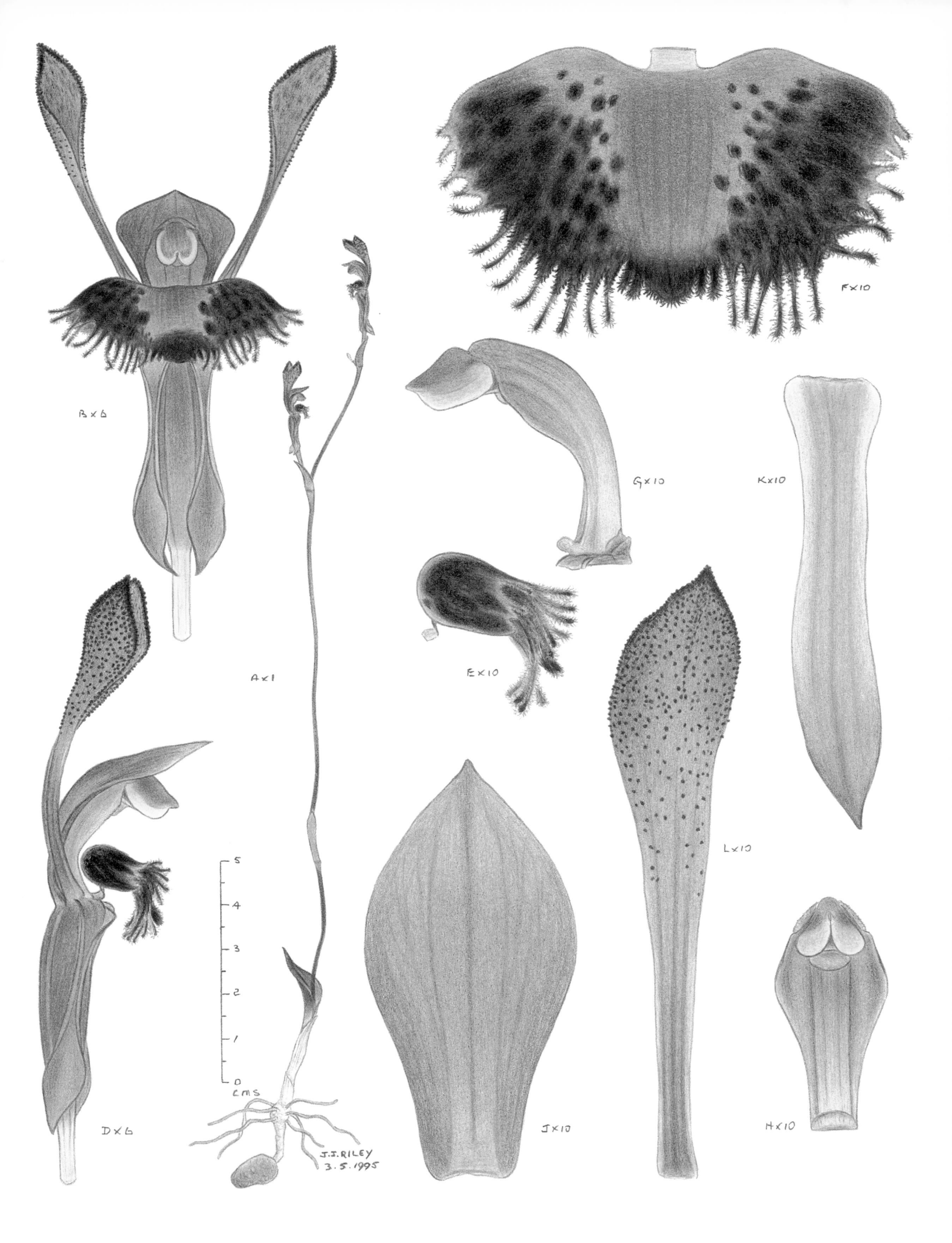
B×6
F×10
G×10
K×10
E×10
A×1
L×10
D×6
J×10
H×10
5
4
3
2
1
0
cms
J.J. RILEY
3.5.1995

Liparis reflexa (R. Br.) Lindl.

1825 | *Edward's Botanical Register* 11:882

TYPE LOCALITY Port Jackson, New South Wales

ETYMOLOGY *reflexa* — bent backwards

FLOWERING TIME February to June

DISTRIBUTION This species is endemic to New South Wales, and is locally common north of Bega to around the catchment area of the Hastings River.

ALTITUDINAL RANGE Up to 1000 m

DISTINGUISHING FEATURES *L. reflexa* is a lithophytic species with a yellowish-green plant and flowers. It has petals and sepals that recurve towards the ovary and a labellum with a blunt apex.

HABITAT This species is usually found on rock outcrops or rock escarpments and cliff lines in wet sclerophyll forest. It is also in sheltered gullies in *Eucalyptus* forest growing on sandstone boulders. Grows on the coast, sometimes within the spray of salt water, and nearby ranges. *L. reflexa* may also grow as an epiphyte, and in some locations, such as Kangaroo Valley, can be quite common on trees.

CONSERVATION STATUS Widespread. Common and conserved in national parks and reserves. Secure.

DISCUSSION *L. reflexa* grows into large clumps that can dominate suitable rock faces. Occasional plants become dislodged from the cliff line and continue to grow on the ground or leaf litter where they have fallen. A particularly robust form occurs at Alum Mountain, Bulahdelah. The flowers of *L. reflexa* have a distinctive, sickly, urine-like odour that is quite unpleasant. This species has been recorded from southern Queensland, but it is likely that these were mistaken for the closely related *L. swenssonii*. Small flies and mosquitoes are known to pollinate this orchid.

Liparis reflexa
Kentlyn, New South Wales

5 May 1999

A plant
B flower from front
C flower from side
D labellum from side
E labellum flattened
F column from side
G column from front
H dorsal sepal
J lateral sepal
K petal

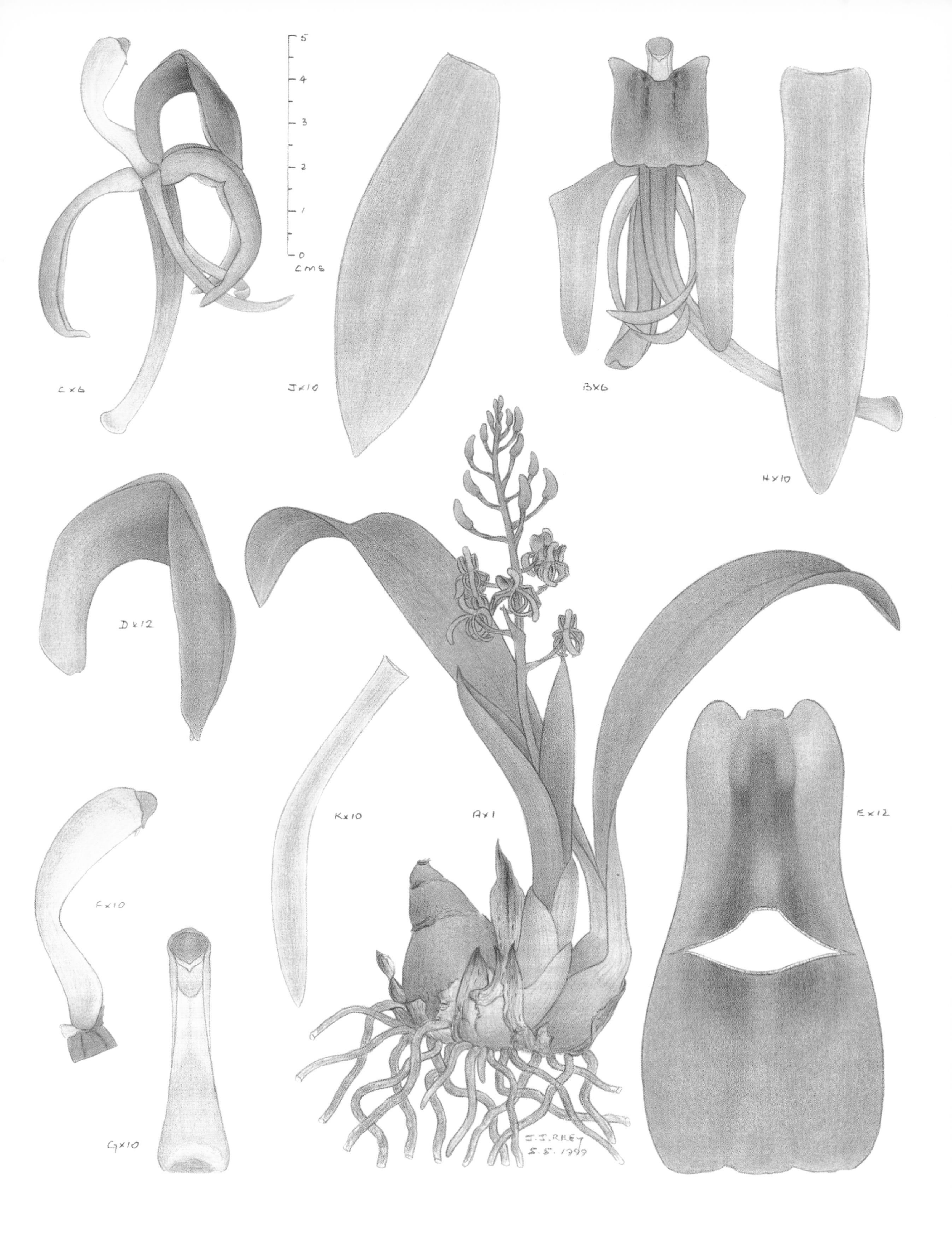
5
4
3
2
1
0
cms
C x 6
J x 10
B x 6
H x 10
D x 12
K x 10
A x 1
E x 12
F x 10
G x 10
J.J. RILEY
5.5.1999

Liparis swenssonii Bailey

1906 | *Queensland Agricultural Journal* 16(9):564

TYPE LOCALITY Emu Vale, Queensland

RECENT SYNONYMS *Liparis reflexa* (R. Br.) Lindl. var. *parviflora* Nicholls

ETYMOLOGY After Carl Swensson

FLOWERING TIME March to July

DISTRIBUTION Found from the Glasshouse Mountains in southern Queensland, south along the ranges and tablelands of New South Wales to Lorne, just north of the Manning River.

ALTITUDINAL RANGE 200 m to 1200 m

DISTINGUISHING FEATURES *L. swenssonii* has smaller flowers than *L. reflexa*, with narrower segments that are spreading and not as reflexed. The labellum also has a rounded apex.

HABITAT This species grows on rocks and cliff lines in wet sclerophyll forest and rainforest. In some locations it can form very large plants. Occasionally it may be found on the trunks of older rainforest trees.

CONSERVATION STATUS Widespread. Common and conserved in national parks and reserves. Secure.

DISCUSSION *L. swenssonii* is very similar to *L. reflexa* and it is very difficult to separate these species out of bloom. There has been confusion over the exact distribution of these taxa, particularly in northern New South Wales, as *L. swenssonii* is poorly known. Like its sister species, *L. swenssonii* also has an unpleasant odour and is pollinated by small flies and mosquitoes.

Liparis swenssonii
Mount Maroon, Queensland

5 May 2001

A plant
B flower from front
C flower from side
D labellum from side
E labellum flattened
F column from side
G column from front
H dorsal sepal
J lateral sepal
K petal

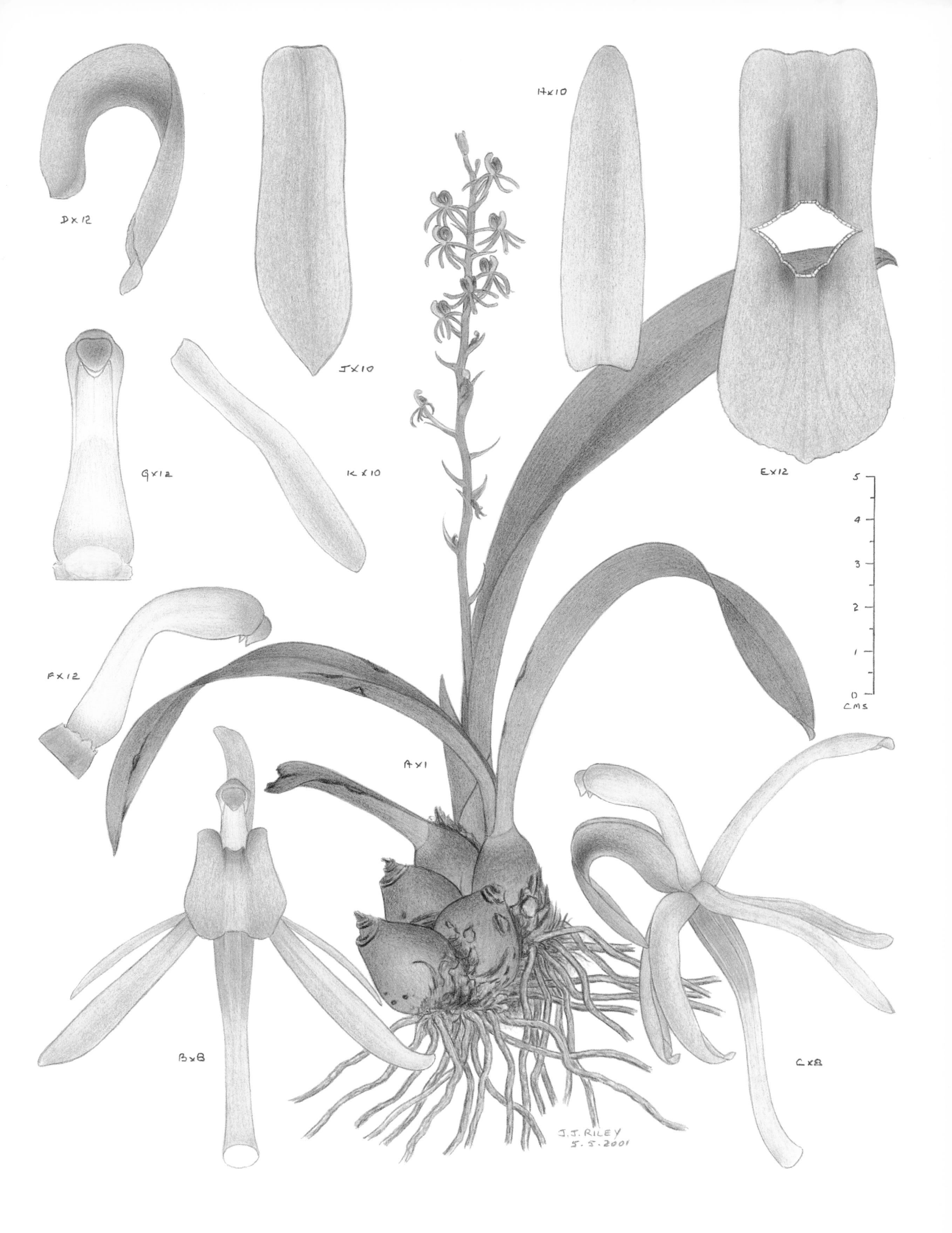
D x 12
J x 10
H x 10
E x 12
G x 12
K x 10
F x 12
A x 1
B x 8
C x 8
5
4
3
2
1
0
CMS
J. J. RILEY
5. 5. 2001

Liparis coelogynoides (F. Muell.) Benth.

1873 | *Flora Australiensis* 6:273

TYPE LOCALITY Bunya Mountains, Queensland

ETYMOLOGY *coelogynoides* — resembling the genus *Coelogyne*

FLOWERING TIME October to March

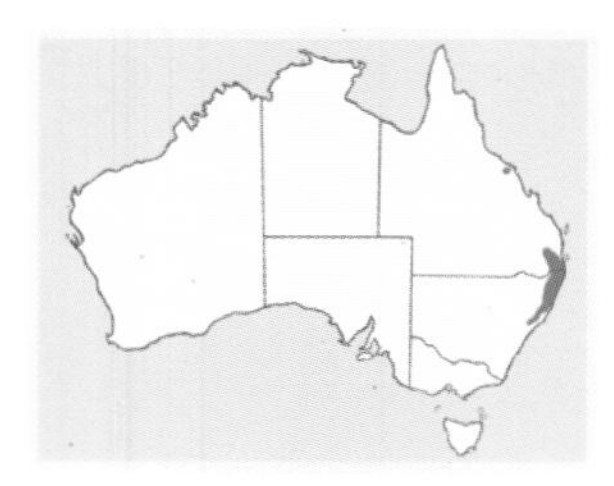

DISTRIBUTION Found from the Bunya Mountains in southern Queensland, south along the ranges and tablelands of New South Wales to the Barrington Tops region and the upper Colo River in the Blue Mountains. There is also a disjunct population at Eungella in Queensland.

ALTITUDINAL RANGE 100 m to 950 m

DISTINGUISHING FEATURES *L. coelogynoides* is a small, clump-forming epiphyte with round, flattened pseudobulbs. The leaves are short-lived and thin-textured. The inflorescence is arching and has numerous, starry blooms with a wedge-shaped labellum.

HABITAT This species grows in wet sclerophyll forest and rainforest. It is mostly found on tree trunks and major limbs, but rarely on rocks.

CONSERVATION STATUS Common. Occurs in national parks and reserves. Secure.

DISCUSSION *L. coelogynoides* is the smallest of the Australian epiphytic *Liparis* species. It is mostly found growing on trees among epiphytic mosses, lichens and *Pyrrosia* fern. Despite not growing into massive plants, it is a very attractive and dainty species when in bloom. *L. coelogynoides* has an unpleasant odour and is pollinated by small flies and mosquitoes.

Liparis coelogynoides
Cunninghams Gap, Queensland

16 February 2000

A plant
B flower from side
C flower from front
D labellum flattened
E labellum from side
F column from side
G column from front
H dorsal sepal
J petal
K lateral sepal

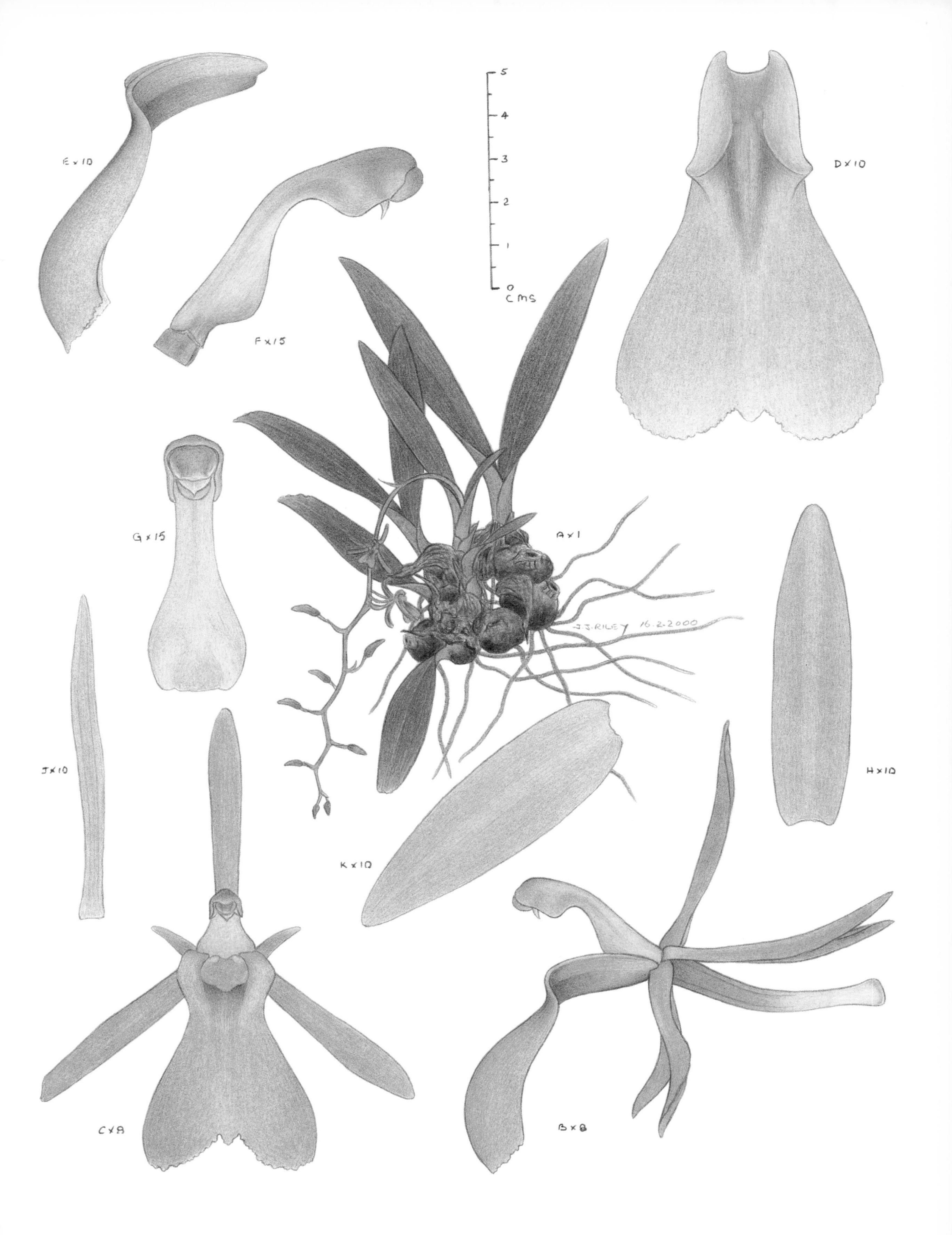

E x 10
F x 15
D x 10
5
4
3
2
1
0
CMS
G x 15
A x 1
J.J. RILEY 16.2.2000
J x 10
H x 10
K x 10
C x 8
B x 8

Liparis habenarina (F. Muell.) Benth.

1873 | *Flora Australiensis* 6:273

TYPE LOCALITY Rockingham Bay, Queensland

ETYMOLOGY *habenarina* — like a *Habenaria*

FLOWERING TIME January to April

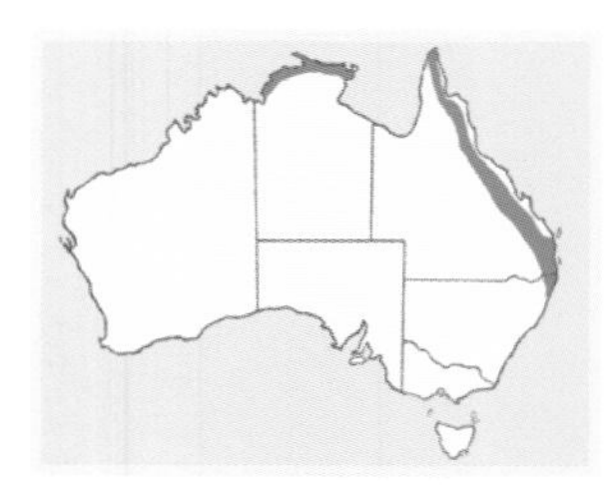

DISTRIBUTION Found from the northern coast of the Northern Territory, the Torres Strait islands, Cape York Peninsula in North Queensland, south along the ranges and coast to Coffs Harbour, New South Wales. In the tropics it is mostly found in the ranges becoming more coastal to the south.

ALTITUDINAL RANGE 50 m to 700 m

DISTINGUISHING FEATURES *L. habenarina* is a terrestrial species with ovoid pseudobulbs that are mostly buried. The leaves are long and slightly pleated. The inflorescence is tall, upright and carries up to 25 yellow-green to reddish-green flowers, with prominent, broad lateral sepals.

HABITAT This species grows in open forest and woodland with a grassy understorey in gravelly clay loams that are usually well-drained. It can also be found in situations that are seasonally wet and swampy.

CONSERVATION STATUS Uncommon. Occurs in national parks. Secure.

DISCUSSION *L. habenarina* is the only deciduous terrestrial species of *Liparis* in Australia. During the winter months the pseudobulbs remain dormant, producing leaves in early summer. It is common in the Queensland tropics where it is usually found away from the coast in the ranges and tablelands where it benefits from summer rains. It is common in coastal, often swampy parts of the Northern Territory. *L. habenarina* is a rare, coastal lowland species in the southern parts of its range, where rainfall is fairly even throughout the year. This orchid reproduces from seed.

Liparis habenarina
Atherton Tableland, Queensland

3 March 1998

A plant
B flower from front
C flower from side
D labellum flattened
E labellum from side
F column from side
G column from front
H dorsal sepal
J lateral sepal
K petal

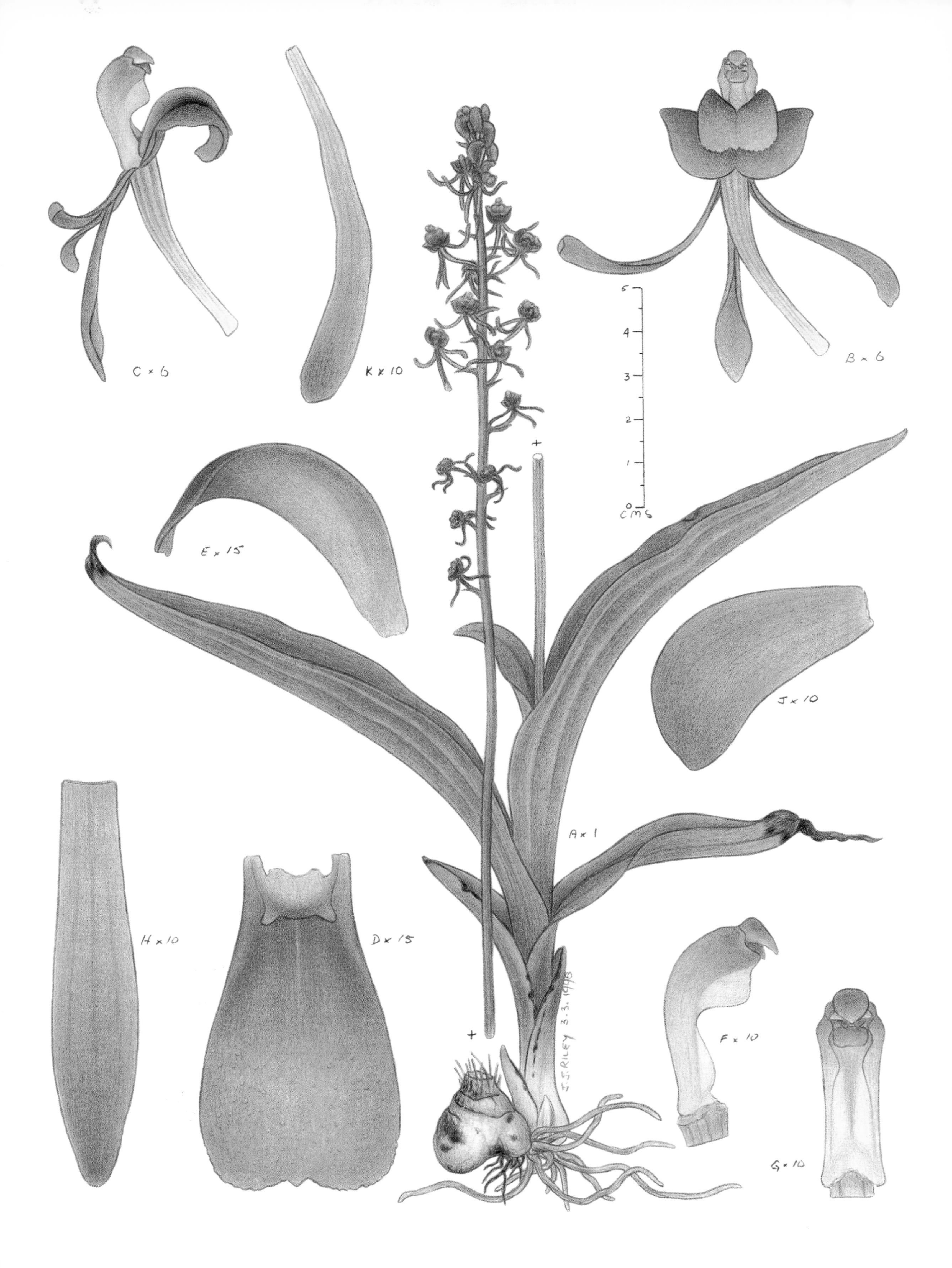
C × 6
K × 10
B × 6
5
4
3
2
1
0
CMS
E × 15
J × 10
A × 1
H × 10
D × 15
F × 10
G × 10
J.J. RILEY 3.3. 1998

Papillilabium beckleri (F. Muell. ex Benth.) Dockr.

1967 | *Australasian Sarcanthinae*: 31

TYPE LOCALITY Clarence River, New South Wales

ETYMOLOGY After Dr H. Beckler

FLOWERING TIME September to November

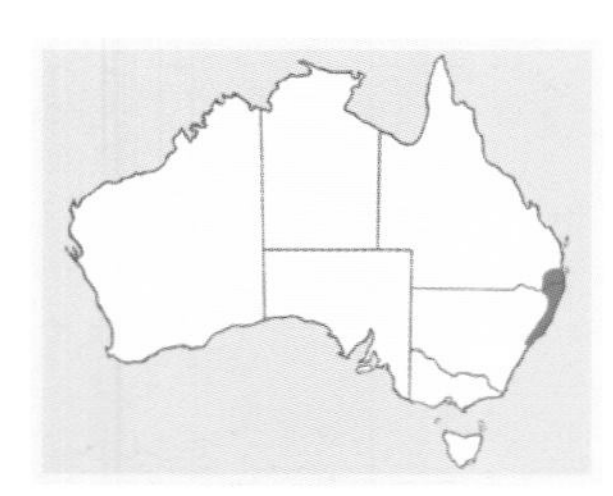

DISTRIBUTION Found from Brisbane Forest Park in south-eastern Queensland, south to the lower Blue Mountains and Royal National Park, New South Wales.

ALTITUDINAL RANGE 60 m to 600 m

DISTINGUISHING FEATURES *P. beckleri* is a small, twig epiphyte with cupped, greenish flowers and some purplish-maroon markings mainly on the petals and column. Labellum is whitish-green with darker, olive green blotches and is covered densely with tiny papillae.

HABITAT This species grows as an epiphyte in wet sclerophyll forest and in or on the fringes of rainforest. It is usually restricted to trees along or very close to flowing creeks and streams, where it can be seen on the outer twiggy branches often over water. Generally the extensive root system is found before the actual plant is located.

CONSERVATION STATUS Widespread. Conserved in national parks and reserves. Secure.

DISCUSSION *P. beckleri* is a very small, monopodial orchid. Because of its preferred habitat, this and similar species are often referred to as twig epiphytes. It can be locally common, with vast colonies in a relatively small area, then further down or upstream there may be few or no plants, despite a similar habitat. It can be difficult to spot, with experience proving that it is easier to look for root systems, and then trace them to locate the plant. This species reproduces from seed and the flowers have a subtle sweet fragrance.

Papillilabium beckleri
Ourimbah, New South Wales

7 September 1999

A plant
B flower from front
C flower from side
D labellum from above
E longitudinal section of labellum
F labellum from front
G labellum and column from side
H column from front
J dorsal sepal
K lateral sepal
L petal

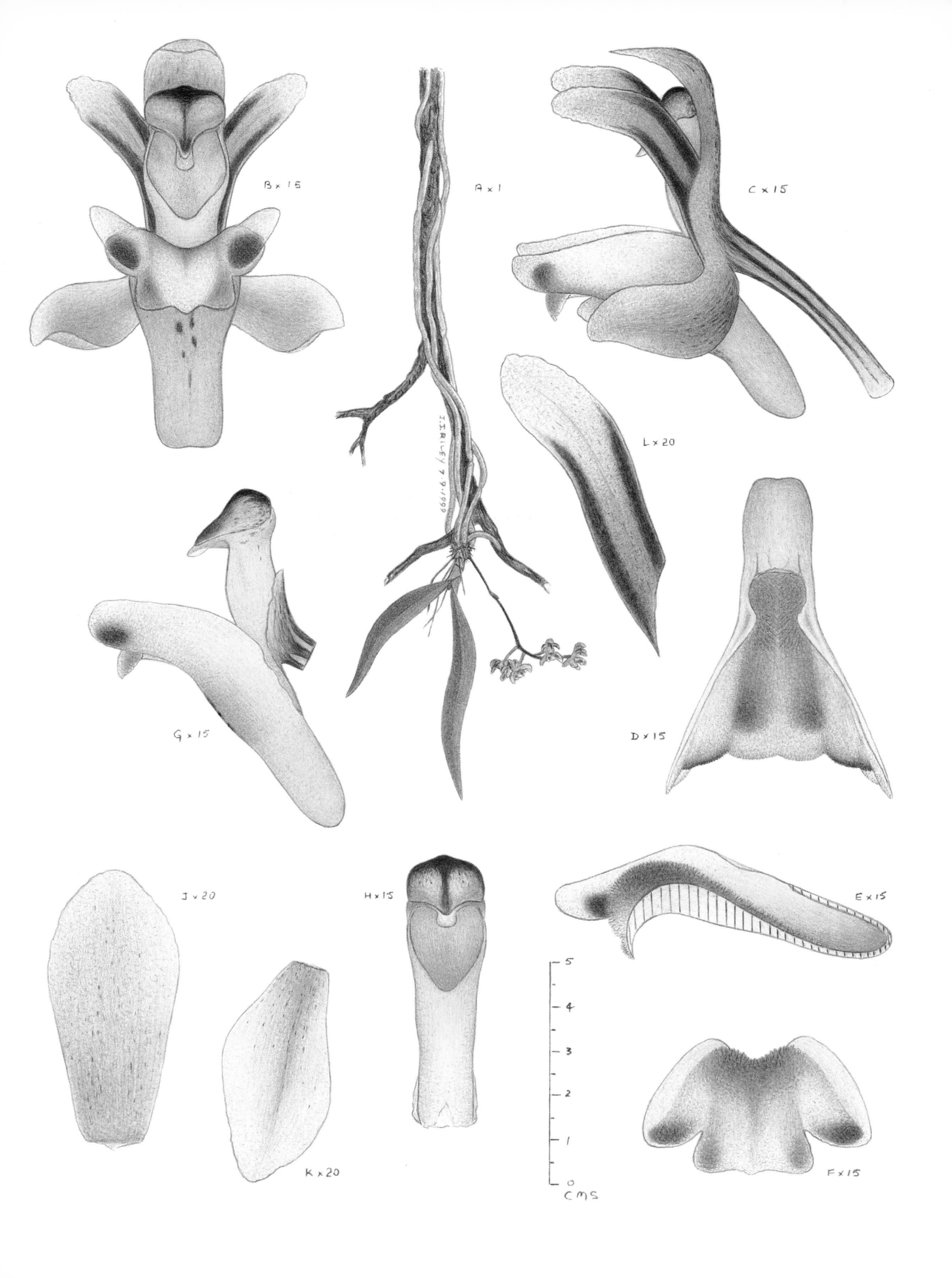
B × 15
A × 1
C × 15
J.J.RILEY 7.9.1999
L × 20
G × 15
D × 15
J × 20
H × 15
E × 15
K × 20
F × 15
5
4
3
2
1
0
CMS

Plectorrhiza tridentata (Lindl.) Dockr.

1967 | *Australasian Sarcanthinae*: 27

TYPE LOCALITY Australia

ETYMOLOGY *tridentata* — with three teeth

FLOWERING TIME September to January

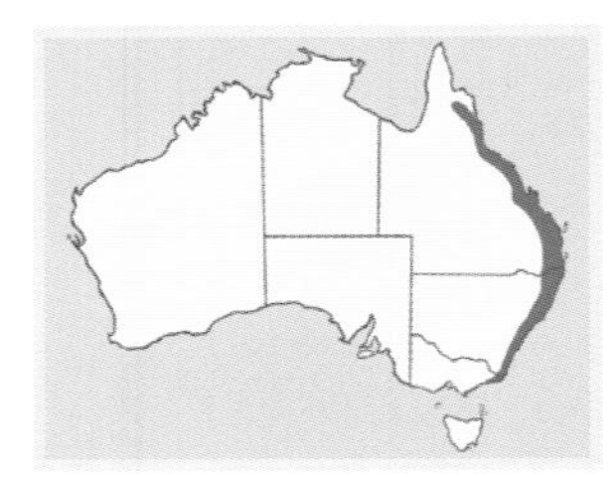

DISTRIBUTION This widespread species is found along the coast and ranges of eastern Australia, from East Gippsland in Victoria, northwards through New South Wales, to the Windsor Tableland in North Queensland.

ALTITUDINAL RANGE 50 m to 900 m

DISTINGUISHING FEATURES *P. tridentata* is a hanging epiphyte with an extensive, tangled, aerial root system, with plants often only clinging by a couple of roots. The labellum is set at right angles to the column and is three-lobed, with the two sidelobes long and narrow and the midlobe wide and blunt.

HABITAT This species grows as an epiphyte in rainforest and wet sclerophyll forests, also in humid sheltered gullies in open woodland. It frequently grows on the thin branches of trees and larger shrubs along or very close to flowing creeks and streams, often overhanging water. Popular hosts include the grey myrtle (*Backhousia myrtifolia*), and the water gum, (*Tristaniopsis laurina*). Sometimes the extensive root system is found before the actual plant is located. It grows on the ranges in the north of its range, becoming more coastal in the south.

CONSERVATION STATUS Common in New South Wales and Queensland. Rare in Victoria. Conserved in national parks and reserves. Secure.

DISCUSSION *P. tridentata* is one of our more common and easily recognised epiphytic orchids. In the tropics, the plants never reach the same dimensions as southern specimens. The flowers have a sweet fragrance, which is quite powerful considering the size of the blooms. Flower colour is variable, from apple green through olive and brown tones. At Mackay, Bundaberg and the Kroombit Tops region of Queensland, there are colonies of plants with dark chocolate brown flowers.

Plectorrhiza tridentata
Paterson, New South Wales

13 January 1999

A plant
B flower from front
C flower from side
D labellum and column from side
E labellum from front
F column from front
G dorsal sepal
H lateral sepal
J petal

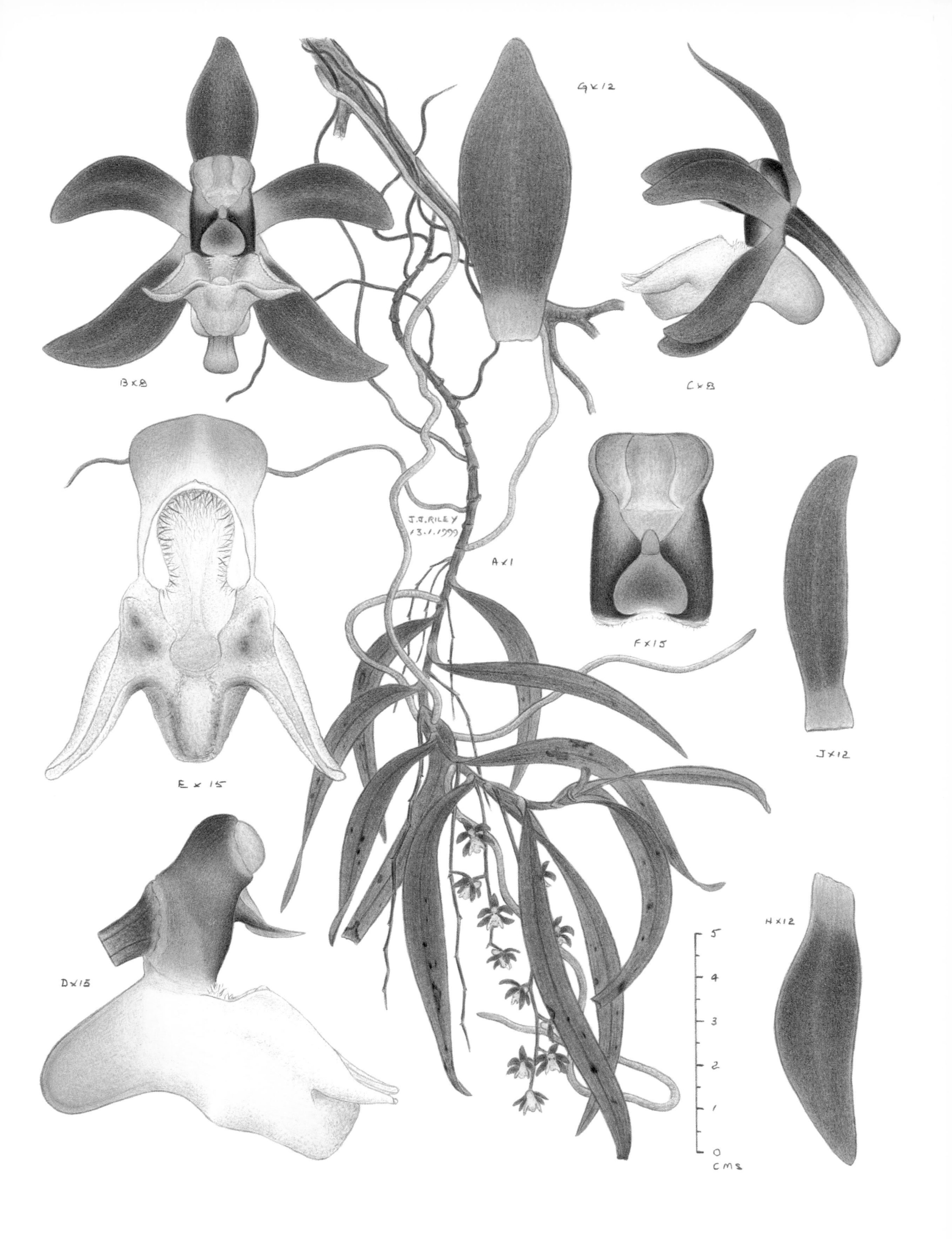
G x 12
B x 8
C x 8
J.J. RILEY
13.1.1999
A x 1
F x 15
J x 12
E x 15
H x 12
D x 15
5
4
3
2
1
0
CMS

Plectorrhiza erecta (Fitzg.) Dockr.

1967 | *Australasian Sarcanthinae*: 28

TYPE LOCALITY Lord Howe Island, off coast of New South Wales

ETYMOLOGY *erecta* — upright, erect

FLOWERING TIME October to January

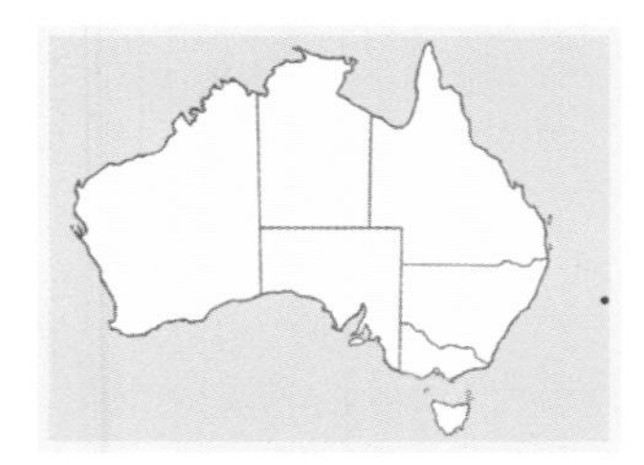

DISTRIBUTION *P. erecta* is endemic to Lord Howe Island, which is about 700 km north-east of Sydney.

ALTITUDINAL RANGE Up to 300 m

DISTINGUISHING FEATURES *P. erecta* is a monopodial epiphyte with erect stems with an untidy, tangled root system. Leaves are short and well spaced with aerial roots produced along their length. It produces small arching inflorescences with up to five flowers that do not open fully and a three-lobed labellum held at right angles to the column.

HABITAT Grows in low, coastal scrub on the north-eastern side of the base of Mount Lidgbird and Mount Gower. It prefers semi-shaded situations, but can also tolerate bright light. Plants are often found in exposed positions.

CONSERVATION STATUS Uncommon. Occurs in a World Heritage Area. Secure.

DISCUSSION This isolated species is related to *P. tridentata*, as the flowers are superficially similar. The plants are always trying to grow upright, even though the stems are not strong enough to support older plants. Unless supported by the root system, these tend to fall over and then bend to continue growing upwards. Mature plants will branch at the base and form small clumps.

Plectorrhiza erecta
Lord Howe Island

20 December 1997

A plant
B flower from front
C flower from side
D labellum and column from side
E labellum from front
F column from front
G dorsal sepal
H lateral sepal
J petal

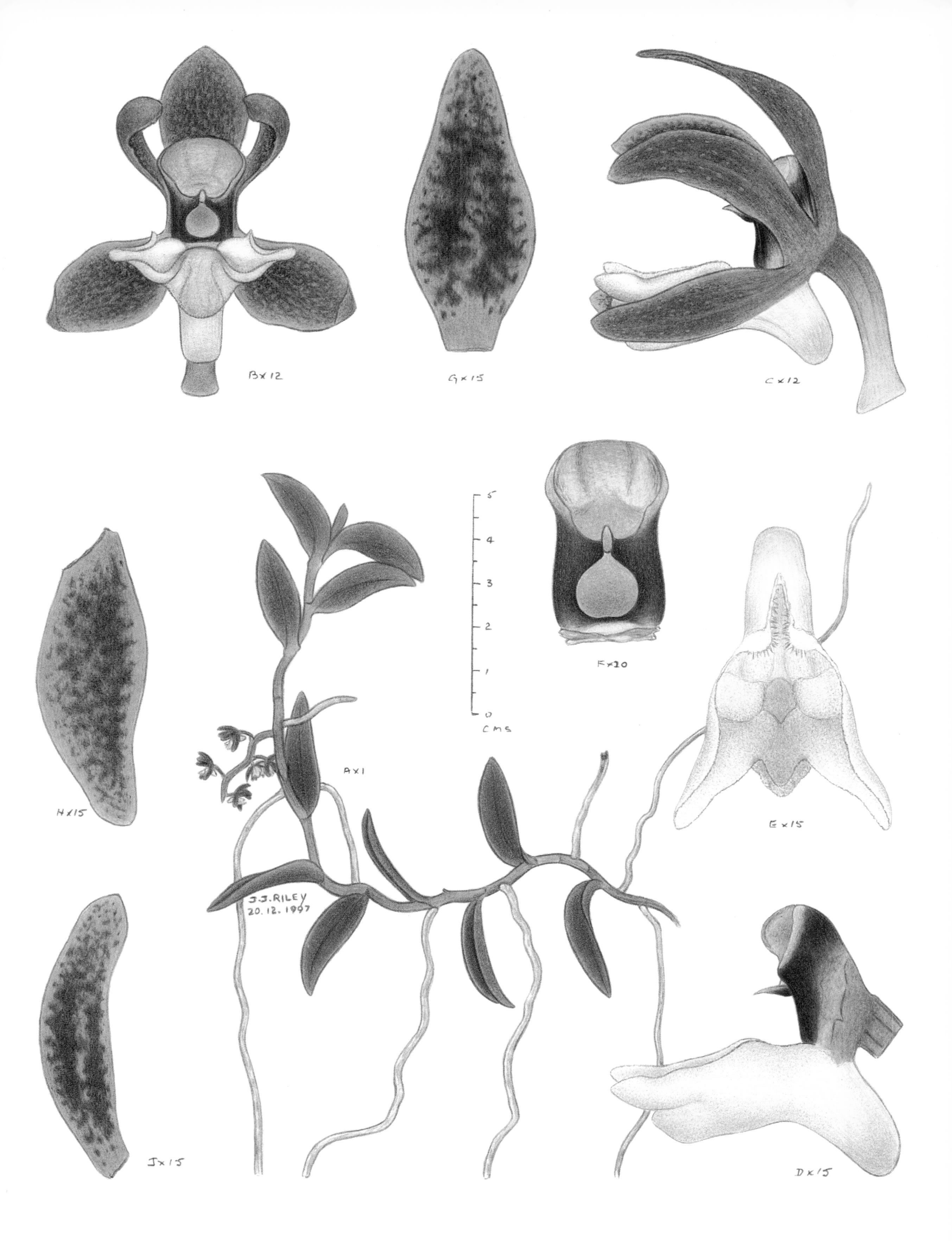

B×12
G×15
C×12
F×20
CMS
A×1
H×15
E×15
J.J.RILEY
20.12.1997
J×15
D×15

Plectorrhiza brevilabris (F. Muell.) Dockr.

1967 | *Australasian Sarcanthinae*: 28

TYPE LOCALITY Proserpine, Queensland

ETYMOLOGY *brevilabris* — with a short lip

FLOWERING TIME November to February

DISTRIBUTION *P. brevilabris* is endemic to Queensland, being found from the McIlwraith Range south to the Noosa River.

ALTITUDINAL RANGE Up to 1200 m

DISTINGUISHING FEATURES *P. brevilabris* is a monopodial epiphyte with a pendent habit producing leaves that have an uneven notch at the apex. The green and brown, sweetly perfumed blooms have a white labellum and distinctive, long, prominent, green spur.

HABITAT This species grows as an epiphyte in wet or dry sclerophyll forest and also in rainforest, growing on small twigs, branches and, less frequently, on the trunks of trees. In North Queensland, it is a plant of the higher ranges, while further south at Noosa it grows in dry rainforest on the coast.

CONSERVATION STATUS Widespread. Occurs in national parks and reserves. Secure.

DISCUSSION *P. brevilabris* is a very attractive species and a tidier plant with broader leaves than *P. tridentata*. The notched leaf apex readily identifies this species even when not in bloom. In the southern part of its range, *P. brevilabris* frequently colonises young hoop pines (*Araucaria cunninghamii*).

Plectorrhiza brevilabris
Wallaman Falls, Queensland

15 January 2000

A plant
B flower from front
C flower from side
D labellum and column from side
E labellum from front
F column from front
G lateral sepal
H dorsal sepal
J petal

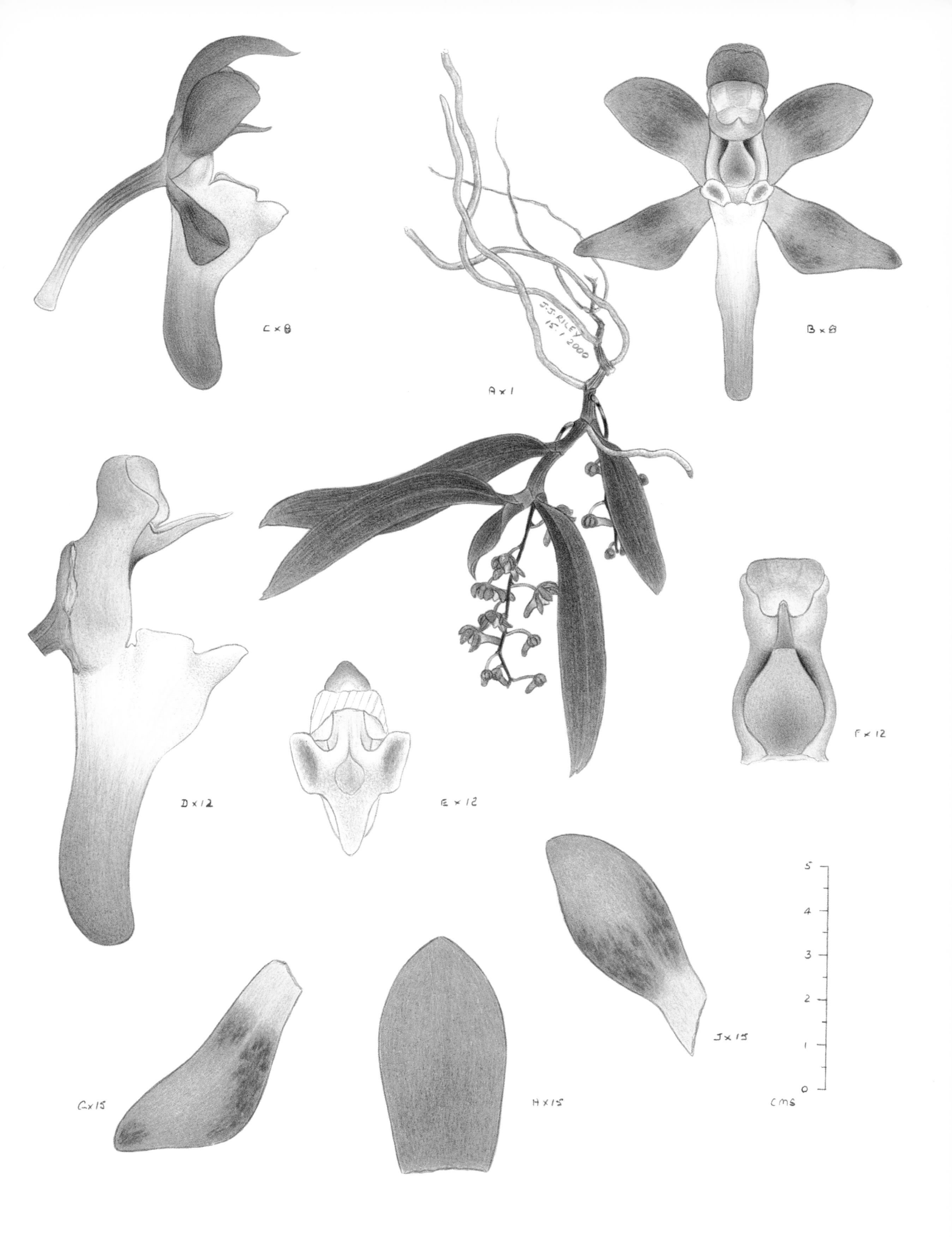
C x 8
J. J. RILEY
15. 1 2000
B x 8
A x 1
D x 12
E x 12
F x 12
5
4
3
2
1
0
J x 15
G x 15
H x 15
CMS

Prasophyllum flavum R. Br.

1810 | *Prodromus Florae Novae Hollandiae*: 318

TYPE LOCALITY Sydney, New South Wales

ETYMOLOGY *flavum* — yellow

FLOWERING TIME October to February

DISTRIBUTION *P. flavum* is of sporadic occurrence and is found from the Moreton District of southern Queensland, south along the coast and tablelands of New South Wales, to the southern half of Victoria, the Bass Strait islands and eastern Tasmania.

ALTITUDINAL RANGE 30 m to 1100 m

DISTINGUISHING FEATURES *P. flavum* appears to be a semi-saprophytic species, with a small, poorly developed, dark purple leaf. The flowers are cup-shaped and yellow to yellow-green in colour with short, wide perianth segments. The lateral sepals are fused and the labellum has a white wavy margin.

HABITAT This species grows in woody heathland, moist, open *Eucalyptus* woodland and forest with a shrubby understorey. Soils are well drained sand, sandy loams and clay loams.

CONSERVATION STATUS Occurs in national parks and reserves. Secure.

DISCUSSION *P. flavum* can be locally common, at times growing as scattered small colonies among shrubs and, more commonly, near the base of old *Eucalyptus* trees. While not requiring the stimulus of fire to bloom, flowering is greatly enhanced the year after a burn. This is an impressive tall-growing species that stands out in the bush when in bloom. This species is pollinated by native bees and wasps, and reproduces from seed.

Prasophyllum flavum
Mount Victoria, New South Wales

22 December 1991

A plant
B flower from front
C flower from above
D flower from side
E labellum from front
F labellum from rear
G labellum from side
H dorsal sepal
J lateral sepals
K petal

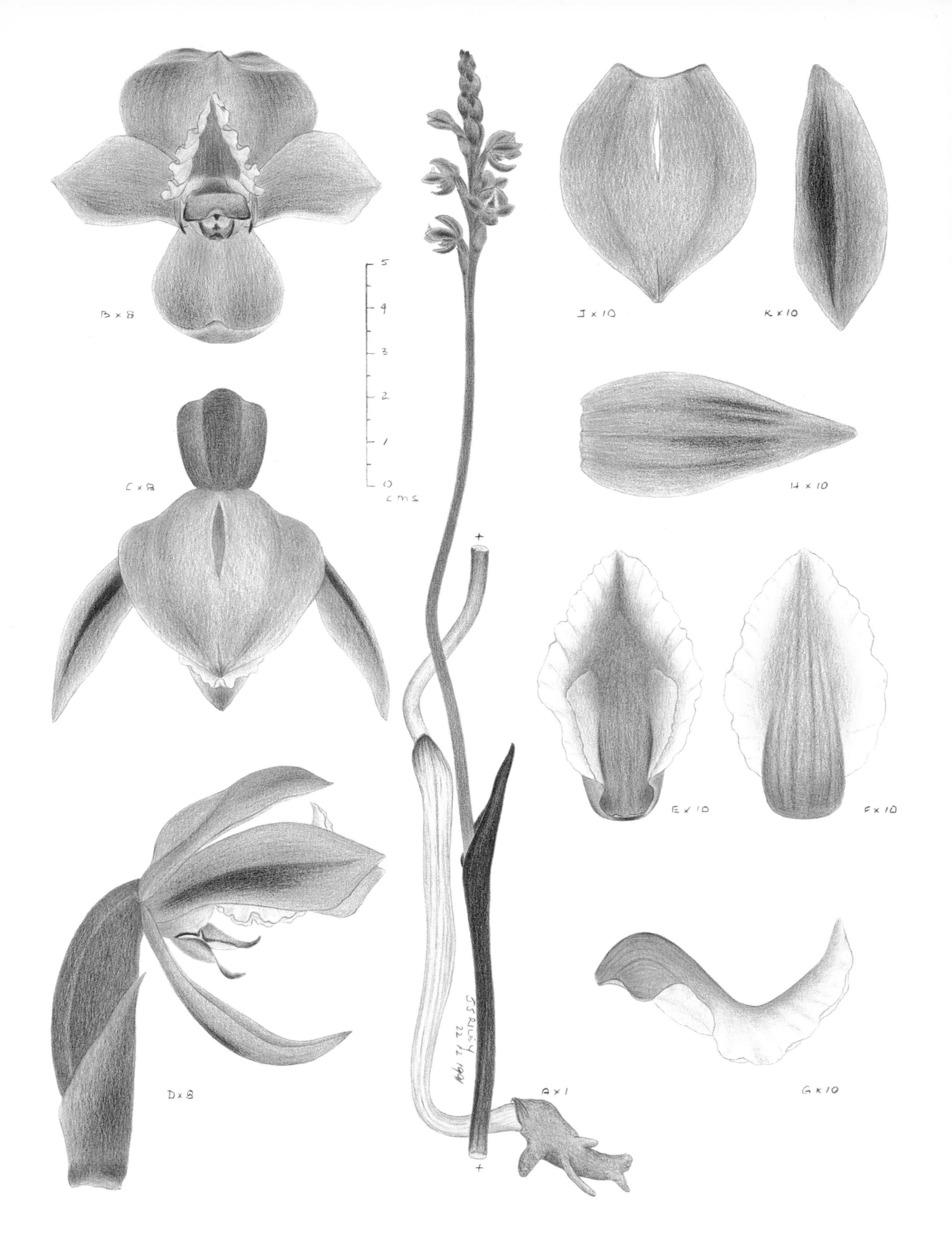
B x 8
C x 8
D x 8
A x 1
5
4
3
2
1
0
cms
J x 10
K x 10
H x 10
E x 10
F x 10
G x 10

Prasophyllum striatum R. Br.

1810 | *Prodromus Florae Novae Hollandiae*: 318

TYPE LOCALITY Sydney, New South Wales

ETYMOLOGY *striatum* — streaked, striped

FLOWERING TIME February to June

DISTRIBUTION *P. striatum* is endemic to New South Wales, being found from the Hawkesbury River along the coast south to around Nowra. Also occurs in the Blue Mountains, west of Sydney.

ALTITUDINAL RANGE 40 m to 1300 m

DISTINGUISHING FEATURES *P. striatum* is the smallest of the eastern states' prasophyllums. The flowering stem always has a distinctive bend, and the predominantly striped blooms have an unpleasant, musky odour with the white labellum having a deeply channelled callus.

HABITAT This species grows in low heathland and areas of shrubby heath in *Eucalyptus* woodland. Soils are moist, shallow sandy loams or clay loams over sandstone and laterites.

CONSERVATION STATUS Occurs in national parks and reserves. Secure.

DISCUSSION *P. striatum* does not occur on the clay loams derived from the Wianamatta Shales of the Sydney Basin. Around Sydney, a lot of its habitat has been lost to development for housing. On the upper Blue Mountains it is often locally common, growing in bare ground among low heath and sedges on the tops of flat sandstone ridges. It also grows on the slopes above hanging swamps. It flowers profusely the year after a fire. This species is pollinated by native bees and wasps, and reproduces from seed.

Prasophyllum striatum
Mount Wilson, New South Wales

20 April 1999

A plant
B flower from front
C flower from side
D labellum from front
E labellum from rear
F labellum from side
G column from side
H column from front
J column from rear
K dorsal sepal
L lateral sepals
M petal

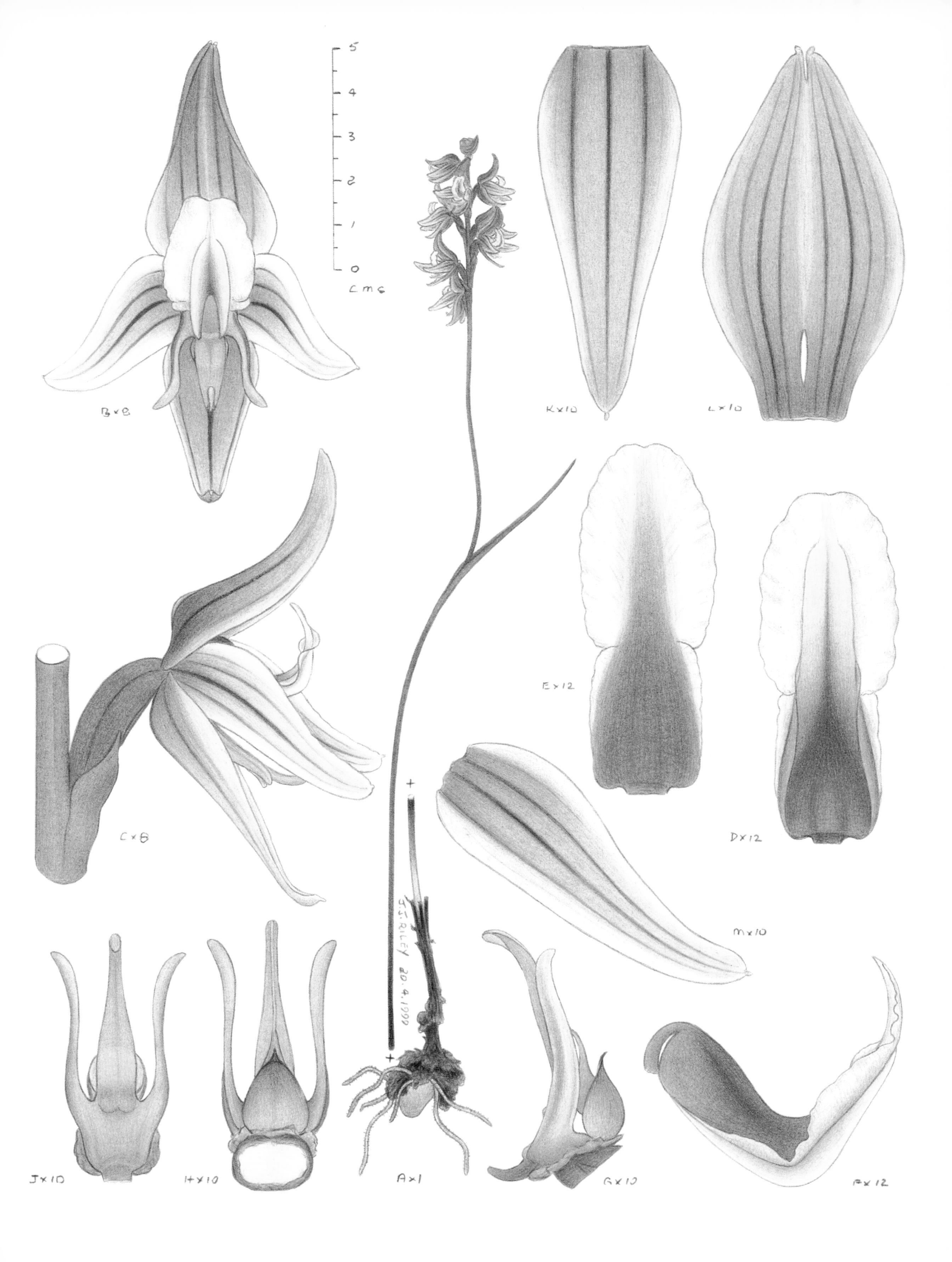
5
4
3
2
1
0
cms
B×8
K×10
L×10
E×12
C×8
D×12
J.J. RILEY 20.4.1990
M×10
J×10
H×10
A×1
G×10
F×12

Prasophyllum parvifolium Lindl.

1840 | *Edward's Botanical Register*, Appendix to Vols 1–23

TYPE LOCALITY Swan River, Western Australia

ETYMOLOGY *parvifolium* — small leaves

FLOWERING TIME June to August

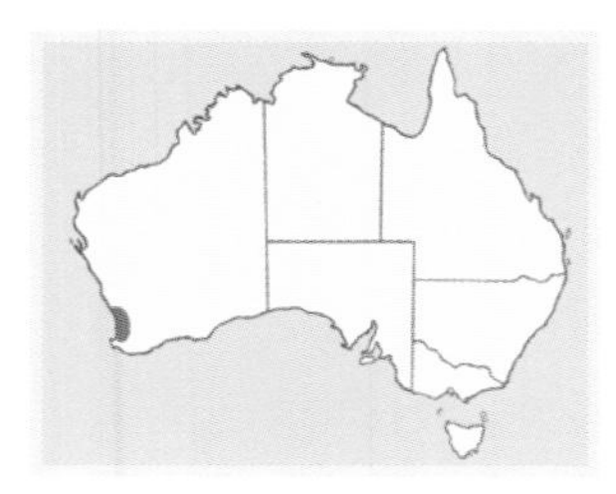

DISTRIBUTION *P. parvifolium* is endemic to Western Australia, being found between Eneabba and Manjimup.

ALTITUDINAL RANGE 30 m to 200 m

DISTINGUISHING FEATURES *P. parvifolium* is a slender species with a distinctive bend where the flower spike emerges from the leaf. The petals and sepals are pale with fine, red-maroon striping and the labellum is predominantly white. The flowers are produced in winter and early spring and are odourless.

HABITAT This species occurs in coastal heath and *Banksia* woodland and can also be seen growing under groves of Casuarina.

CONSERVATION STATUS Occurs in national parks and reserves. Secure.

DISCUSSION Two similar taxa have been known in Western Australia as *P. parvifolium*. One is an early, green-flowered species with a southern, and more coastal, distribution, while the other flowers later and has red-maroon markings and a more northern distribution. Only the later-flowered, darker coloured species grows around the Swan River area, the type-site, making it true *P. parvifolium*. This small and attractive orchid is very similar to *P. striatum* from New South Wales. This species is pollinated by native bees and wasps, and reproduces from seed.

Prasophyllum parvifolium
Canning Vale, Western Australia

13 June 1991

A plant
B flower from side
C flower from above
D flower from front
E labellum from front — flattened
F labellum from rear — flattened
G labellum from side
H column from side
J column from rear
K column from front

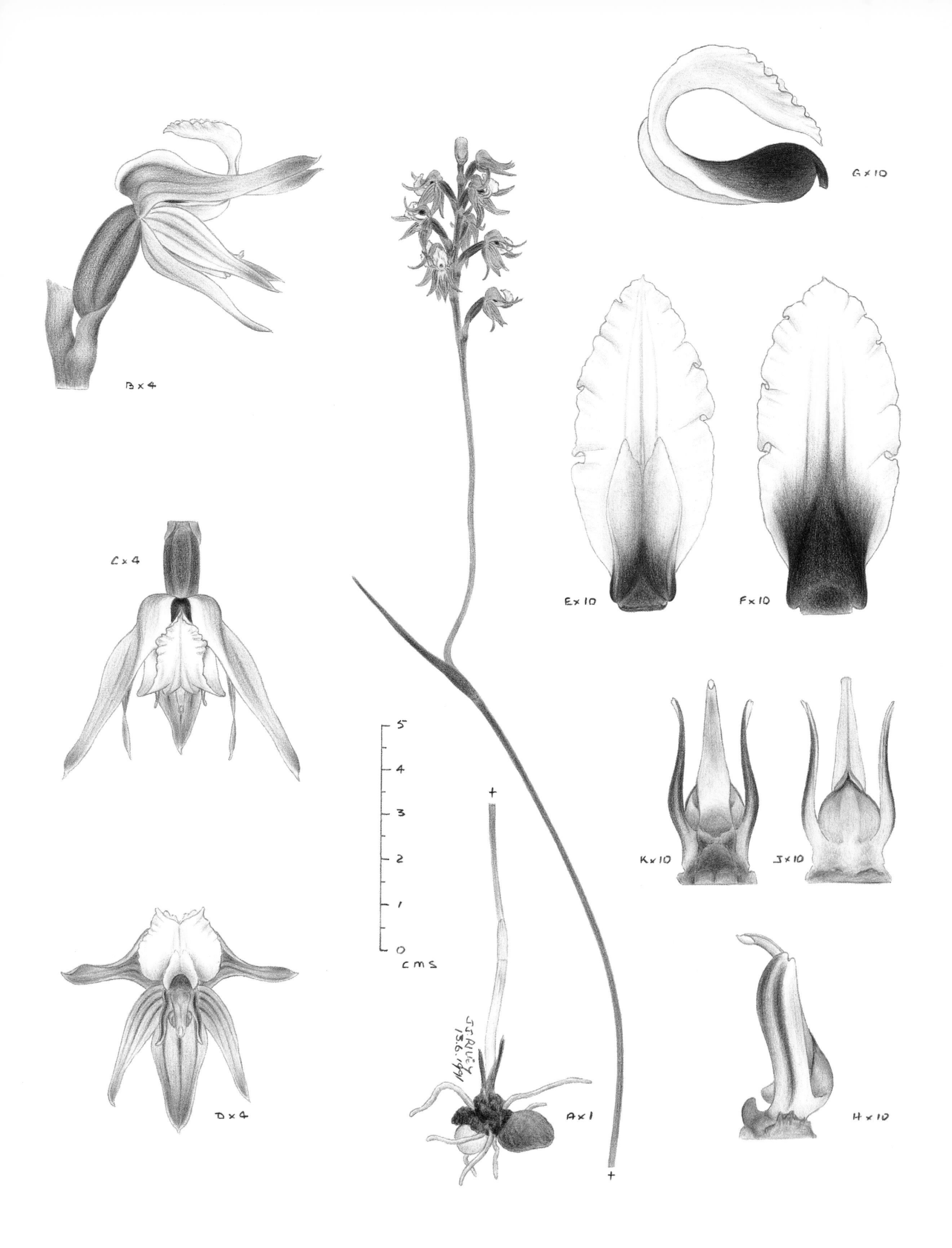
G x 10
B x 4
E x 10
F x 10
C x 4
5
4
3
2
1
0
cms
K x 10
J x 10
J.S. Riley
13.6.1991
D x 4
A x 1
H x 10

Prasophyllum alpestre D.L. Jones

1998 | *Australian Orchid Research* 3:98

TYPE LOCALITY Charlotte Pass, New South Wales

ETYMOLOGY *alpestre* — of high mountains

FLOWERING TIME January to March

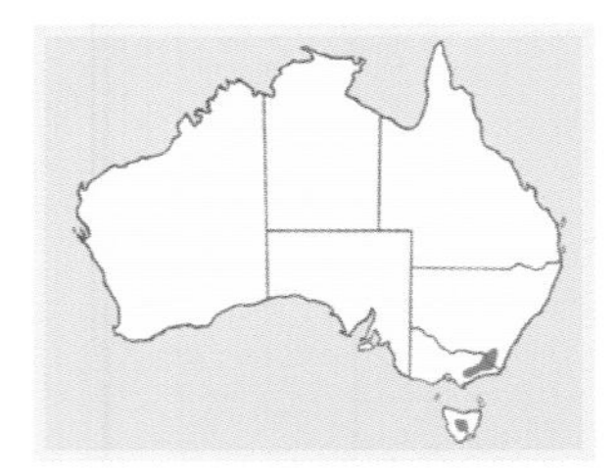

DISTRIBUTION *P. alpestre* is an alpine species found from the Snowy Mountains of New South Wales and the Brindabella Range of the Australian Capital Territory, to the Eastern Highlands of Victoria, the central plateau of Tasmania and Mount Wellington.

ALTITUDINAL RANGE 1100 m to 2000 m

DISTINGUISHING FEATURES *P. alpestre* has green flowers with mauve suffusions and lateral sepals that are usually fused. The broad labellum is white with a green callus.

HABITAT This attractive, terrestrial species occurs in sub-alpine to alpine frost hollows and herbfields and in the grassy clearings among snow gums (*Eucalyptus pauciflora*). Plants may be locally common and grow together with dense tussocky grasses and sparse low shrubs. Soils are moist peaty loams to moist, gravelly clay loams.

CONSERVATION STATUS Well represented in national parks and reserves. Secure.

DISCUSSION *P. alpestre* is a recently named species that was previously misinterpreted as *P. suttonii*. A study of the type material of the latter species showed that two distinct taxa were involved. It appears that *P. suttonii* is a rare Victorian endemic, which flowers earlier and is poorly known. The familiar and widespread orchid commonly called *P. suttonii* was in fact an unnamed species that was subsequently described in 1998 as *P. alpestre*. This species is pollinated by native bees and wasps, and reproduces from seed.

Prasophyllum alpestre
Kiandra, New South Wales

4 February 1991

A plant
B flower from front
C flower from side
D flower from above
E labellum from front — flattened
F column from side
G column from front
H column from rear

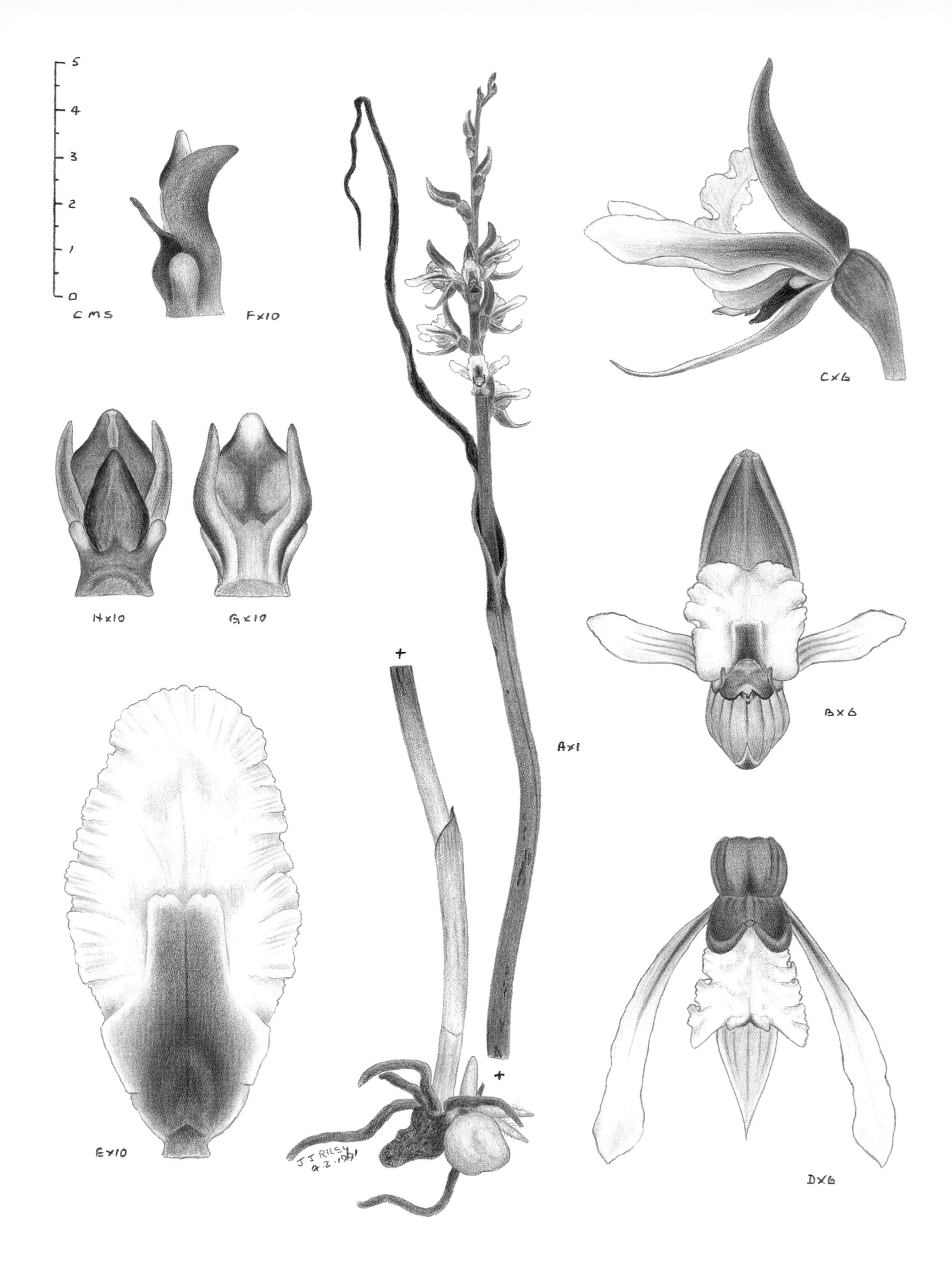
5
4
3
2
1
0
CMS
F×10
H×10
G×10
E×10
A×1
C×6
B×6
D×6
J.J. RILEY
4.2.1991

Prasophyllum rogersii Rupp

1928 | *Proceedings of the Linnean Society of New South Wales* 53:340

TYPE LOCALITY Barrington Tops, New South Wales

ETYMOLOGY After Richard Sanders Rogers

FLOWERING TIME December to February

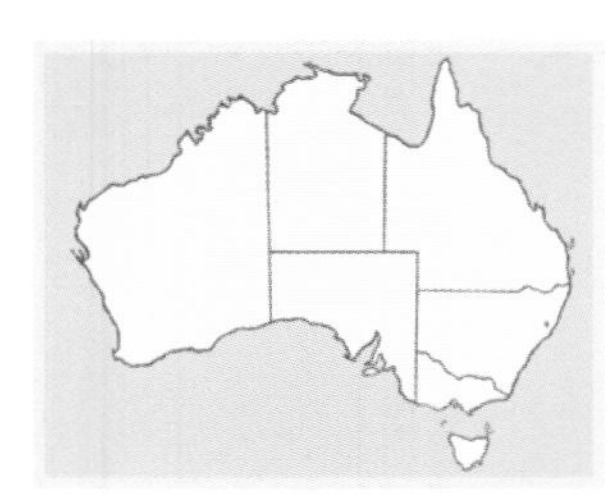

DISTRIBUTION This species is confined to the Barrington Tops, New South Wales.

ALTITUDINAL RANGE 1400 m to 1550 m

DISTINGUISHING FEATURES *P. rogersii* has green flowers with short, wide perianth segments. The labellum is short and wide, green with a prominent, darker green callus.

HABITAT This species can be found in open, sub-alpine herbfields bordering swamps. Usually seen as scattered individuals among tussocky snow grasses and small herbs. Soils are peaty clay loams.

CONSERVATION STATUS Uncommon and restricted. Occurs in a national park. Vulnerable.

DISCUSSION *P. rogersii* is a poorly known orchid, despite growing in the sub-alpine area of the Barrington Tops National Park, a popular and well visited area for campers, bushwalkers and native plant enthusiasts. There is a similar taxon in the Eastern Highlands of Victoria, which has been referred to as *P. rogersii*. This Victorian endemic species was described in 2000 as *P. niphopedium*. *P. rogersii* is pollinated by native bees and wasps, and reproduces from seed.

Prasophyllum rogersii
Barrington Tops, New South Wales

22 December 1993

A plant
B flower from front
C flower from side
D labellum from front — flattened
E labellum from rear — flattened
F labellum from side
G column from side
H column from front
J column from rear
K dorsal sepal
L lateral sepal
M petal

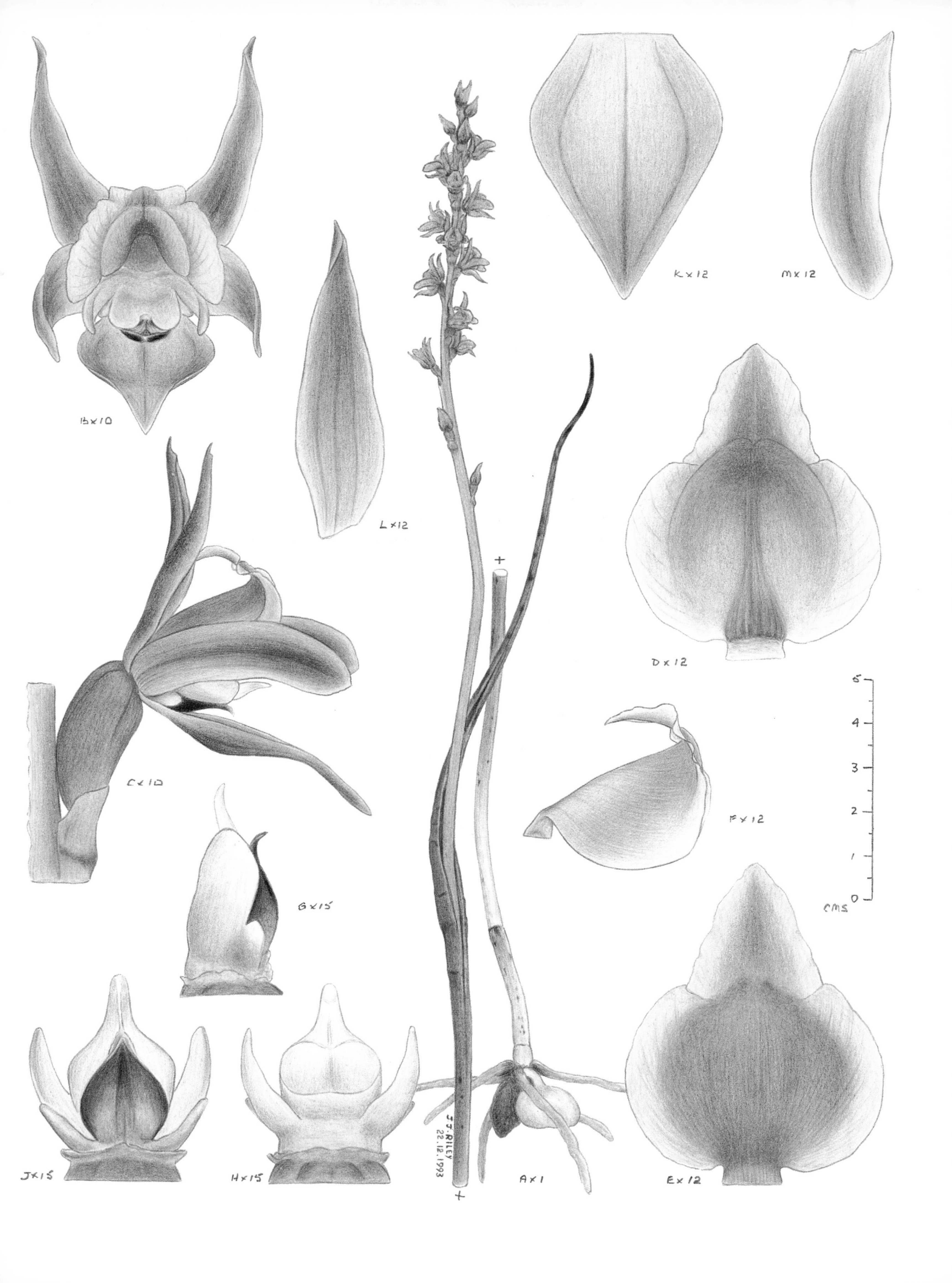

B x 10
L x 12
K x 12
M x 12
D x 12
C x 10
F x 12
G x 15
J x 15
H x 15
J.J. RILEY
22.12.1993
A x 1
E x 12
5
4
3
2
1
0
CMS

Prasophyllum pallens D.L. Jones

2000 | *Orchadian* 13:162

TYPE LOCALITY Mount Banks, New South Wales

ETYMOLOGY *pallens* — pale

FLOWERING TIME October to December

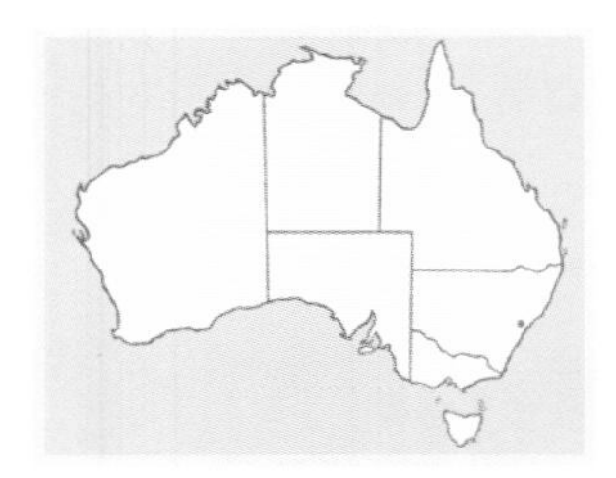

DISTRIBUTION This species is endemic to the upper Blue Mountains, New South Wales.

ALTITUDINAL RANGE 850 m to 1100 m

DISTINGUISHING FEATURES *P. pallens* has pale green flowers with faint tawny brown to dark green suffusions. The petals and sepals are narrow and taper to a distinct point. The white labellum, which has a prominent, yellow-green callus, is long, narrow and pointed at the apex. The flowers have a musky scent.

HABITAT This species occurs in dense, sedgy heath, often on slopes where there is water seepage. These areas are mostly at the tops of the numerous cliffs or along the creeks feeding the waterfalls of the upper Blue Mountains. These are known locally as hanging swamps. Soils are sandy peat to sandy loams on a sandstone base.

CONSERVATION STATUS Occurs in a national park. Secure.

DISCUSSION In the past, *P. pallens* was misidentified as *P. fuscum*, which is an earlier-flowering, sweetly scented, lowland species. *P. pallens* is surprisingly easy to locate in the field, despite growing mostly in the sedgy heath. Plants can be scattered or in small groups. It is also found growing in the wet, more open heath above the hanging swamps. This species is pollinated by native bees, and reproduces from seed.

Prasophyllum pallens
Blackheath, New South Wales

10 December 1992

A plant
B flower from side
C flower from front
D flower from rear
E labellum from front — flattened
F labellum from rear — flattened
G labellum from side
H column from side
J column from front
K column from rear
L dorsal sepal
M lateral sepals
N petal
O pollinia

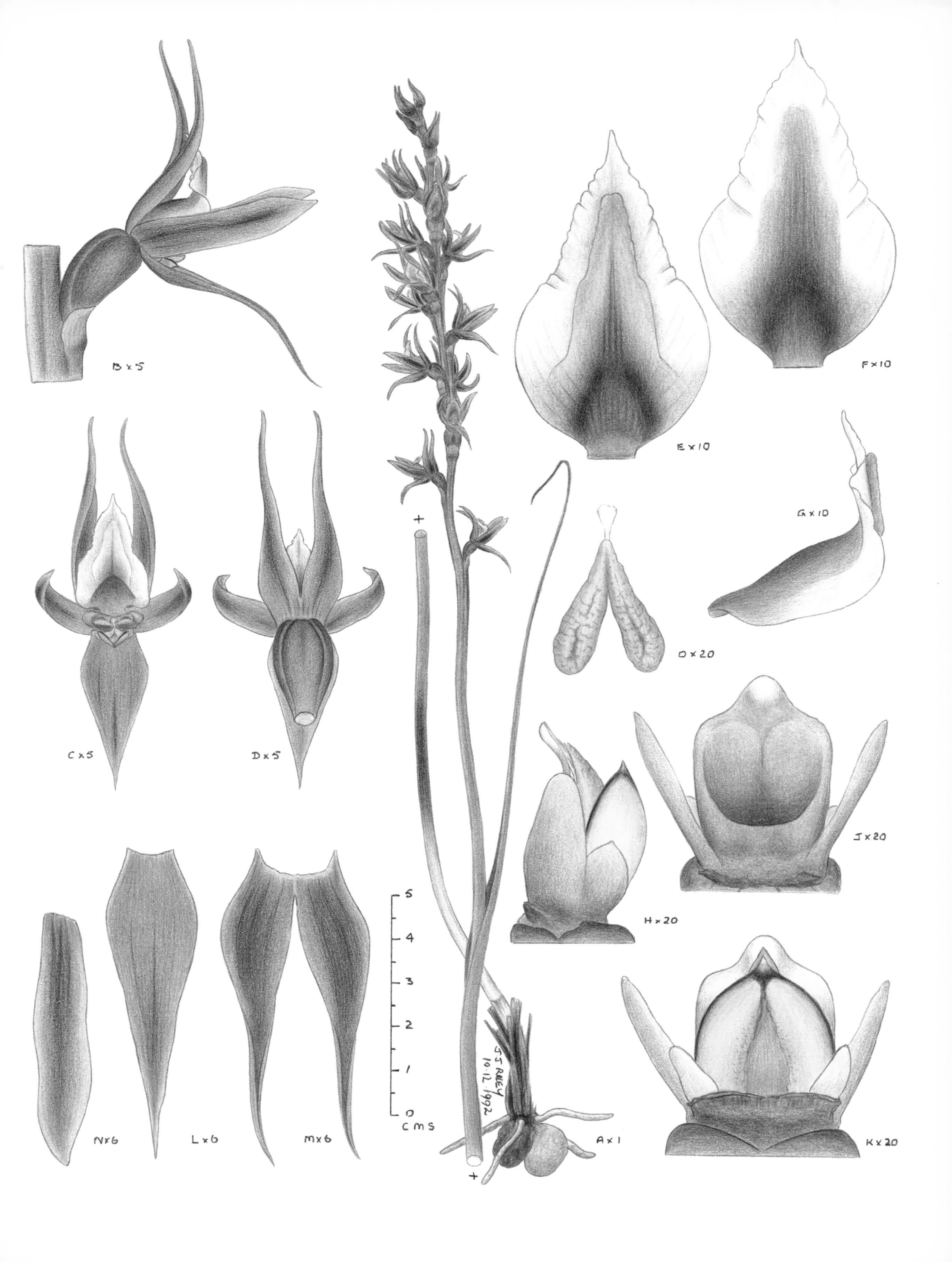
B × 5
C × 5
D × 5
E × 10
F × 10
G × 10
O × 20
H × 20
J × 20
K × 20
N × 6
L × 6
M × 6
A × 1
5
4
3
2
1
0
CMS
J.J. RILEY
10.12.1992

Prasophyllum caricetum D.L. Jones

2000 | *Orchadian* 13:151

TYPE LOCALITY Cathcart, New South Wales

ETYMOLOGY *caricetum* — growing with sedges

FLOWERING TIME December to February

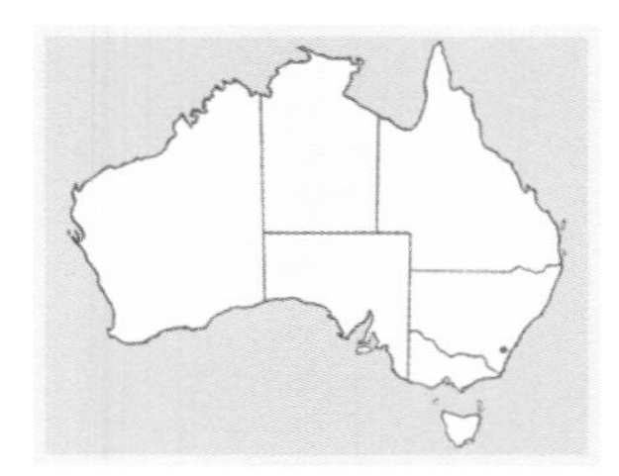

DISTRIBUTION This species is endemic to New South Wales, being found in the Cathcart district.

ALTITUDINAL RANGE 700 m to 800 m

DISTINGUISHING FEATURES *P. caricetum* is a robust species with white petals and green sepals. The mostly white labellum is long and narrow with a pale yellowish-green callus with the apex papillate.

HABITAT This species occurs in montane swamps dominated by sedges and low heathy shrubs and herbs. It grows in brown to black peaty loams.

CONSERVATION STATUS Uncommon and restricted. Does not occur in a national park. Vulnerable.

DISCUSSION *P. caricetum* is a sister species to *P. dossenum*, which is from the New England Tableland. The later species has larger flowers that are densely packed on the inflorescence and a different floral structure. This species is pollinated by native bees and wasps, and reproduces from seed.

Prasophyllum caricetum
Hains Swamp, New South Wales

20 January 1995

A plant
B flower from side
C flower from front
D labellum flattened
E labellum from side
F column from side
G column from front
H dorsal sepal
J lateral sepals
K petal

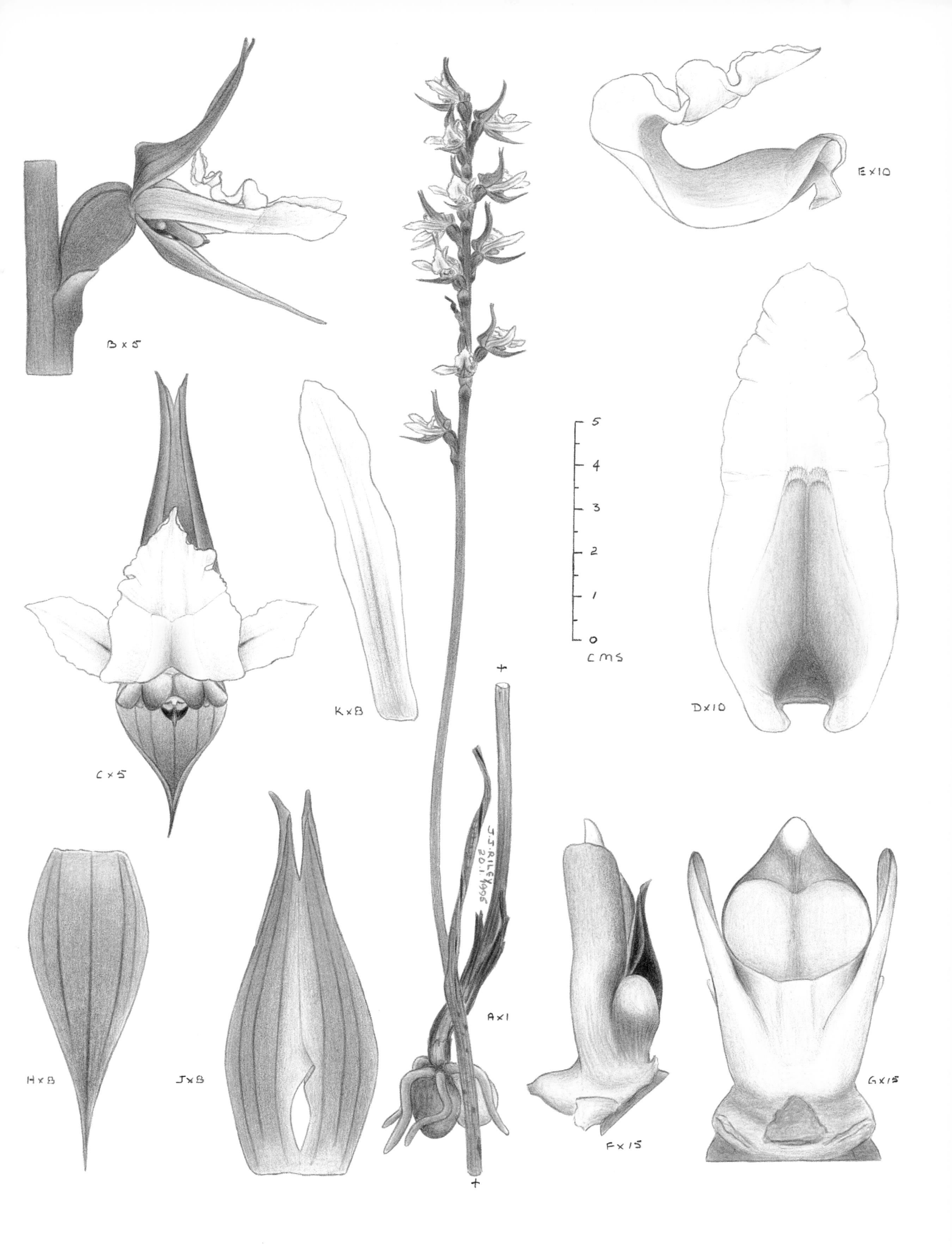

B x 5
E x 10
C x 5
K x 8
D x 10
5
4
3
2
1
0
CMS
J.J. RILEY
20.1.1996
A x 1
H x 8
J x 8
F x 15
G x 15

Pterostylis curta R. Br.

1810 | *Prodromus Florae Novae Hollandiae*: 326

TYPE LOCALITY Port Jackson, New South Wales

ETYMOLOGY *curta* — short

FLOWERING TIME July to November

DISTRIBUTION This common and widespread species is found from south-eastern Queensland, along the coast and tablelands of New South Wales, to Victoria, eastern South Australia and eastern Tasmania.

ALTITUDINAL RANGE 20 m to 1000 m

DISTINGUISHING FEATURES *P. curta* has a basal rosette of leaves when flowering and a broad, brownish labellum with a very pronounced twist to the apex.

HABITAT This robust species can be found in many different plant communities. These range from coastal heath, open and closed woodland and forest to moist grassland. Mostly growing in sheltered, moist, humus-rich habitats, however, on occasions it can be seen in drier, more exposed situations. Often forms colonies among ferns and shrubs, sometimes close to streams.

CONSERVATION STATUS Common and widespread. Conserved in national parks and reserves. Secure.

DISCUSSION *P. curta* is one of the most widespread and best known of the greenhoods in Australia. It is a free-flowering species with large populations providing a spectacular display. Robust specimens may produce two blooms. It can form large colonies, reproducing vegetatively from daughter tubers formed on the end of stolonoid roots. Also reproduces from seed to form new colonies, as fungus gnats, and sometimes mosquitoes, pollinate the flowers. The labellum is highly irritable and closes when the pollinator triggers the appendage at the base of the lip.

Pterostylis curta
Menangle, New South Wales

25 July 1992

A plant
B flower from front
C flower from side
D flower from rear
E flower from above
F labellum flattened
G labellum and column from side
H column from front

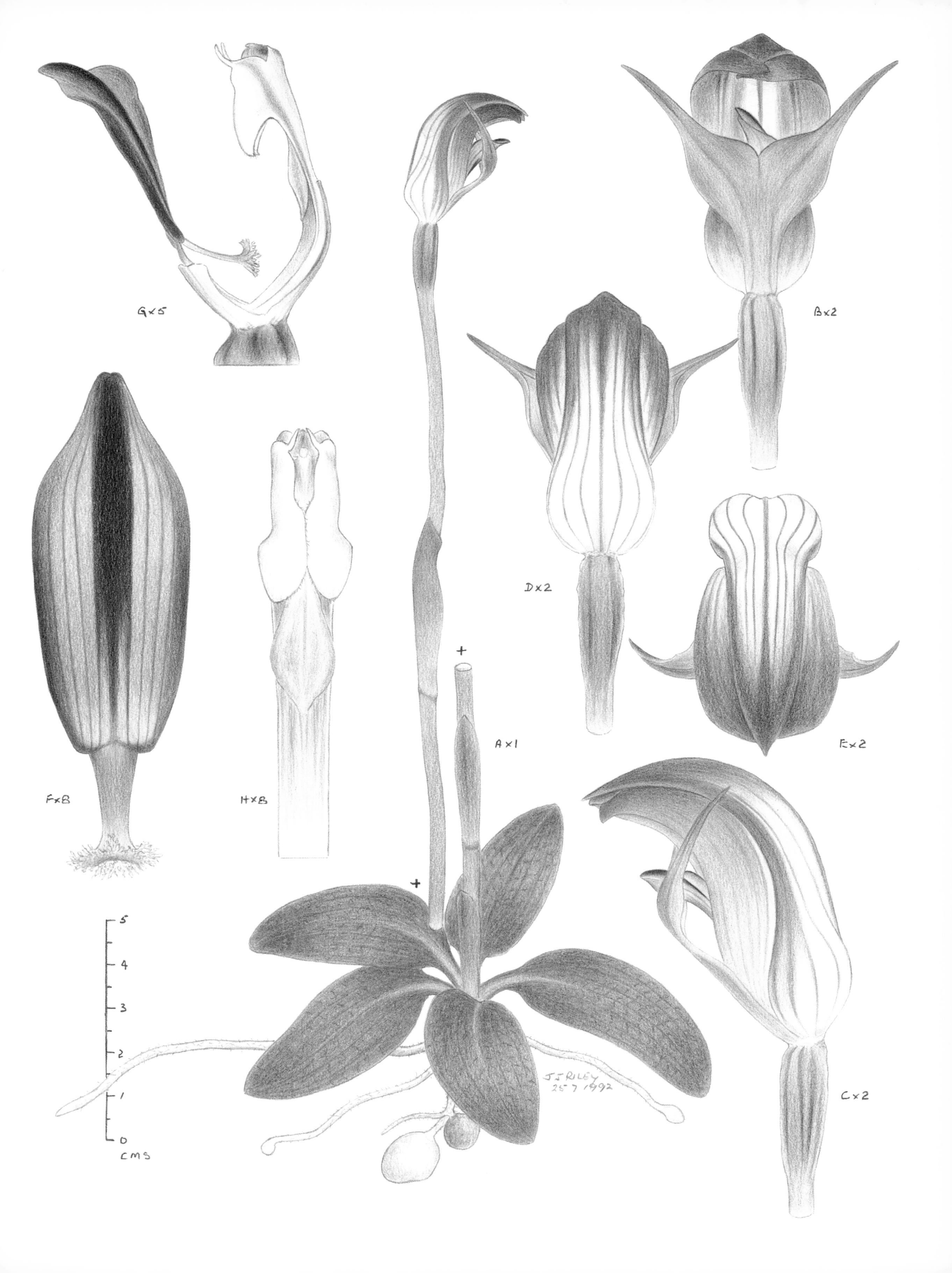
G×5
B×2
D×2
F×8
H×8
A×1
E×2
C×2
5
4
3
2
1
0
CMS
J J RILEY
25 7 1992

Pterostylis hildae Nicholls

1937 | *Queensland Naturalist* 10:39

TYPE LOCALITY Mount Tamborine, Queensland

ETYMOLOGY After Hilda Curtis

FLOWERING TIME March to September

DISTRIBUTION This species is found from the Macquarie Pass, south-west of Wollongong, north along the coast and nearby ranges of New South Wales, to Eungella and the Atherton Tableland in North Queensland.

ALTITUDINAL RANGE 50 m to 900 m

DISTINGUISHING FEATURES *P. hildae* has smaller and narrower flowers than *P. curta*. It also differs from the latter species by having an untwisted, green labellum with a blunt apex. *P. hildae* has an open, basal rosette when flowering, often with only three or four leaves.

HABITAT This species is found in humus-rich soils in wet sclerophyll forest. It prefers moist, sheltered, shrubby areas along streams and drainage areas in *Eucalyptus* woodland and on the margins of rain-forest.

CONSERVATION STATUS Widespread. Occurs in national parks and reserves. Secure.

DISCUSSION *P. hildae* is of sporadic occurrence in the south of its range, becoming increasingly common to the north. It forms small colonies, reproducing vegetatively from daughter tubers formed on the end of stolonoid roots. Also reproduces from seed to form new colonies, as fungus gnats pollinate the flowers. The labellum is highly irritable and closes when the pollinator triggers the appendage at the base of the lip.

Pterostylis hildae
Springvale, Queensland

5 May 1998

A plant
B flower from side
C flower from front
D flower from rear
E labellum flattened
F labellum and column from side
G column from front
H dorsal sepal
J lateral sepals
K petal

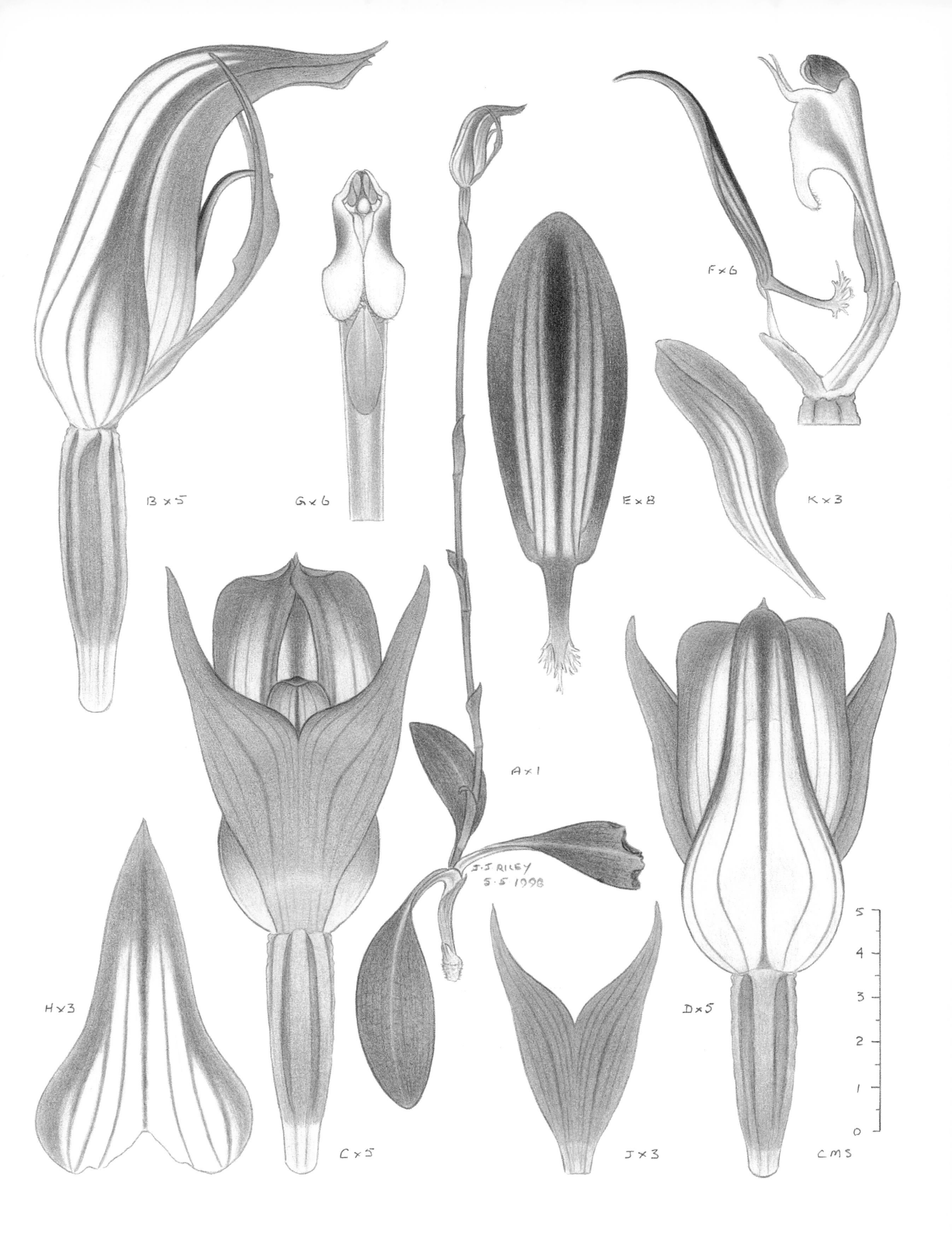
B x 5
G x 6
E x 8
K x 3
F x 6
A x 1
J. J. RILEY
5.5.1998
H x 3
C x 5
J x 3
D x 5
CMS
5
4
3
2
1
0

Pterostylis stricta Clemesha & B. Gray

1972 | *Orchadian* 4:18

TYPE LOCALITY Ravenshoe, Queensland

ETYMOLOGY *stricta* — stiff, rigid

FLOWERING TIME March to August

DISTRIBUTION *P. stricta* is endemic to North Queensland, being found between Eungella and Windsor Tablelands.

ALTITUDINAL RANGE 750 m to 1100 m

DISTINGUISHING FEATURES *P. stricta* is more robust than *P. hildae* and lacks the twisted labellum of *P. curta*. This species is easily identified by its curved labellum that tapers to a point and has fine, ciliate hairs on the margins.

HABITAT This species is found in moist, protected sites in dry, open, grassy woodland. It grows around the base of grass trees, (*Xanthorrhoea* spp.), sheltered among granite boulders and along the banks of creeks and rivers. It may also be seen among grasses and ferns in wet sclerophyll forest and on rainforest margins, often seen near walking tracks.

CONSERVATION STATUS Occurs in national parks and reserves. Secure.

DISCUSSION *P. stricta* can be locally common, forming scattered colonies. It is sometimes seen growing with the related *P. procera*, yet hybrids between the two are unknown. It reproduces vegetatively from daughter tubers formed on the end of stolonoid roots. Also reproduces from seed to form new colonies, as fungus gnats pollinate the flowers. The labellum is highly irritable and snaps back towards the reproductive parts of the bloom when touched.

Pterostylis stricta
Atherton Tableland, Queensland

13 June 1995

A plant
B flower from front
C flower from side
D flower from rear
E flower from above
F labellum flattened
G labellum and column from side
H column from front
J lateral sepals
K petal

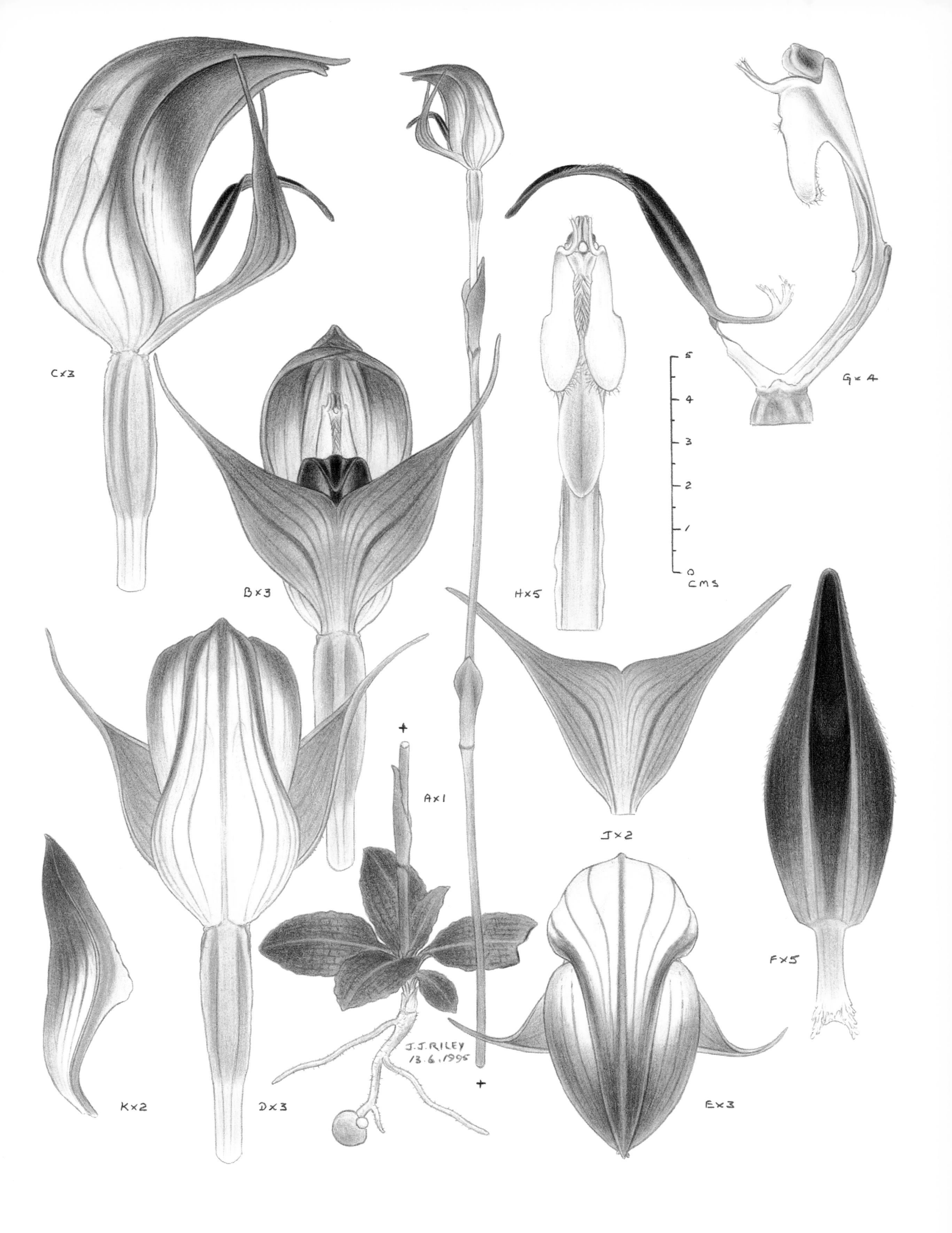
C×3
B×3
H×5
5
4
3
2
1
0
CMS
G×4
A×1
J×2
F×5
K×2
D×3
E×3
J.J. RILEY
13.6.1995

Pterostylis baptistii Fitzg.

1875 | *Australian Orchids* 1(1)

TYPE LOCALITY Sydney, New South Wales

ETYMOLOGY After John Baptist

FLOWERING TIME July to October

DISTRIBUTION *P. baptistii* is found from near Coffs Harbour in northern New South Wales, south to just over the Victorian border. It is a coastal species that may be encountered on the adjacent ranges.

ALTITUDINAL RANGE 10 m to 700 m

DISTINGUISHING FEATURES *P. baptistii* is a tall-growing and large-flowered species that has a basal rosette of leaves, with a distinctive wavy edge. The labellum, when flattened, is long and narrow, tapering to an extended, blunt point in the apical third. The flower, among the largest of the greenhoods, is upright with a prominent, bulging sinus when viewed from the side.

HABITAT This majestic species can be found growing close to creeks and swamps in open and closed woodland and wet sclerophyll forest. It generally favours damp situations close to water, but can sometimes be seen growing in sheltered dry sites. *P. baptistii* can be common in low lying, seasonally wet woodland dominated by *Melaleuca quinquenervia*.

CONSERVATION STATUS Common in New South Wales. Occurs in national parks and reserves. Rare in Victoria.

DISCUSSION *P. baptistii* is part of a complex of related taxa. Two of these have been named, being *P. procera* and *P. anatona*. There are also spring, summer and early autumn-flowering plants in southern Queensland (including Fraser Island) and northern New South Wales that probably need further study. *P. baptistii* has the most southerly distribution and is the latest member of the complex to flower. It was once common around Sydney, but is now very rare due to development for housing. It reproduces vegetatively from daughter tubers that are formed on the end of stolonoid roots about six weeks after blooming. Also reproduces from seed to form new colonies, and is pollinated by fungus gnats. The labellum is highly irritable and closes when the pollinator triggers the appendage at the base of the lip.

Pterostylis baptistii
Hallidays Point, New South Wales

1 September 2000

A plant
B flower from side
C flower from front
D flower from rear
E flower from above
F labellum flattened
G labellum and column from side
H column from front

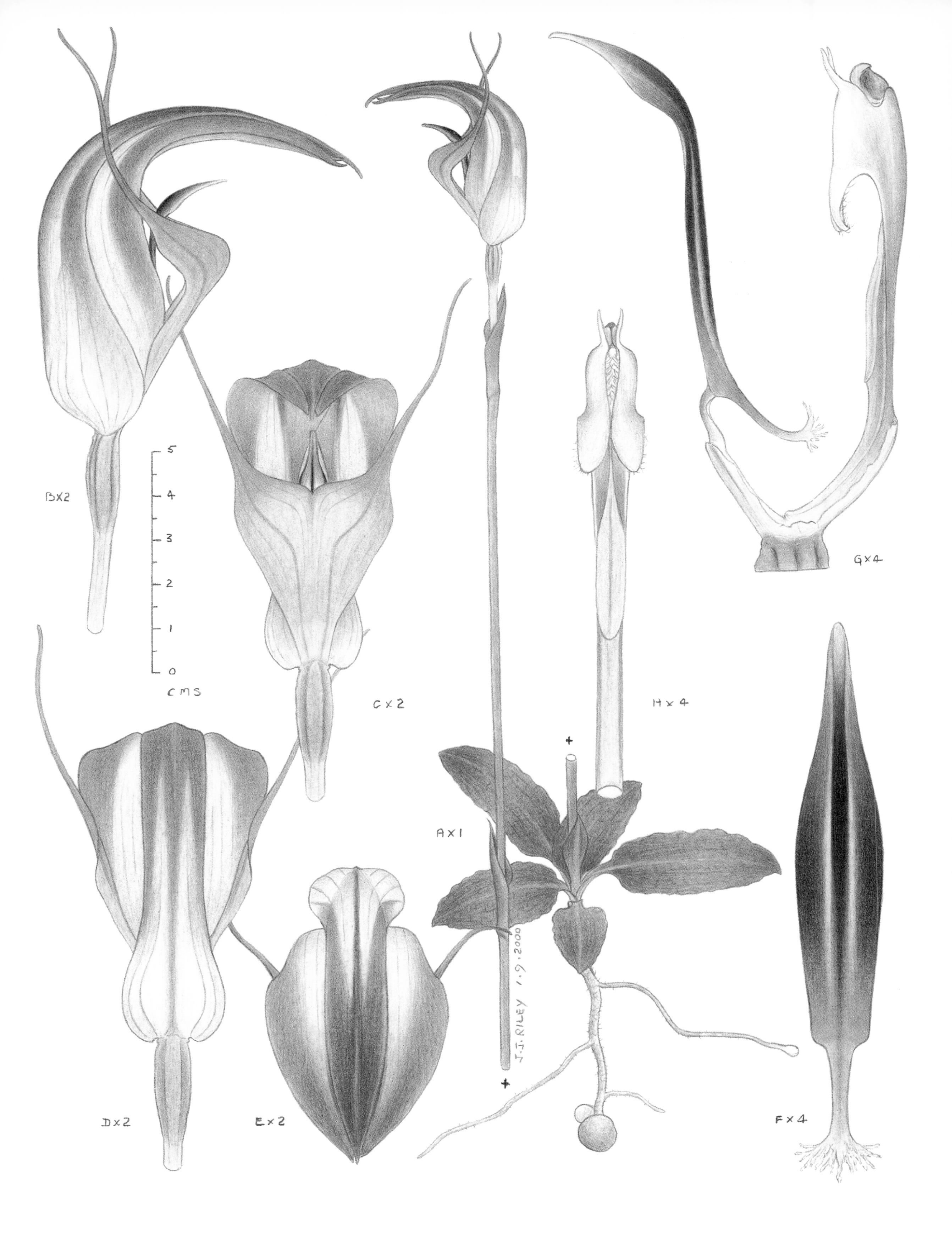
B×2
5
4
3
2
1
0
CMS
C×2
G×4
H×4
A×1
J.J. RILEY 1.9.2000
D×2
E×2
F×4

Pterostylis procera D.L. Jones & M.A. Clem.

1989 | *Australian Orchid Research* 1:125

TYPE LOCALITY Herberton Range, Queensland

ETYMOLOGY *procera* — slim, erect

FLOWERING TIME March to June

DISTRIBUTION *P. procera* is endemic to North Queensland, being found between the Seaview Range and the Atherton Tableland.

ALTITUDINAL RANGE 800 m to 1100 m

DISTINGUISHING FEATURES *P. procera* has a basal rosette of leaves when flowering. It has smaller, narrower flowers than *P. baptistii* and blooms earlier than that species. The labellum, when flattened, is much shorter and wider than that of *P. baptistii* and is constricted in the apical quarter.

HABITAT This species is found among grasses and ferns in wet sclerophyll forest and sheltered areas in open woodland, often beside rivers and streams. Away from creeks, it is often found growing near the tops of ridges, sheltered among granite boulders or around the base of grass trees (*Xanthorrhoea* spp.).

CONSERVATION STATUS Occurs in national parks and reserves. Secure.

DISCUSSION *P. procera* is a free-flowering species and one of the northern members of the *P. baptistii* complex that grows in loose, scattered colonies. It reproduces vegetatively from daughter tubers formed on the end of stolonoid roots. Also reproduces from seed to form new colonies, as fungus gnats pollinate the flowers. The labellum is highly irritable and closes when the pollinator triggers the appendage at the base of the lip.

Pterostylis procera
Mount Banksia, Queensland

27 March 1996

A plant
B flower from front
C flower from side
D flower from rear
E flower from above
F labellum flattened
G labellum and column from side
H column from front

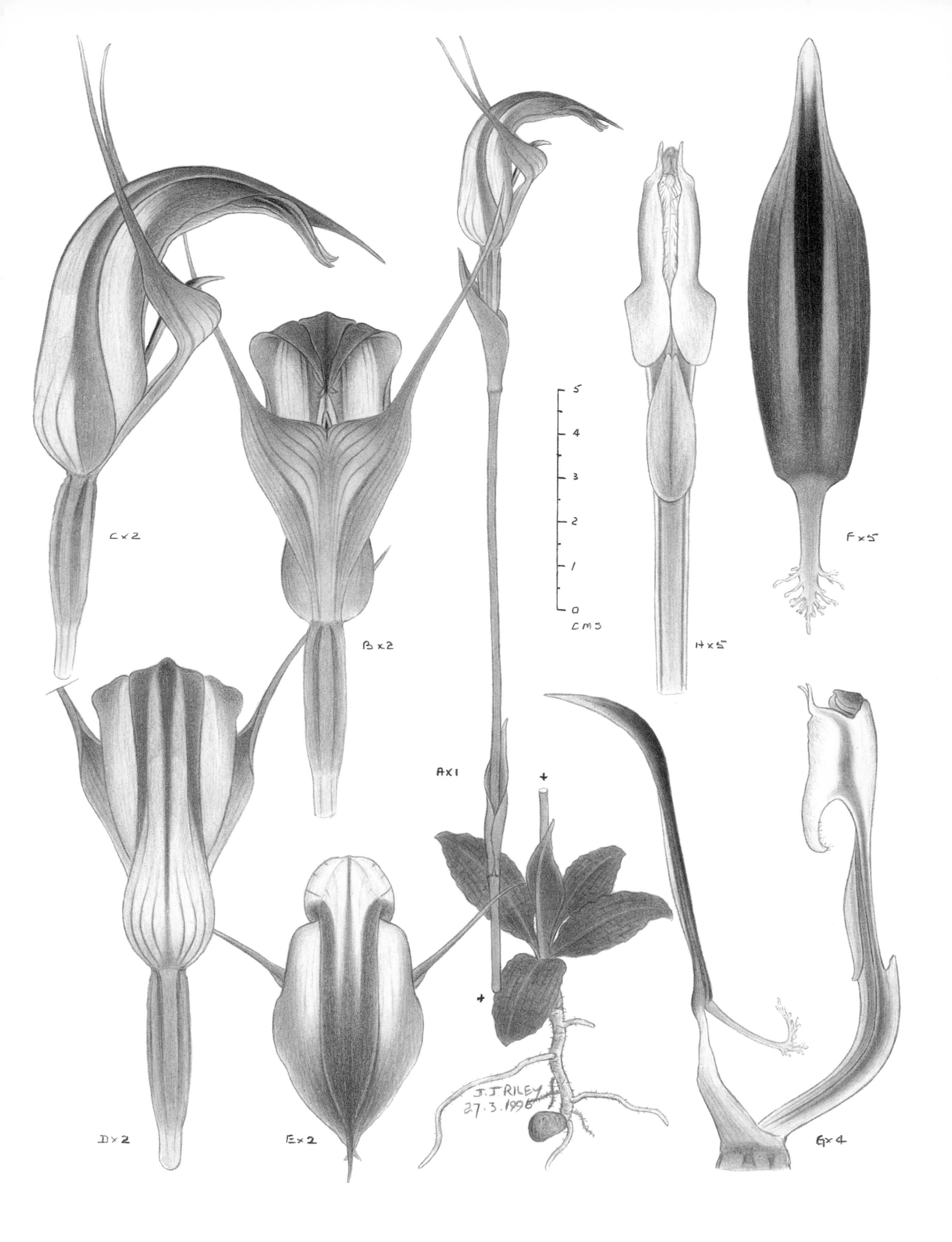
C×2
B×2
5
4
3
2
1
0
CMS
H×5
F×5
A×1
D×2
E×2
J.J. RILEY
27.3.1996
G×4

Pterostylis anatona D.L. Jones

1997 | *Orchadian* 12:245

TYPE LOCALITY Eungella, Queensland

ETYMOLOGY *anatona* — stretching upward

FLOWERING TIME June to August

DISTRIBUTION *P. anatona* is endemic to North Queensland, being found on the Clarke Range, north-west of Mackay and Eungella.

ALTITUDINAL RANGE 800 m to 1000 m

DISTINGUISHING FEATURES *P. anatona* has a basal rosette of leaves when flowering and is similar to *P. procera* but has smaller flowers with shorter lateral sepals. The galea is more upright at the apex while the galea apex of *P. procera* curves downwards. The labellum, when in the set position, protrudes noticeably through the sinus and, when flattened, is widest at half its length, tapering to a blunt point.

HABITAT This species is found in moist, sheltered sites in dry sclerophyll woodland with a grassy and scattered, shrubby understorey. *P. anatona* more commonly grows near the small creeks and washes that form the drainage patterns. Occasional, small colonies are encountered in the more open areas around the margins of rainforest.

CONSERVATION STATUS Common. Occurs in a national park. Secure.

DISCUSSION *P. anatona* is another member of the *P. baptistii* complex. It can form loose, scattered colonies, reproducing vegetatively from daughter tubers formed on the end of stolonoid roots. Also reproduces from seed to form new colonies, as fungus gnats pollinate the flowers. The labellum is highly irritable and closes when the pollinator triggers the appendage at the base of the lip.

Pterostylis anatona
Eungella, Queensland

13 May 1998

A plant
B flower from front
C flower from side
D flower from rear
E flower from above
F labellum flattened
G labellum and column from side
H column from front

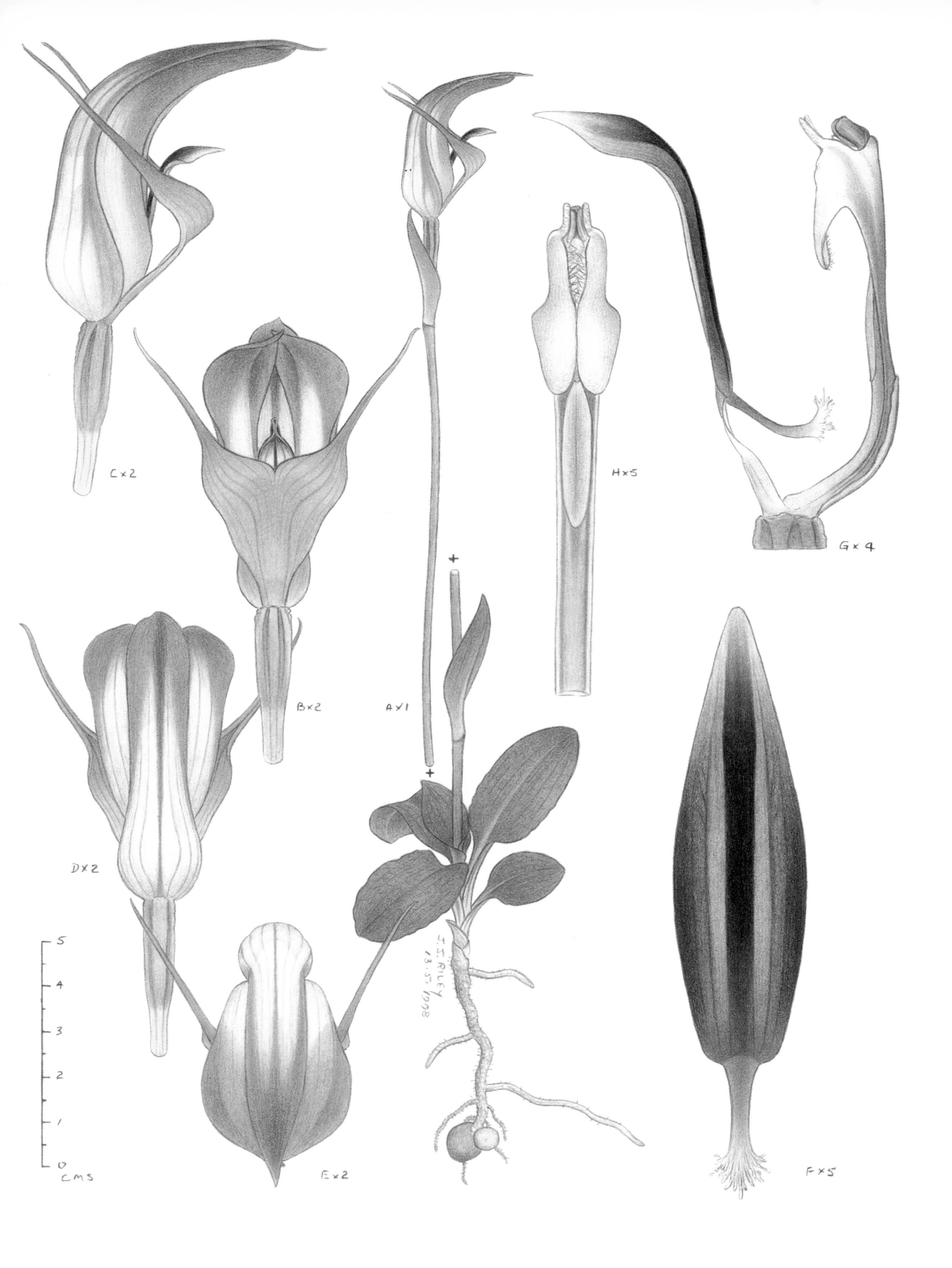
C×2
B×2
A×1
H×5
G×4
D×2
E×2
F×5
5
4
3
2
1
0
CMS
J.J. RILEY
13.5.1998

Pterostylis ophioglossa R. Br.

1810 | *Prodromus Florae Novae Hollandiae*: 326

TYPE LOCALITY Port Jackson, New South Wales

ETYMOLOGY *ophioglossa* — with a snake-like tongue

FLOWERING TIME April to July

DISTRIBUTION *P. ophioglossa* is found from around the Campbelltown district of New South Wales, north along the coast to south-eastern Queensland as far north as Fraser Island and the adjacent mainland.

ALTITUDINAL RANGE 20 m to 300 m

DISTINGUISHING FEATURES *P. ophioglossa* has a ground-hugging, basal rosette of leaves when flowering. It has large, brownish-green to pale green flowers and a distinctive labellum that is forked, with the points parallel.

HABITAT This species grows in the coastal lowlands in open woodland and forest with an understorey of grass and small shrubs, and occasionally in littoral rainforest. Also found in more exposed situations among rocks and ridge tops and slopes. Soils are heavy clay loams. *P. ophioglossa* occasionally will colonise small areas of coastal heath dominated by *Banksia* spp. and *Leptospermum* spp., growing in sand.

CONSERVATION STATUS Common and widespread. Occurs in national parks and reserves. Secure.

DISCUSSION *P. ophioglossa* is a robust species with impressive blooms that are quite large considering the size of the plant. The exact northern distribution is unclear due to confusion with *P. collina*. *P. ophioglossa* is involved in at least two natural hybrids which have been formally named, with *P. alveata* to produce *P.* X *furcillata* and with *P. concinna* to make *P.* X *conoglossa*. *P. ophioglossa* can form large colonies, reproducing vegetatively from daughter tubers formed on the end of stolonoid roots. Also reproduces from seed to form new colonies, as fungus gnats pollinate the flowers. The labellum is highly irritable and closes when the pollinator triggers the appendage at the base of the lip.

Pterostylis ophioglossa
Casula, New South Wales

15 May 1994

A plant
B flower from front
C flower from side
D flower from rear
E flower from above
F labellum flattened
G labellum and column from side
H column from front
J dorsal sepal
K lateral sepals
L petal

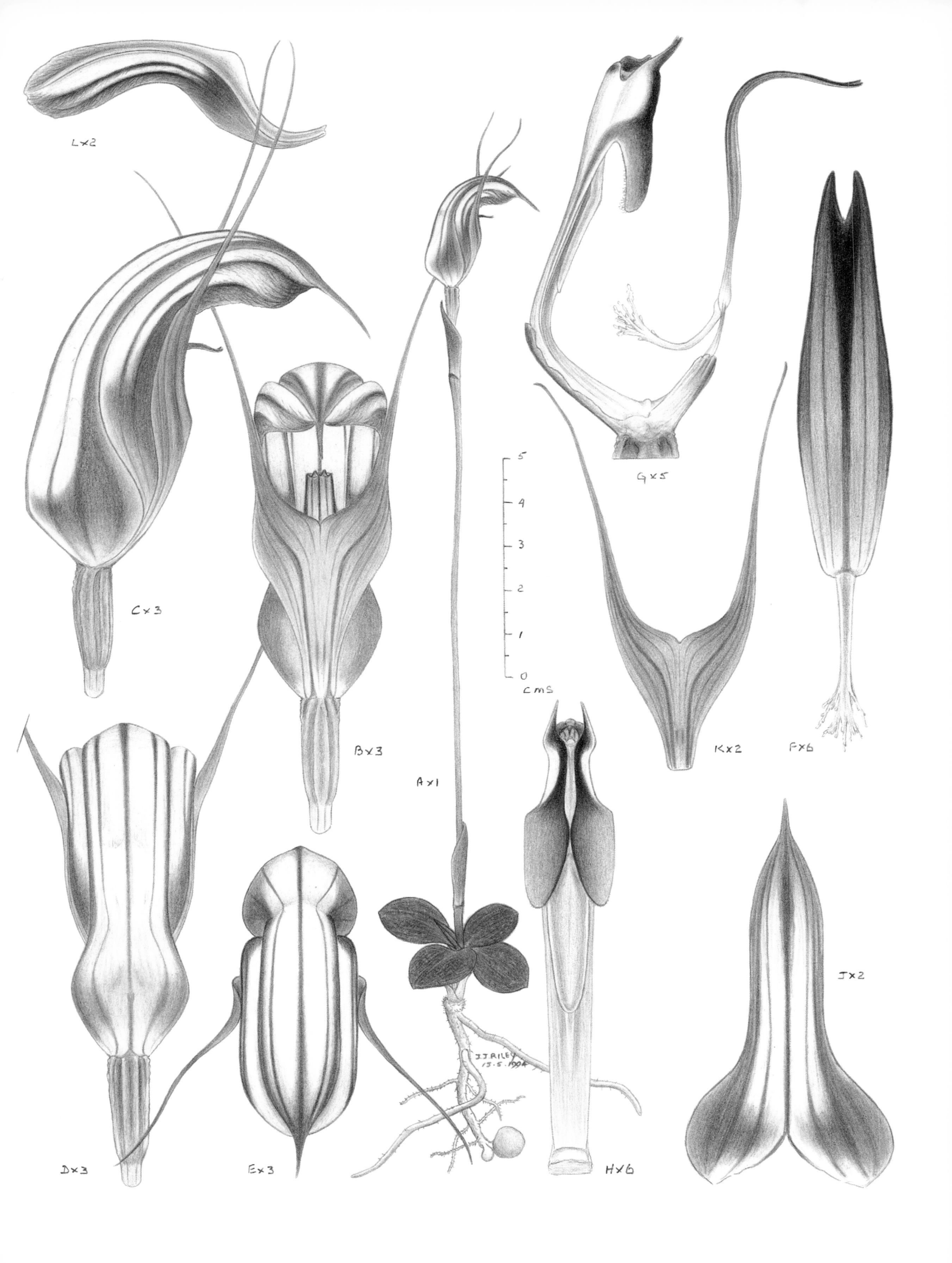

L×2
C×3
B×3
A×1
G×5
K×2
F×6
5
4
3
2
1
0
cms
D×3
E×3
H×6
J×2
J.J.RILEY
13.5.1994

Pterostylis collina (Rupp) M.A. Clem. & D.L. Jones

1989 | *Australian Orchid Research* 1:121

TYPE LOCALITY Paterson, New South Wales

RECENT SYNONYMS *Pterostylis ophioglossa* R. Br. var. *collina* Rupp

ETYMOLOGY *collina* — living in the hills

FLOWERING TIME May to August

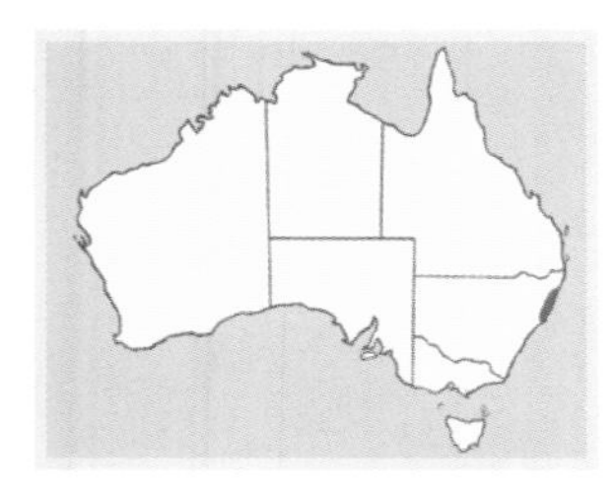

DISTRIBUTION *P. collina* is endemic to New South Wales, being found from Paterson and the Port Stephens district, north along the coast and adjacent ranges to around Coffs Harbour.

ALTITUDINAL RANGE 20 m to 500 m

DISTINGUISHING FEATURES *P. collina* has a basal rosette of leaves when flowering and has smaller blooms than *P. ophioglossa*. The labellum is deeply forked with the free points widely spread. This characteristic becomes obvious when the labellum is flattened. *P. collina* has significantly darker blooms than *P. ophioglossa*, being an attractive reddish-brown.

HABITAT *P. collina* mostly favours wet sclerophyll forest and the margins of rainforest. It can colonise disturbed ground along roads and tracks. It is usually found in the low foothills away from the coast. *P. collina* grows in sand under *Banksia* spp. and *Eucalyptus* spp. on the Tomaree Peninsula at Port Stephens.

CONSERVATION STATUS Occurs in national parks and reserves. Secure.

DISCUSSION The dark colouration and compact habit make *P. collina* an attractive species. Occasionally clones that are striped green will appear in normal coloured populations. These are not to be confused with an entirely green-striped, undescribed species from northern New South Wales and southern Queensland. *P. collina* is a free-flowering species that forms colonies, reproducing vegetatively from daughter tubers formed on the end of stolonoid roots. Also reproduces from seed to form new colonies, as fungus gnats pollinate the flowers. The labellum is highly irritable and closes when the pollinator triggers the appendage at the base of the lip.

Pterostylis collina
Copeland, New South Wales

16 June 1993

A plant
B flower from front
C flower from side
D flower from rear
E flower from above
F labellum flattened
G labellum and column from side
H column from front

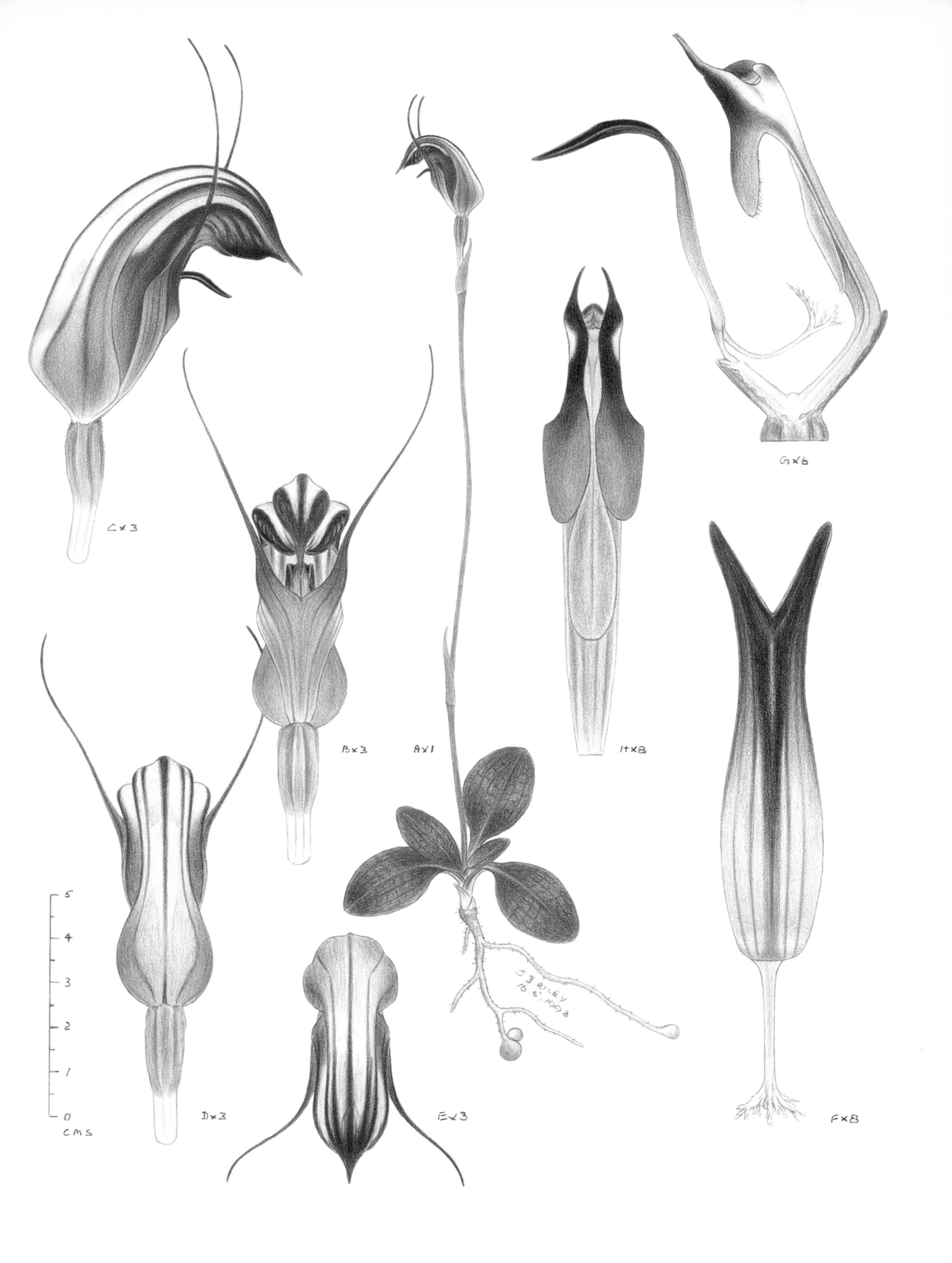
C×3
B×3
A×1
H×8
G×6
F×8
D×3
E×3
5
4
3
2
1
0
CMS
J J RILEY
10.8.1993

Pterostylis taurus M.A. Clem. & D.L. Jones

1989 | *Australian Orchid Research* 1:127

TYPE LOCALITY Ravenshoe, Queensland

RECENT SYNONYMS *Pterostylis ophioglossa* R. Br. subsp. *fusca* Clemesha

ETYMOLOGY *taurus* — like a bull

FLOWERING TIME May to August

DISTRIBUTION This species is endemic to North Queensland, being found from Mackay to the Windsor Tableland.

ALTITUDINAL RANGE 60 m to 1000 m

DISTINGUISHING FEATURES *P. taurus* has a flat, basal rosette of leaves when blooming and has slightly nodding flowers that are smaller than *P. collina*. *P. taurus* is an intense reddish-brown in colour with no green stripes. The labellum has a distinct fork at the apex, which is narrower in *P. taurus* and spreading in the case of *P. collina*.

HABITAT Found in a range of habitats, from wet forest close to rainforest, among stands of grass trees (*Xanthorrhoea* spp.) and in exposed locations to the dry open forest of the inland ranges. Colonies have also been observed on the dry, steep and bare slopes of the Herbert River and Blencoe Creek gorges. It forms small, scattered colonies, occurring in almost full sun to quite shaded situations. This region has a dry winter and a wet summer.

CONSERVATION STATUS Occurs in national parks and state forests. Secure.

DISCUSSION *P. taurus* is the smallest species in the *P. ophioglossa* complex in Australia. It is a very colourful, free-flowering and attractive orchid. The specific name refers to the blooms resemblance to a bull with its head down ready to charge. This is a colony-forming species, which multiplies by producing daughter tubers off the end of stolonoid roots. Also reproduces from seed to form new colonies, as fungus gnats pollinate the flowers. The labellum is highly irritable and closes when the pollinator triggers the appendage at the base of the lip.

Pterostylis taurus
Eungella, Queensland

1 August 1997

A plant
B flower from front
C flower from side
D flower from rear
E flower from above
F labellum flattened
G labellum and column from side
H column from front
J dorsal sepal
K lateral sepals
L petal

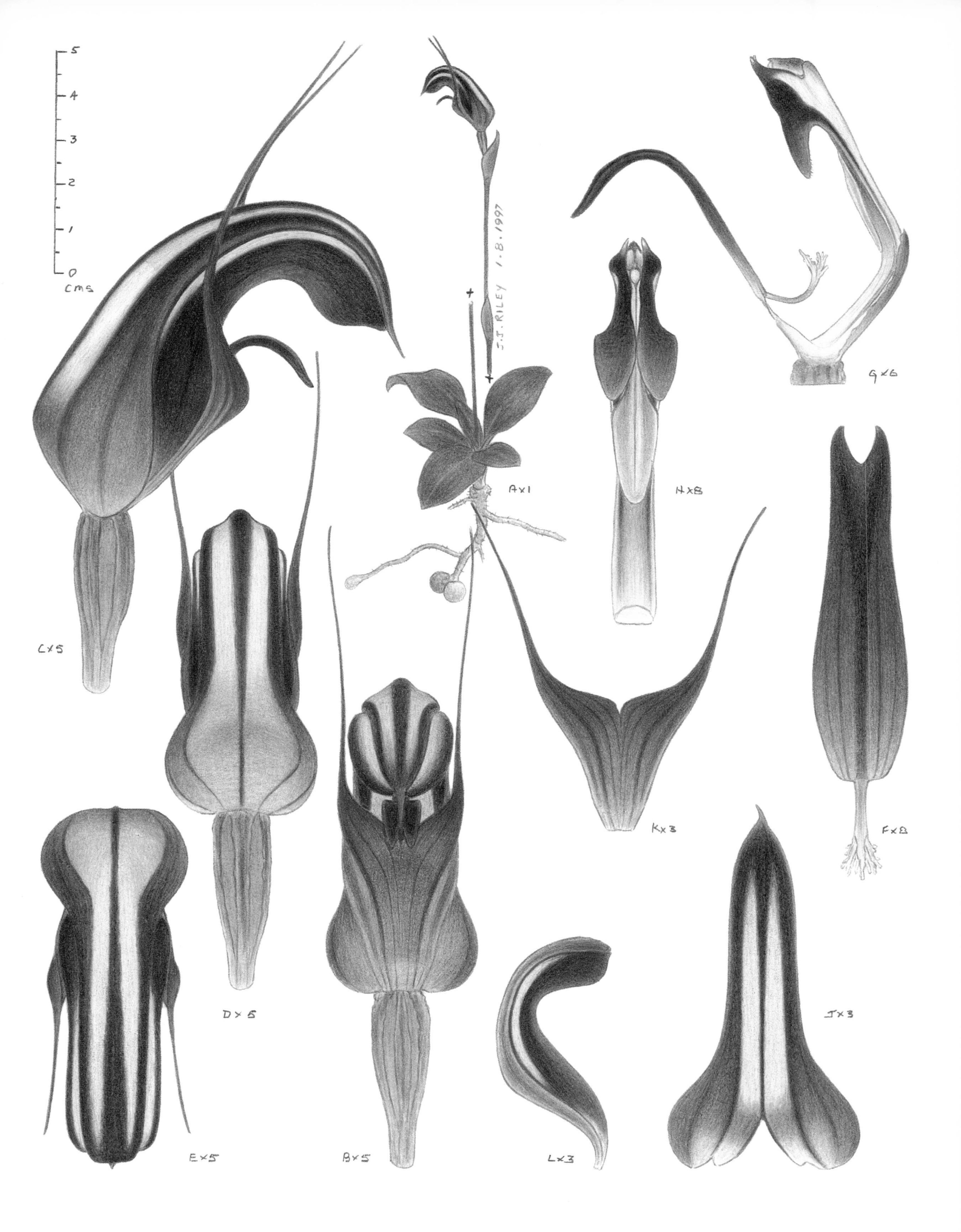
5
4
3
2
1
0
cms
J.J. RILEY 1.8.1997
A×1
B×5
C×5
D×5
E×5
F×8
G×6
H×8
J×3
K×3
L×3

Pterostylis pedoglossa Fitzg.

1877 | *Australian Orchids* 1(3)

TYPE LOCALITY Long Bay, Sydney, New South Wales

ETYMOLOGY *pedoglossa* — foot-like tongue

FLOWERING TIME March to July

DISTRIBUTION This distinctive species occurs on the north coast and East Coast of Tasmania and Cape Barren Island, across to coastal Victoria from around Melbourne, northwards along the New South Wales coast to Port Stephens. There are two disjunct inland populations in New South Wales at Sassafras and Capertee. There is also an isolated population at Daves Creek in the Lamington National Park, south-eastern Queensland.

ALTITUDINAL RANGE 10 m to 900 m

DISTINGUISHING FEATURES *P. pedoglossa* has a basal rosette of leaves when flowering. It has small olive green and white flowers with sepals ending in extremely long, filiform points. When viewed from the side, the flower looks like a prawn. The labellum is short, wide and not visible through the sinus.

HABITAT It is found growing in sand and sandy peat loams around the base of shrubs in coastal heathland and heathy woodland. At higher elevations inland it grows in different plant communities. At Sassafras *P. pedoglossa* grows at the base of large boulders in open *Eucalyptus* forest, while at Capertee it grows in sandy loams under *Callitris* spp.

CONSERVATION STATUS Uncommon. Occurs in national parks and reserves. Secure.

DISCUSSION *P. pedoglossa* is not a common or free-flowering species and has a sporadic distribution. The leaf rosettes are small and are a bluish-green colour. It forms scattered colonies, reproducing vegetatively from daughter tubers formed on the end of stolonoid roots. Also reproduces from seed to form new colonies, as fungus gnats pollinate the flowers. The labellum is highly irritable and closes when the pollinator triggers the appendage at the base of the lip.

Pterostylis pedoglossa
Capertee, New South Wales

8 May 1996

A plant
B flower from side
C flower from front
D flower from rear
E flower from above
F petal
G lateral sepals
H dorsal sepal
J labellum flattened
K labellum and column from side
L column from front

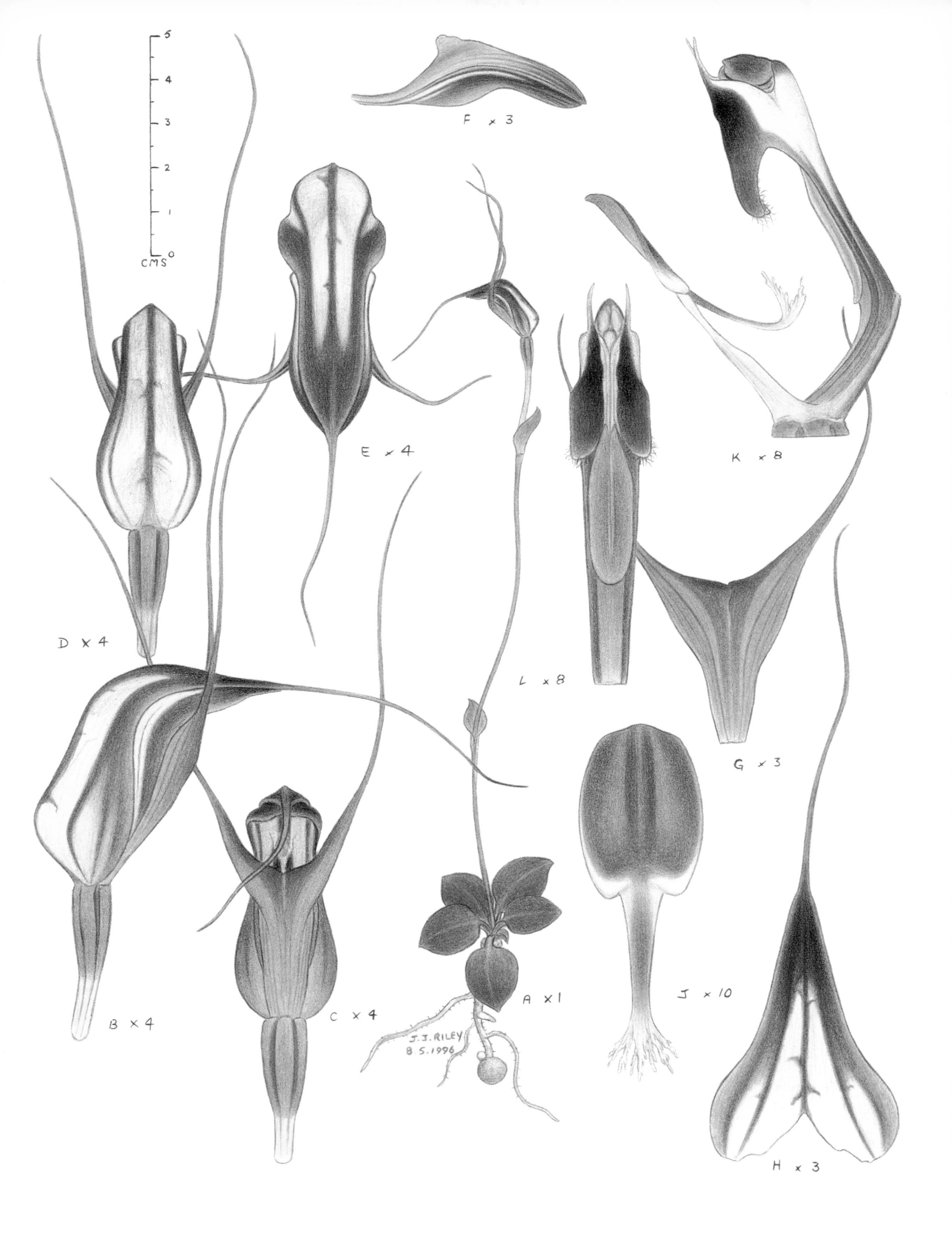

CMS
0
1
2
3
4
5
F x 3
E x 4
K x 8
D x 4
L x 8
G x 3
B x 4
C x 4
A x 1
J x 10
H x 3
J.J. RILEY
8.5.1996

Pterostylis concinna R. Br.

1810 | *Prodromus Florae Novae Hollandiae*: 326

TYPE LOCALITY Sydney, New South Wales

ETYMOLOGY *concinna* — neat, trim

FLOWERING TIME May to October

DISTRIBUTION This widespread species occurs on the north coast and East Coast of Tasmania, and Furneaux Islands, across to south-eastern South Australia, the southern half of Victoria, and north along the coast and tablelands of New South Wales to Taree.

ALTITUDINAL RANGE Up to 900 m

DISTINGUISHING FEATURES *P. concinna* has a basal rosette of leaves when flowering. The white flowers have dark green stripes and reddish-brown suffusions on the galea, petals and most noticeably around the edge of the sinus. The short, wide, forked labellum is just visible through the sinus.

HABITAT *P. concinna* is a common, widespread and well known species which grows in a variety of plant communities. These include stabilised, heathy coastal dunes dominated by *Banksia* spp., heathland and open *Eucalyptus* woodland and forest. Although sometimes seen growing among dense vegetation, it prefers a more open and drier habitat. Soils vary from sand, sandy loams to heavy clay loams.

CONSERVATION STATUS Common and widespread. Occurs in national parks and reserves. Secure.

DISCUSSION *P. concinna* has bluish-green leaves and is a parent in two named natural hybrids, with *P. alata* to produce *P.* X *toveyana* and with *P. ophioglossa* to make *P.* X *conoglossa*. It is a very free-flowering orchid that can form large colonies. It reproduces vegetatively from daughter tubers formed on the end of stolonoid roots. Also reproduces from seed to form new colonies, as fungus gnats pollinate the flowers. The labellum is highly irritable and closes when the pollinator triggers the appendage at the base of the lip. There is also a rare aberrant form with pale yellow flowers.

Pterostylis concinna
South Arm Peninsula, Tasmania

22 June 1996

A plant
B flower from front
C flower from side
D flower from rear
E flower from above
F labellum flattened
G labellum and column from side
H column from front
J dorsal sepal
K lateral sepals
L petal

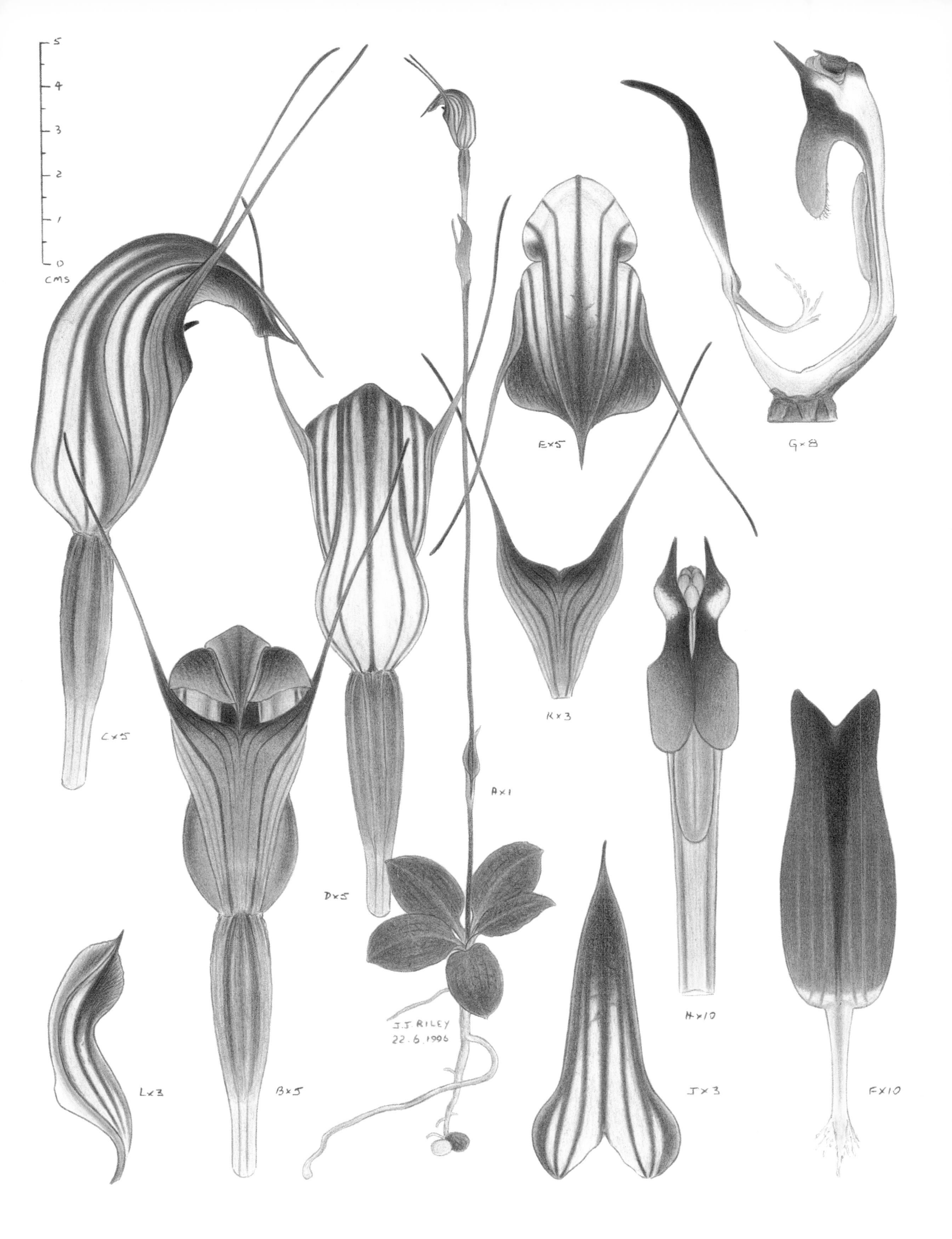

5
4
3
2
1
0
CMS
C×5
D×5
B×5
L×3
A×1
E×5
K×3
J×3
G×8
H×10
F×10
J.J. RILEY
22.6.1996

Pterostylis pulchella Messmer

1933 | *Proceedings of the Linnean Society of New South Wales* 58:429

TYPE LOCALITY Fitzroy Falls, New South Wales

ETYMOLOGY *pulchella* — beautiful

FLOWERING TIME February to May

DISTRIBUTION This species is endemic to New South Wales, being restricted to the precipitous valleys of the Illawarra escarpment between Fitzroy Falls and Barren Grounds.

ALTITUDINAL RANGE 500 m to 700 m

DISTINGUISHING FEATURES *P. pulchella* is one of the cauline-leafed species, with non-blooming plants having a basal rosette of leaves, and flowering plants without a rosette, instead having stem-clasping leaves up the stalk. It has large reddish-brown flowers and a prominent sinus that protrudes forward. The labellum is long, wide and fleshy with the apex deeply notched.

HABITAT This free-flowering species is found near several waterfalls, where it grows on mossy rocks and in moist humus-rich soil. It makes its home among vegetation that has managed to become established on the steep sandstone escarpments.

CONSERVATION STATUS Rare. Occurs in a national park. Vulnerable.

DISCUSSION *P. pulchella* is one of Australia's most beautiful species. It forms small groups and scattered colonies, reproducing vegetatively from daughter tubers formed on the end of stolonoid roots, mostly off the non-flowering plants. Also reproduces from seed to form new colonies, as fungus gnats pollinate the flowers. The labellum is highly irritable and closes when the pollinator triggers the appendage at the base of the lip. At present it is only known from several high waterfalls where there is limited access. The rugged and vertical cliffs of the escarpment present an insurmountable barrier to finding additional populations.

Pterostylis pulchella
Fitzroy Falls, New South Wales

15 March 1996

A plant
B flower from front
C flower from side
D flower from rear
E flower from above
F labellum flattened
G labellum and column from side
H column from front
J dorsal sepal
K lateral sepals
L petal

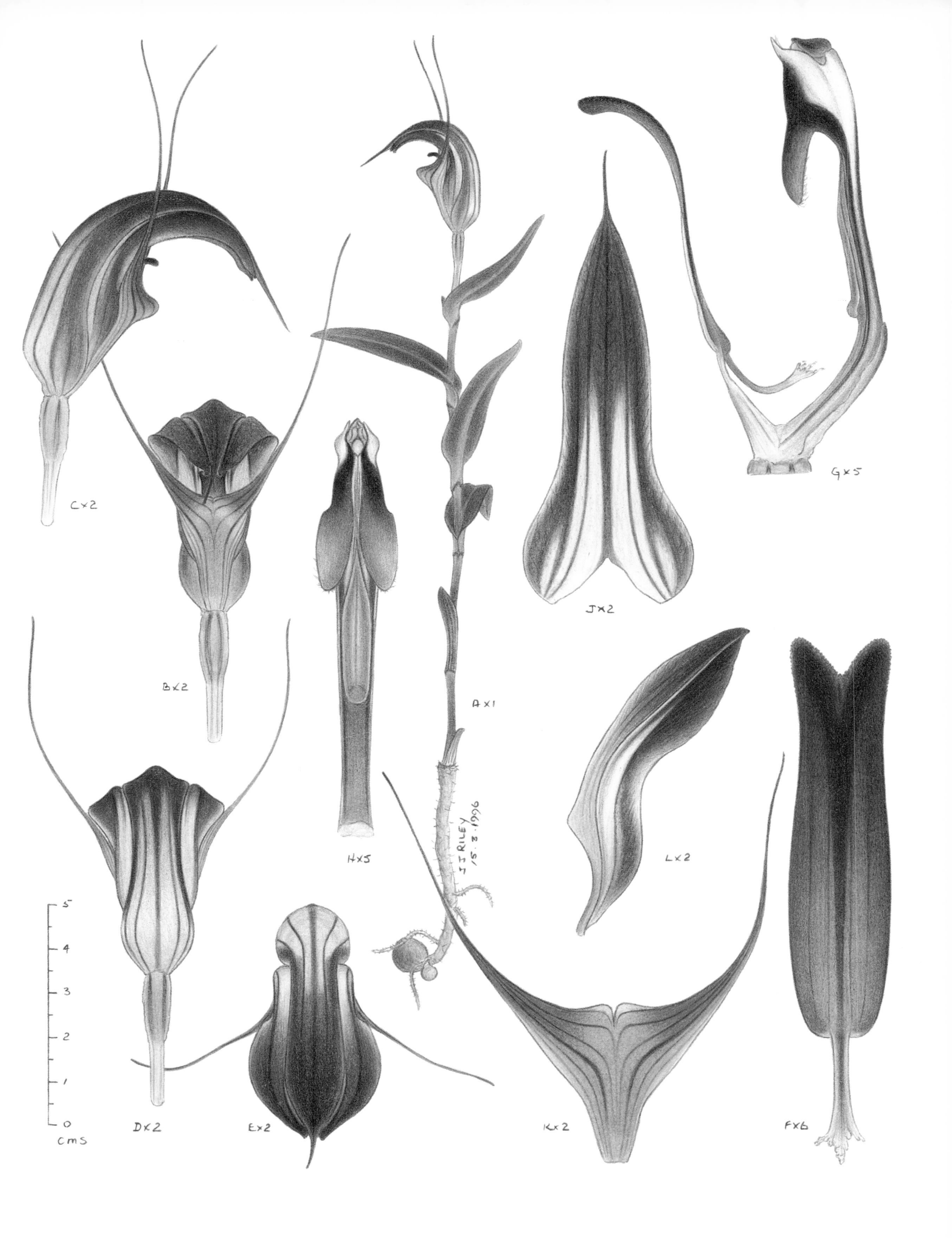

C×2
B×2
A×1
H×5
J×2
G×5
L×2
D×2
E×2
K×2
F×6
cms
0
1
2
3
4
5
J J RILEY
15.3.1996

Pterostylis truncata Fitzg.

1878 | *Australian Orchids* 1(4)

TYPE LOCALITY Mittagong, New South Wales

ETYMOLOGY *truncata* — cut off

FLOWERING TIME February to May

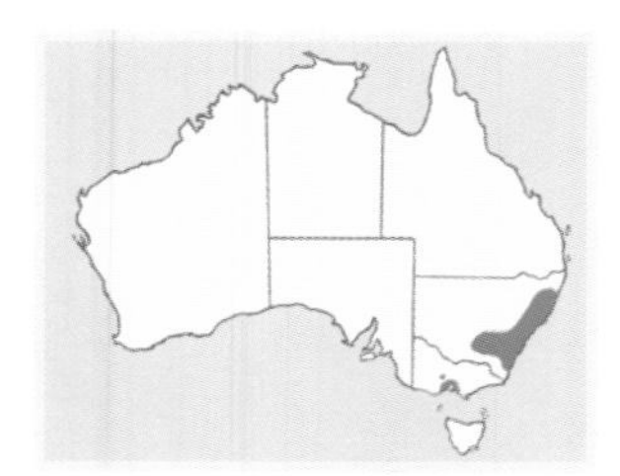

DISTRIBUTION This dumpy species is found in New South Wales, south from Guyra to the Batemans Bay area on the coast, extending inland to Griffith and into Victoria, where it is known from sites around Melbourne, Geelong and Maldon.

ALTITUDINAL RANGE 10 m to 1000 m

DISTINGUISHING FEATURES *P. truncata* is a distinctive, short-stemmed, cauline-leafed species, with non-blooming plants having a ground-hugging basal rosette of leaves, and flowering plants without a rosette, instead having stem-clasping leaves up the stalk. The wide, bulbous petals create a blunt, rounded apex to the flower. The long narrow labellum, with its entire apex, readily distinguishes this orchid from similar species.

HABITAT Found in open heath, open woodland and forest, often in moderately exposed areas. Commonly seen growing on stony ridges and slopes. Soils are always well drained, and vary from sandy loams to gravelly, clay loams.

CONSERVATION STATUS Common and secure in New South Wales. Conserved in national parks and reserves. Rare and vulnerable in Victoria.

DISCUSSION *P. truncata* has large green and brown striped flowers that lean forward. It can form extensive colonies, reproducing vegetatively from daughter tubers formed on the end of stolonoid roots, mostly off the non-flowering plants. Although there may be many hundreds of individuals in a colony, *P. truncata* seems shy, with few, and sometimes no, plants blooming. It also reproduces from seed to form new colonies, as fungus gnats pollinate the flowers. The labellum is highly irritable and closes when the pollinator triggers the appendage at the base of the lip.

Pterostylis truncata
Duncans Creek, New South Wales

9 May 1997

A flowering plant
B non-flowering plant
C flower from front
D flower from side
E flower from rear
F flower from above
G labellum flattened
H labellum and column from side
J column from front
K dorsal sepal
L lateral sepals
M petal

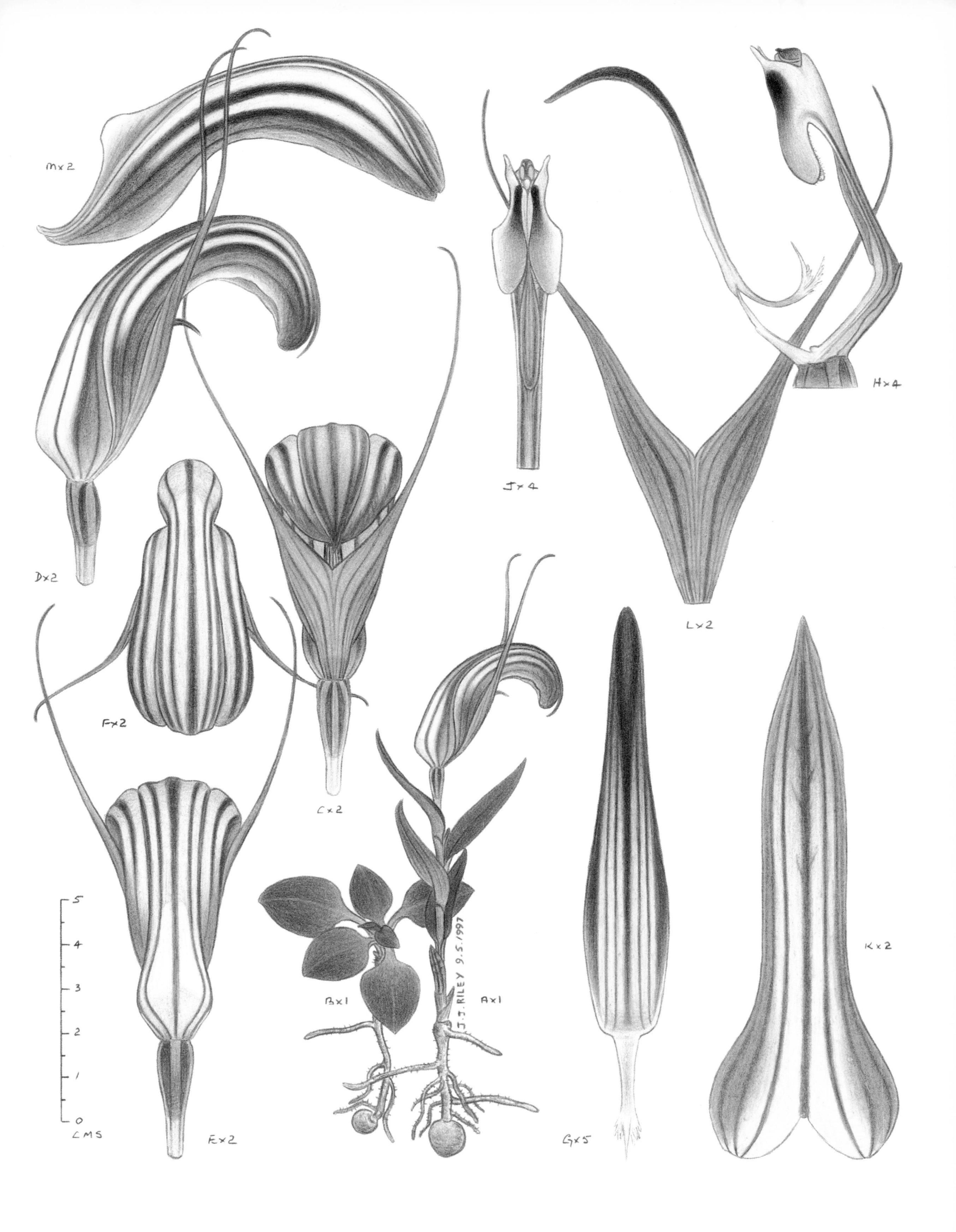
M×2
D×2
F×2
C×2
E×2
J×4
L×2
H×4
G×5
K×2
B×1
A×1
J.J. RILEY 9.5.1997
5
4
3
2
1
0
CMS

Pterostylis fischii Nicholls

1950 | *Victorian Naturalist* 67:45

TYPE LOCALITY Woodside, Victoria

ETYMOLOGY After Paul Fisch

FLOWERING TIME February to May

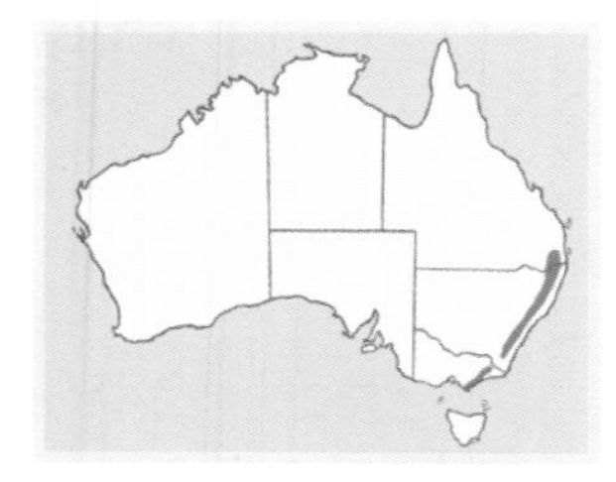

DISTRIBUTION This species is found from south-east Queensland, south through the ranges of New South Wales to eastern Victoria. It is mostly found on the ranges and tablelands, but occurs near the coast in the Gippsland region of Victoria.

ALTITUDINAL RANGE 100 m to 1200 m

DISTINGUISHING FEATURES *P. fischii* is a tall-growing, cauline-leafed species, with non-blooming plants having a rosette of leaves, and flowering plants without a rosette, instead having a few small leaves up the stem. The narrow flowers have clear green and brown stripes. The short, wide labellum is not visible through the sinus.

HABITAT *P. fischii* is a plant of the ranges and tablelands. It occurs in montane forest among grasses and shrubs, often close to large boulders. On the lower ranges and close to the coast, it prefers cool woodland with a grassy and bracken fern understorey. Soils are mostly gravelly loams.

CONSERVATION STATUS Uncommon. Occurs in national parks and reserves. Secure.

DISCUSSION *P. fischii* is a tall, elegant species of sporadic occurrence and is never locally common. It is considered rare in Victoria. It forms loose colonies, reproducing vegetatively from daughter tubers formed on the end of stolonoid roots, mostly off the non-flowering plants. Also reproduces from seed to form new colonies, as fungus gnats pollinate the flowers. The labellum is highly irritable and closes when the pollinator triggers the appendage at the base of the lip.

Pterostylis fischii
Mount Hamilton, Victoria

12 February 1996

A plant
B flower from front
C flower from side
D flower from rear
E flower from above
F labellum flattened
G labellum and column from side
H column from front
J dorsal sepal
K lateral sepals
L petal

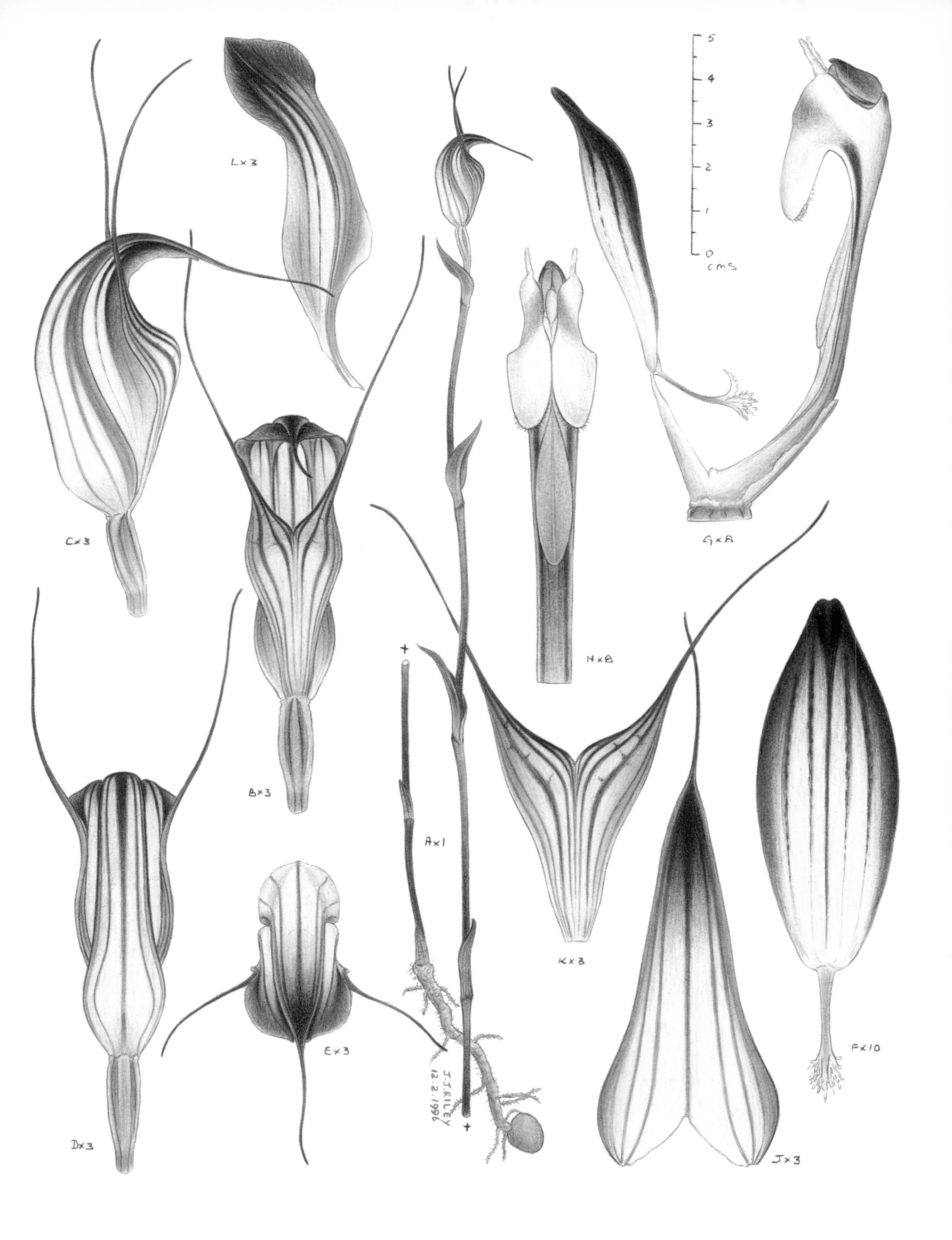
L×3
C×3
B×3
A×1
D×3
E×3
H×8
K×3
J×3
F×10
G×8
5
4
3
2
1
0
cms
J.J.RILEY
12.2.1996

Pterostylis angusta A.S. George

1971 | *Nuytsia* 1(2):164

TYPE LOCALITY Stirling Range, Western Australia

ETYMOLOGY *angusta* — narrow

FLOWERING TIME May to June

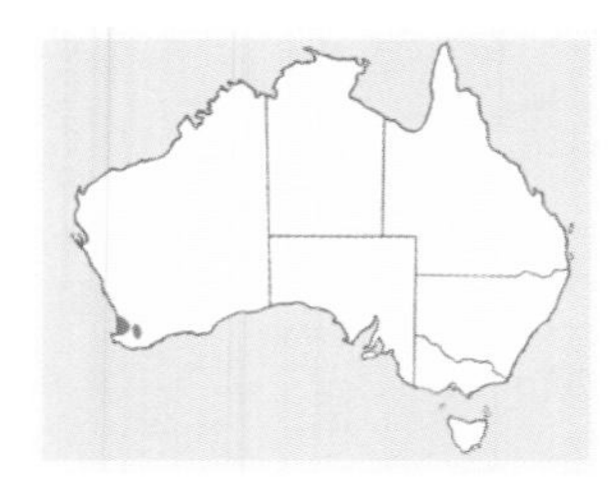

DISTRIBUTION This species is endemic to Western Australia, occurring in the south-western corner around Bunbury and also further east between Brookton and the Stirling Ranges.

ALTITUDINAL RANGE 50 m to 400 m

DISTINGUISHING FEATURES *P. angusta* is one of the cauline-leafed *Pterostylis*. In this species the non-blooming plants have small, ground-hugging, bluish-green, basal rosettes of leaves, and flowering plants have leaves up the stem. *P. angusta* has a very narrow galea and a large bulbous base when viewed from the front. The galea is long and narrow, tapering to a point when viewed from above. The labellum is long and narrow, and protrudes well beyond the sinus.

HABITAT The coastal populations are found in *Banksia* heathland, with the inland populations preferring mallee woodland. Soils range from deep sand to sandy clay loams.

CONSERVATION STATUS Occurs in national parks and reserves. Secure.

DISCUSSION *P. angusta* is from a distinct group of *Pterostylis* that are colloquially known as shell orchids. It can form massive colonies in the wild, reproducing vegetatively from daughter tubers formed on the end of stolonoid roots, mostly off the non-flowering plants. Also reproduces from seed to form new colonies, as fungus gnats pollinate the flowers. The labellum is highly irritable and closes when the pollinator triggers the appendage at the base of the lip.

Pterostylis angusta
Stirling Ranges, Western Australia

20 June 1992

A flowering plant
B non-flowering plant
C flower from front
D flower from side
E flower from rear
F flower from above
G labellum flattened
H labellum and column from side
J column from front

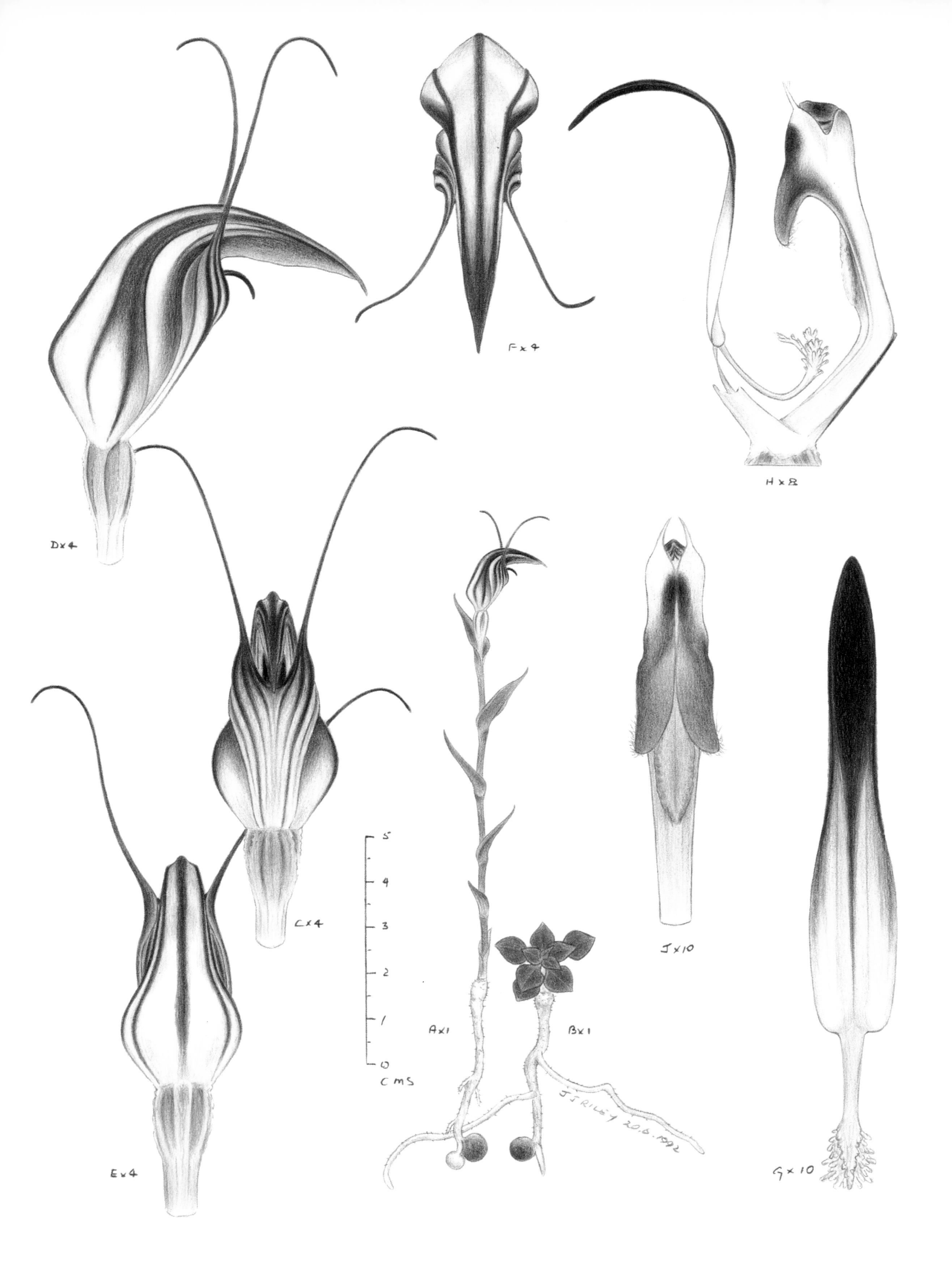

F x 4
H x 8
D x 4
J x 10
C x 4
A x 1
B x 1
E x 4
G x 10
CMS
J J RILEY 20.6.1992

Pterostylis hamiltonii Nicholls

1933 | *Victorian Naturalist* 50:89–91

TYPE LOCALITY Boyup Brook, Western Australia

ETYMOLOGY After A.G. Hamilton

FLOWERING TIME May to August

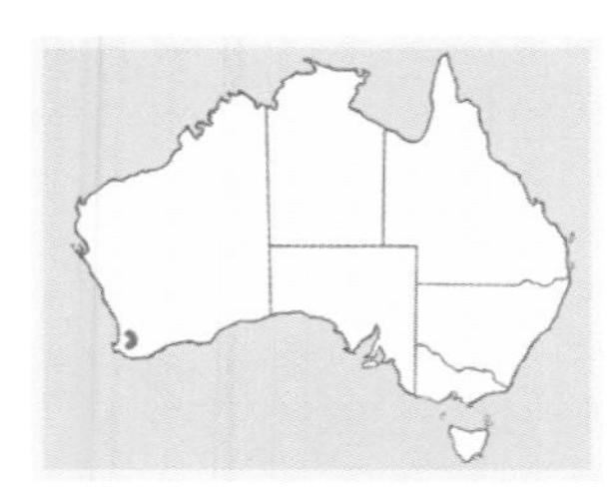

DISTRIBUTION This species is endemic to Western Australia, restricted to the area between Broomehill and Williams.

ALTITUDINAL RANGE 200 m to 350 m

DISTINGUISHING FEATURES *P. hamiltonii* is another cauline-leafed species, with non-blooming plants having a basal rosette of ground-hugging, bluish-green leaves, and flowering plants without a rosette, instead having stem-clasping leaves up the robust stalk. The large flower is striped reddish-brown, with the labellum long and narrowing in the upper third.

HABITAT This species is commonly found in open woodland and under Casuarina spp., growing around the edges of low granite sheets and outcrops. The soils are well drained, gravelly loams.

CONSERVATION STATUS Common. Occurs in national parks and reserves. Secure.

DISCUSSION *P. hamiltonii* is another shell orchid, that is restricted in distribution. It is a beautiful species with large flowers that are striped a rich red-brown. Large colonies have a low ratio of flowering plants. It forms healthy colonies, reproducing vegetatively from daughter tubers formed on the end of stolonoid roots, mostly off the non-flowering plants. Also reproduces from seed to form new colonies, as fungus gnats pollinate the flowers. The labellum is highly irritable and closes when the pollinator triggers the appendage at the base of the lip.

Pterostylis hamiltonii
Wagin, Western Australia

2 June 1991

A flowering plant
B non-flowering plant
C flower from front
D flower from side
E flower from rear
F flower from above
G labellum flattened
H labellum and column from side
J column from front

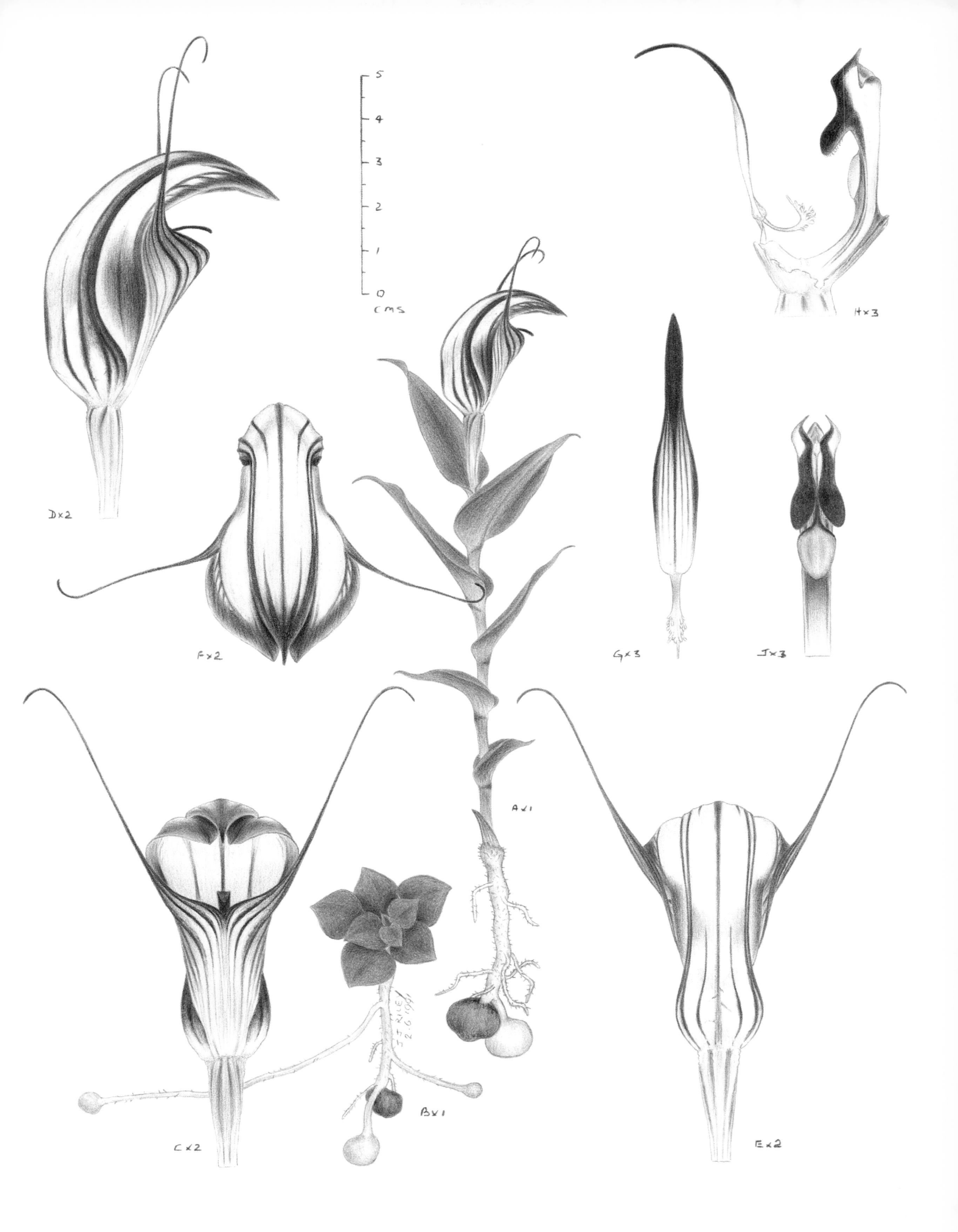

D×2
5
4
3
2
1
0
CMS
H×3
F×2
G×3
J×3
A×1
J.J. RILEY
2.6.1991
B×1
C×2
E×2

Pterostylis erythroconcha M.A. Clem. & D.L. Jones

1986 | *Flora of South Australia*, Part IV:2118

TYPE LOCALITY Corny Point, South Australia

ETYMOLOGY *erythroconcha* — like a red shell

FLOWERING TIME June to September

DISTRIBUTION This striking species is endemic to South Australia, and occurs in the south and west of the State, and on the Eyre and York Peninsulas and Kangaroo Island.

ALTITUDINAL RANGE 30 m to 300 m

DISTINGUISHING FEATURES *P. erythroconcha* is one of the cauline-leafed species, with non-blooming plants having a basal rosette of ground-hugging, bluish-green leaves, and flowering plants without a rosette, instead having stem-clasping leaves up the stem. *P. erythroconcha* has a large, pale reddish-white flower with bright dark red-brown stripes, produced on a short stem. The labellum is fleshy, widest in the basal third and tapering to a blunt point.

HABITAT *P. erythroconcha* is found in scrubby woodland and scrubby mallee communities that grow on red sandy soils, deep sand and the shallow terra rossa soils of the low, flat limestone outcrops. It often forms extensive colonies around the edges or under shrubs where there has been an accumulation of humus. Large colonies have a low ratio of flowering plants.

CONSERVATION STATUS Occurs in national parks and reserves. Secure.

DISCUSSION *P. erythroconcha* is one of the most beautiful and impressive species in the shell orchid group. It forms scattered colonies, reproducing vegetatively from daughter tubers formed on the end of stolonoid roots, mostly off the non-flowering plants. Also reproduces from seed to form new colonies, as fungus gnats pollinate the flowers. The labellum is highly irritable and closes when the pollinator triggers the appendage at the base of the lip.

Pterostylis erythroconcha
Tintinara, South Australia

20 September 1994

A plant
B flower from front
C flower from side
D flower from rear
E flower from above
F labellum flattened
G labellum and column from side
H column from front
J dorsal sepal
K lateral sepals
L petal

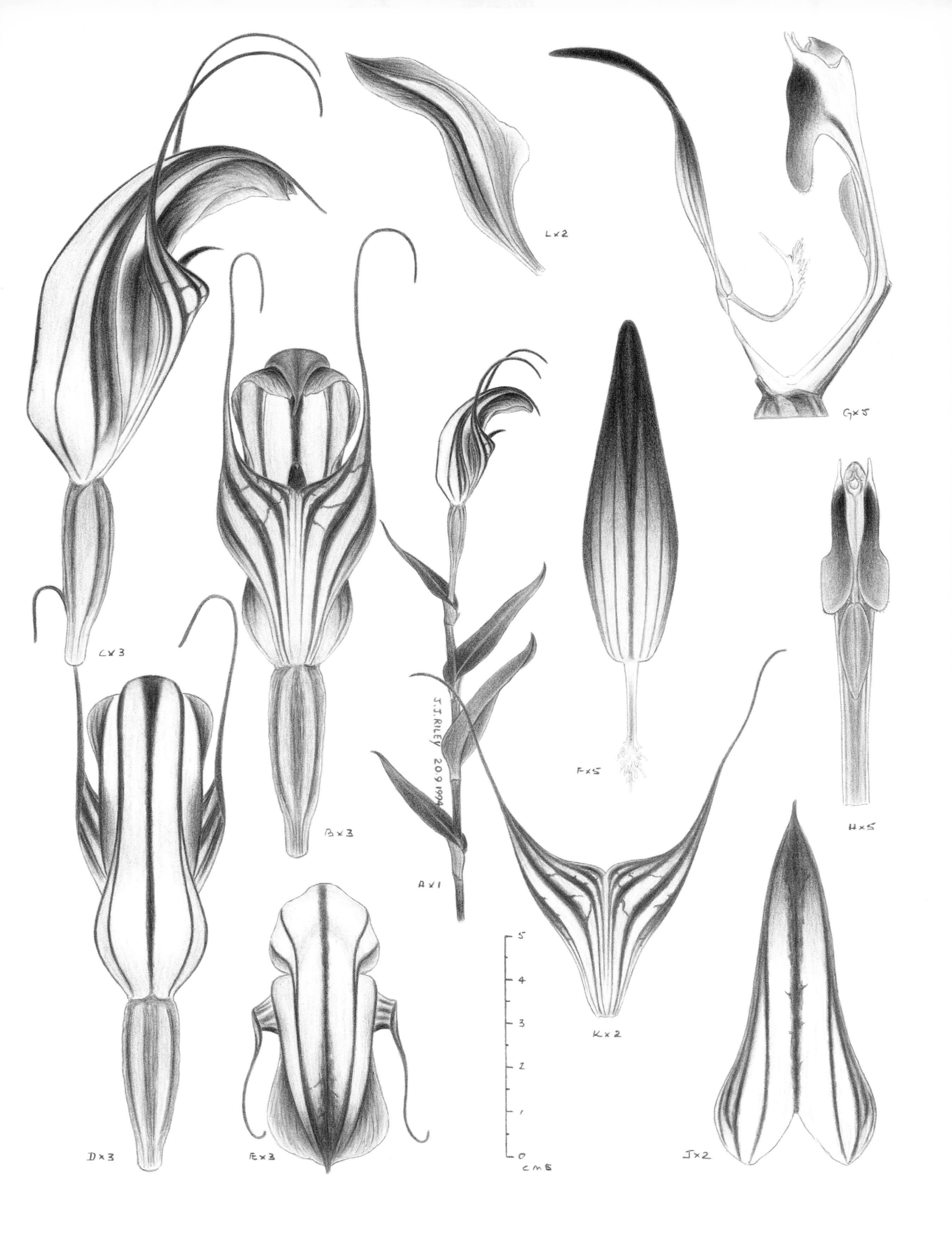
L×2
G×5
C×3
B×3
A×1
J.I. RILEY 20.9.1994
F×5
H×5
K×2
D×3
E×3
J×2
5
4
3
2
1
0
cms

Pterostylis obtusa R. Br.

1810 | *Prodromus Florae Novae Hollandiae*: 327

TYPE LOCALITY Port Jackson, New South Wales

ETYMOLOGY *obtusa* — blunt

FLOWERING TIME February to June

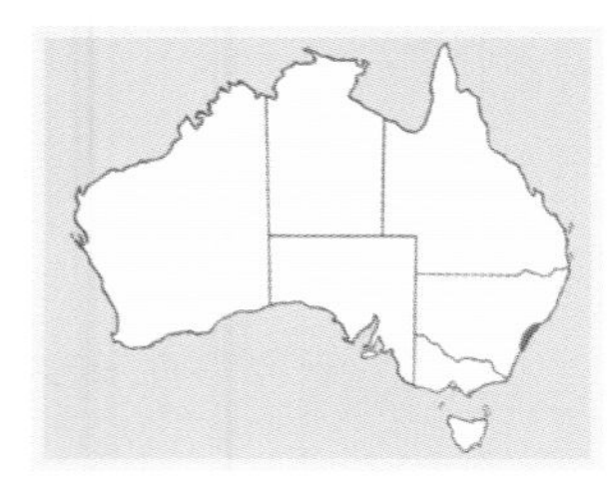

DISTRIBUTION This species is endemic to New South Wales, and occurs along the coast and adjacent ranges from around Gosford south to Nowra.

ALTITUDINAL RANGE 10 m to 400 m

DISTINGUISHING FEATURES *P. obtusa* is one of the cauline-leafed species, with non-blooming plants having a basal rosette of leaves, and flowering plants without a rosette, instead having leaves up the stem. *P. obtusa* is readily recognised by the prominent, flat sinus that projects well forward. The green labellum is short, wide and has a blunt apex.

HABITAT This species is found in a range of habitats, from open woodland to stabilised scrubby sand dunes where it is commonly seen growing among scattered ferns and small heathy plants. Occasionally seen in more exposed situations among grass and growing in the shallow soils of the moss gardens on sandstone outcrops.

CONSERVATION STATUS Common. Grows in national parks and reserves. Secure.

DISCUSSION In the past, *P. obtusa* was considered a widespread and very variable species. Recent research has shown that a complex of unnamed taxa were involved. Most of these have been formally named. As well, the name *P. alveata*, previously regarded as a synonym of *P. obtusa*, has been reinstated. *P. obtusa* forms small to large colonies, reproducing vegetatively from daughter tubers formed on the end of stolonoid roots, mostly off the non-flowering plants. Also reproduces from seed to form new colonies, as fungus gnats pollinate the flowers. The labellum is highly irritable and closes when the pollinator triggers the appendage at the base of the lip. Large colonies in full bloom are an impressive sight.

Pterostylis obtusa
Kentlyn, New South Wales

26 April 2000

A flowering plant
B non-flowering plant
C flower from side
D flower from front
E flower from rear
F flower from above
G labellum and column from side
H labellum flattened
J column from front
K dorsal sepal
L lateral sepals
M petal

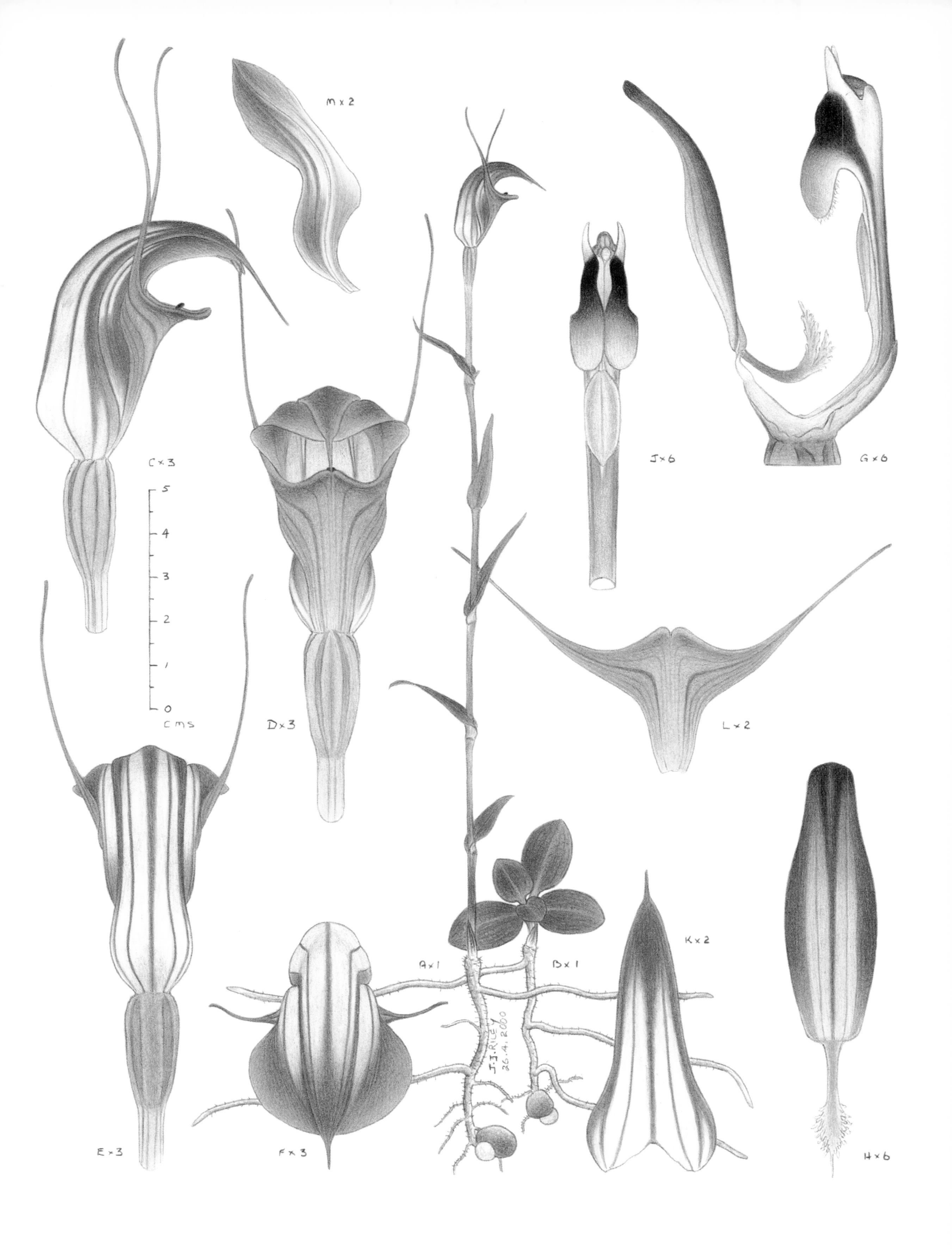
M x 2
C x 3
J x 6
G x 6
5
4
3
2
1
0
cms
D x 3
L x 2
A x 1
B x 1
K x 2
J.J. RILEY
26.4.2000
E x 3
F x 3
H x 6

Pterostylis alveata Garnet

1939 | *Victorian Naturalist* 59:91–94

TYPE LOCALITY Snake Island, Victoria

ETYMOLOGY *alveata* — having a cavity

FLOWERING TIME March to June

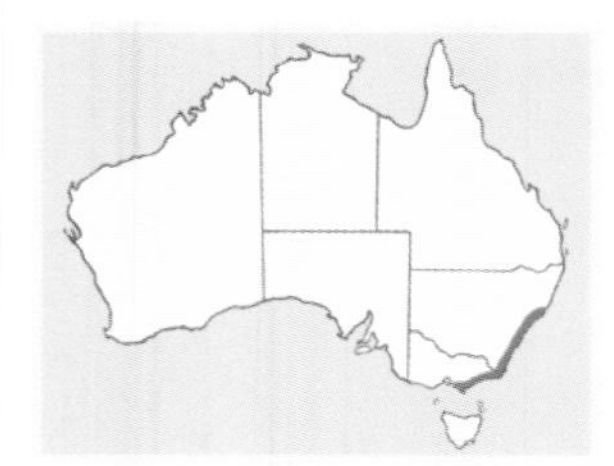

DISTRIBUTION This species is found in coastal Victoria from around Melbourne, north along the coast and nearby ranges to Taree, New South Wales.

ALTITUDINAL RANGE 10 m to 500 m

DISTINGUISHING FEATURES *P. alveata* is one of the cauline-leafed species, with non-blooming plants having a basal rosette of leaves, and flowering plants without a rosette, instead having leaves up the stem. *P. alveata* has smaller flowers on a shorter stem than its sister species, *P. obtusa*. The sinus is less prominent, and the blunt, dark brownish labellum can be seen through the sinus.

HABITAT *P. alveata* is commonly found growing in coastal heathy woodland and scrubby stabilised sand dunes. Away from the coast in New South Wales, it grows in moist open woodland and on the lower ranges in heathy forest. Soils are sand, sandy loams and clay loams.

CONSERVATION STATUS Common. Occurs in national parks, state forests and reserves. Secure.

DISCUSSION *P. alveata* is a very free-flowering species that is part of the *P. obtusa* complex, and in the past was included as a synonym of that species. It prefers to grow in humus-rich soils under shrubs. A rare petaloid form with a rigid fixed labellum has formed small colonies in a couple of locations in Victoria. These were once known as *P. crypta*. *P. alveata* hybridises with *P. ophioglossa* to form *P.* X *furcillata*. It forms scattered colonies, reproducing vegetatively from daughter tubers formed on the end of stolonoid roots, mostly off the non-flowering plants. Also reproduces from seed to form new colonies, as fungus gnats pollinate the flowers. The labellum is highly irritable and closes when the pollinator triggers the appendage at the base of the lip.

Pterostylis alveata
Currawong, New South Wales

16 May 1993

A flowering plant
B non-flowering plant
C flower from front
D flower from side
E flower from rear
F flower from above
G labellum flattened
H labellum and column from side
J column from front
K dorsal sepal
L lateral sepals
M petal

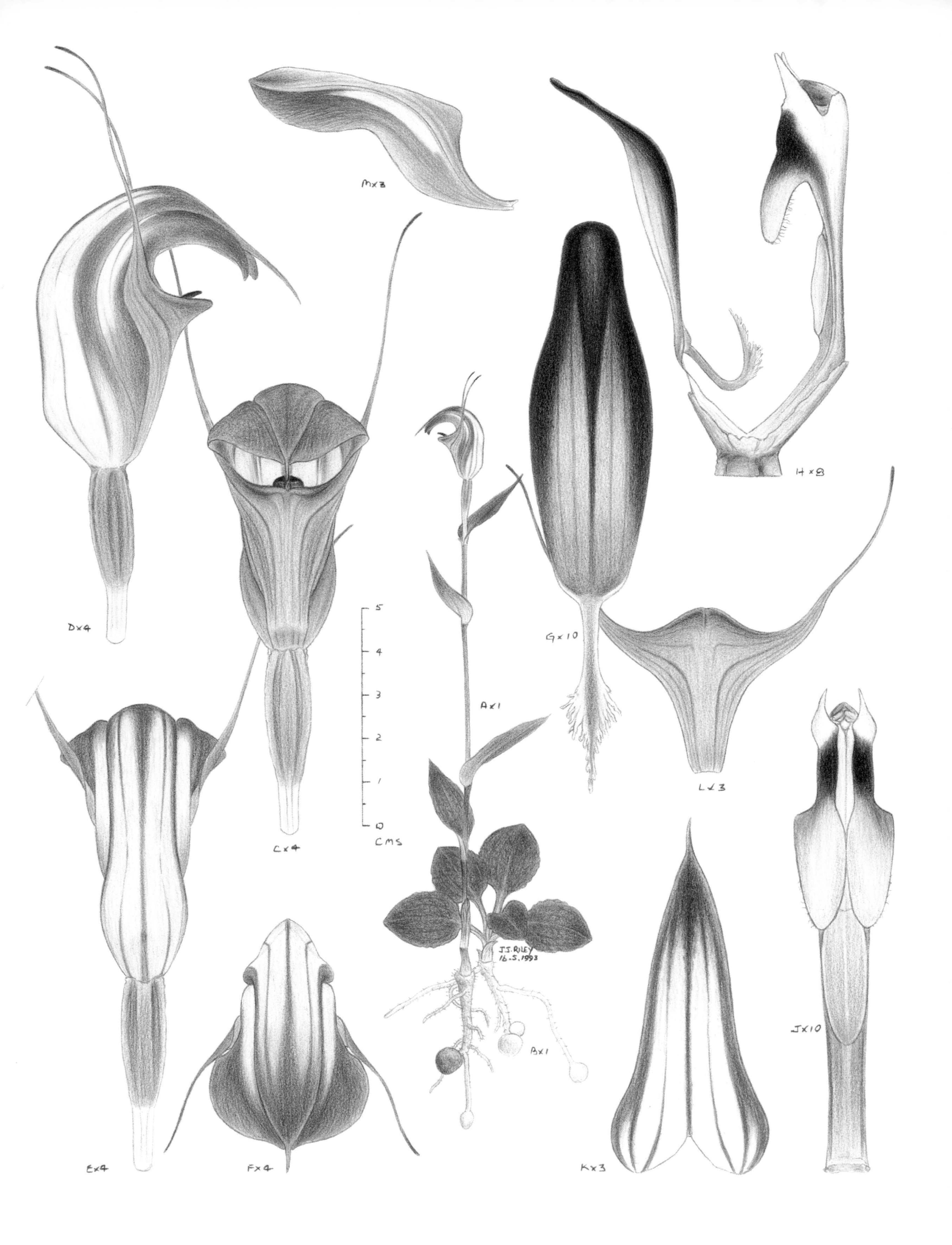

M×3
H×8
D×4
G×10
A×1
C×4
CMS
L×3
J.J. RILEY
16.5.1993
B×1
J×10
E×4
F×4
K×3

Pterostylis bryophila D.L. Jones

1997 | *Orchadian* 12:180

TYPE LOCALITY Hindmarsh Valley, South Australia

ETYMOLOGY *bryophila* — moss loving

FLOWERING TIME April to July

DISTRIBUTION This is a rare species that is only known from the Hindmarsh Valley of South Australia.

ALTITUDINAL RANGE 50 m to 150 m

DISTINGUISHING FEATURES *P. bryophila* is one of the cauline-leafed species, with non-blooming plants having a basal rosette of leaves, and flowering plants without a rosette, instead having leaves up the stem. It has a larger flower than *P. alveata* and the labellum has a rounded apex with a distinct notch.

HABITAT Found in open grassy woodland, growing in moss among grass and rocks.

CONSERVATION STATUS Rare and restricted. Occurs in a nature reserve. Vulnerable. The invasion of introduced weeds into its habitat is posing a serious threat to the survival of this species.

DISCUSSION *P. bryophila* is part of the *P. obtusa* complex. Historical reports of *P. obtusa* from South Australia refer to this species. *P. bryophila* is poorly known, being a rare plant with a very limited distribution. It forms scattered colonies, reproducing vegetatively from daughter tubers formed on the end of stolonoid roots, mostly off the non-flowering plants. Also reproduces from seed to form new colonies, as fungus gnats pollinate the flowers. The labellum is highly irritable and closes when the pollinator triggers the appendage at the base of the lip.

Pterostylis bryophila
Mount Billy, South Australia

12 April 1998

A plant
B flower from front
C flower from side
D flower from rear
E flower from above
F labellum flattened
G labellum and column from side
H column from front
J dorsal sepal
K lateral sepals
L petal

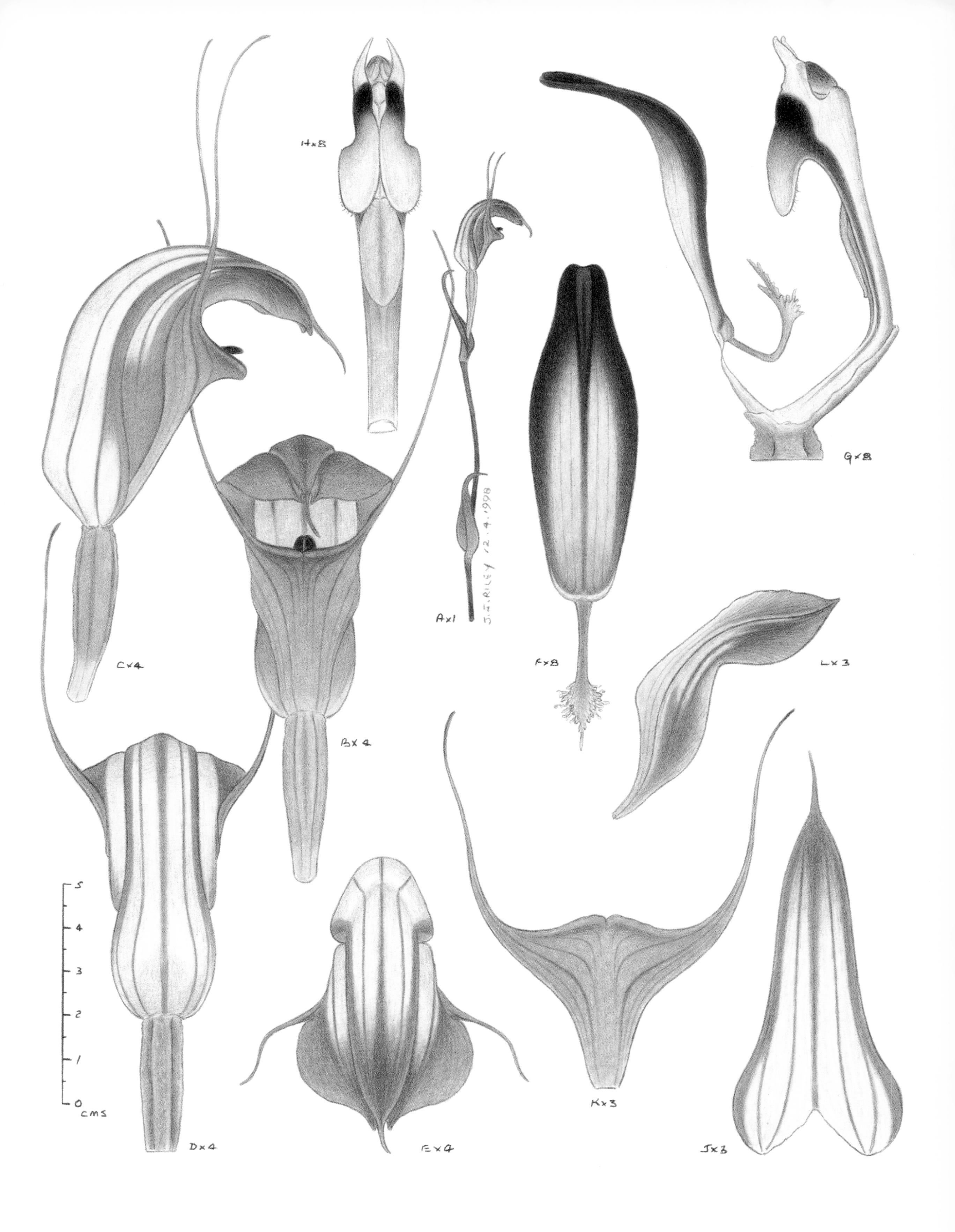
H×8
G×8
A×1
J.E. RILEY 12.4.1998
C×4
F×8
L×3
B×4
5
4
3
2
1
0
CMS
D×4
E×4
K×3
J×3

Pterostylis hians D.L. Jones

1997 | *Orchadian* 12:185

TYPE LOCALITY Manyana, New South Wales

ETYMOLOGY *hians* — yawning, gaping

FLOWERING TIME March to May

DISTRIBUTION Endemic to New South Wales, known only from the type-site near Manyana.

ALTITUDINAL RANGE Up to 50 m

DISTINGUISHING FEATURES *P. hians* is a diminutive cauline-leafed species, with non-blooming plants having a loose rosette of about three, small, wrinkly, kidney-shaped leaves, and flowering plants without a rosette, instead having reduced leaves up the stem. It has small, bottle green and white flowers with an upward pointing galea on a very short stem. The green labellum is oval-shaped with purple edging. When viewed from the front, the flower appears to be open and gaping. Before it was formally named, it was referred to as the Opera House Greenhood, on account of its floral shape when viewed from the side, resembling the Sydney landmark.

HABITAT Found growing near a creek in open forest with a bushy understorey of *Baeckea* and *Leptospermum*.

CONSERVATION STATUS Very rare. Not conserved. Vulnerable. Illegal collecting, followed by a devastating hot bushfire, have placed *P. hians* in danger of extinction in the wild.

DISCUSSION *P. hians* is known only from the type locality, where it is extremely rare. Other species of *Pterostylis* occur in the area. Growing under thick tea tree scrub, this tiny species is very difficult to see and may be easily overlooked. This species slowly reproduces vegetatively from daughter tubers formed at the end of stolonoid roots. Also reproduces from seed, as fungus gnats pollinate the flowers. The labellum is highly irritable and closes when the pollinator triggers the appendage at the base of the lip. Recent extensive searches of the surrounding bush, and similar areas, to locate this species have been unsuccessful.

Pterostylis hians
Manyana, New South Wales

8 May 1992

A flowering and non-flowering plant
B flower from side
C flower from front
D flower from rear
E flower from above
F labellum flattened
G labellum and column from side
H column from front
J dorsal sepal
K petal
L lateral sepals

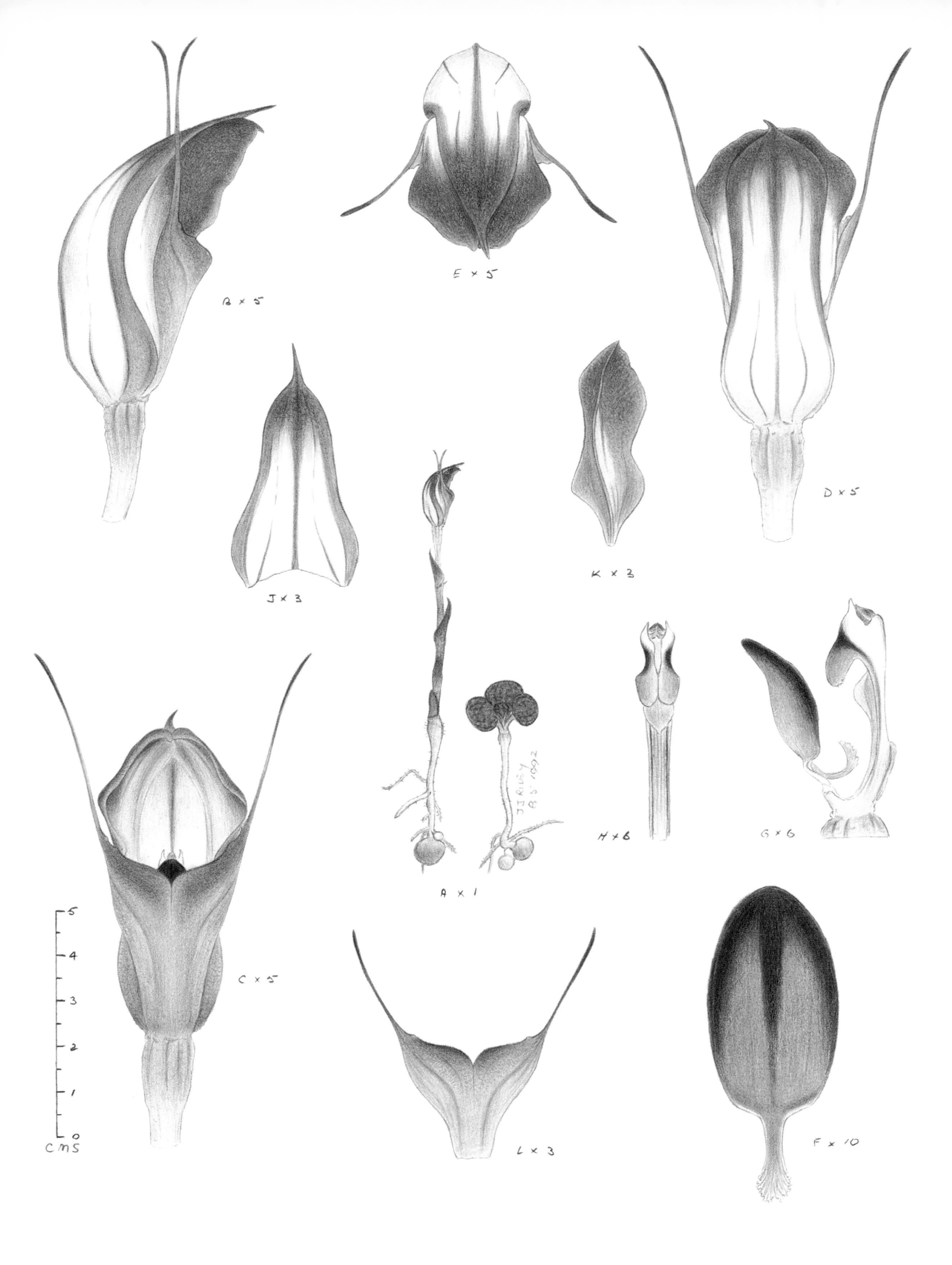
B × 5
E × 5
D × 5
J × 3
K × 3
H × 6
G × 6
A × 1
C × 5
L × 3
F × 10
5
4
3
2
1
0
CMS

Pterostylis longifolia R. Br.

1810 | *Prodromus Florae Novae Hollandiae*: 327

TYPE LOCALITY Port Jackson, New South Wales

ETYMOLOGY *longifolia* — long leaves

FLOWERING TIME April to September

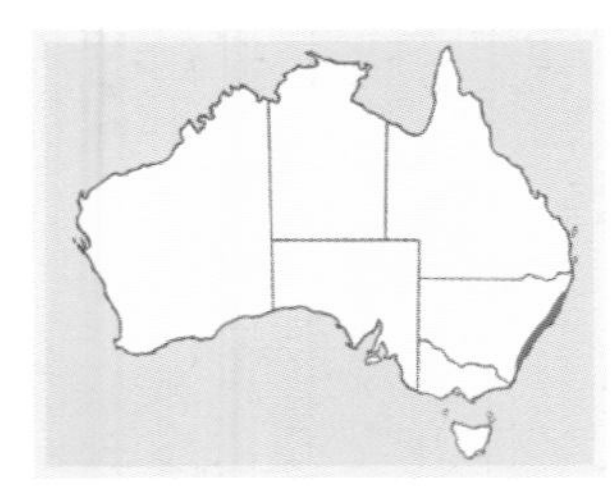

DISTRIBUTION Endemic to New South Wales, found from the Coffs Harbour area, south along the coast and lower parts of the nearby ranges to Narooma.

ALTITUDINAL RANGE Up to 400 m

DISTINGUISHING FEATURES *P. longifolia* is part of a complex of species that are tall-growing and have multiple flowers and several stem-clasping leaves without a rosette. Non-blooming plants have an untidy loose rosette of leaves that are held off the ground on a small stem. It has up to a dozen flowers and a yellow-green labellum with a dark longitudinal stripe down the centre, which is densely fringed on the base and sides with long white cilia.

HABITAT Found in a wide range of habitats, from coastal heath, scrubby woodland, wet sclerophyll forest and moist sheltered areas in open forest. Grows in leaf litter and the humus-rich soils among ferns and shrubs. It is also common in the stabilised coastal sand dunes growing in leaf litter and sand under *Banksia integrifolia*.

CONSERVATION STATUS Common and widespread. Occurs in national parks and reserves. Secure.

DISCUSSION Until recently, *P. longifolia* was treated as a single wide-spread and highly variable species occurring in all the eastern states. Research has shown a group of taxa was involved; some of these have recently been named. Studies are continuing into the status of the rest of the complex. It occurs as scattered individuals or in small groups. It reproduces from seed, with fungus gnats being the main pollinating agents. The labellum is highly sensitive and snaps back towards the reproductive parts of the bloom when touched.

Pterostylis longifolia
Kentlyn, New South Wales

2 June 1995

A flowering plant
B flower from front
C flower from side
D flower from rear
E flower from above
F labellum flattened
G labellum and column from side
H column from front
K lateral sepals
L petal

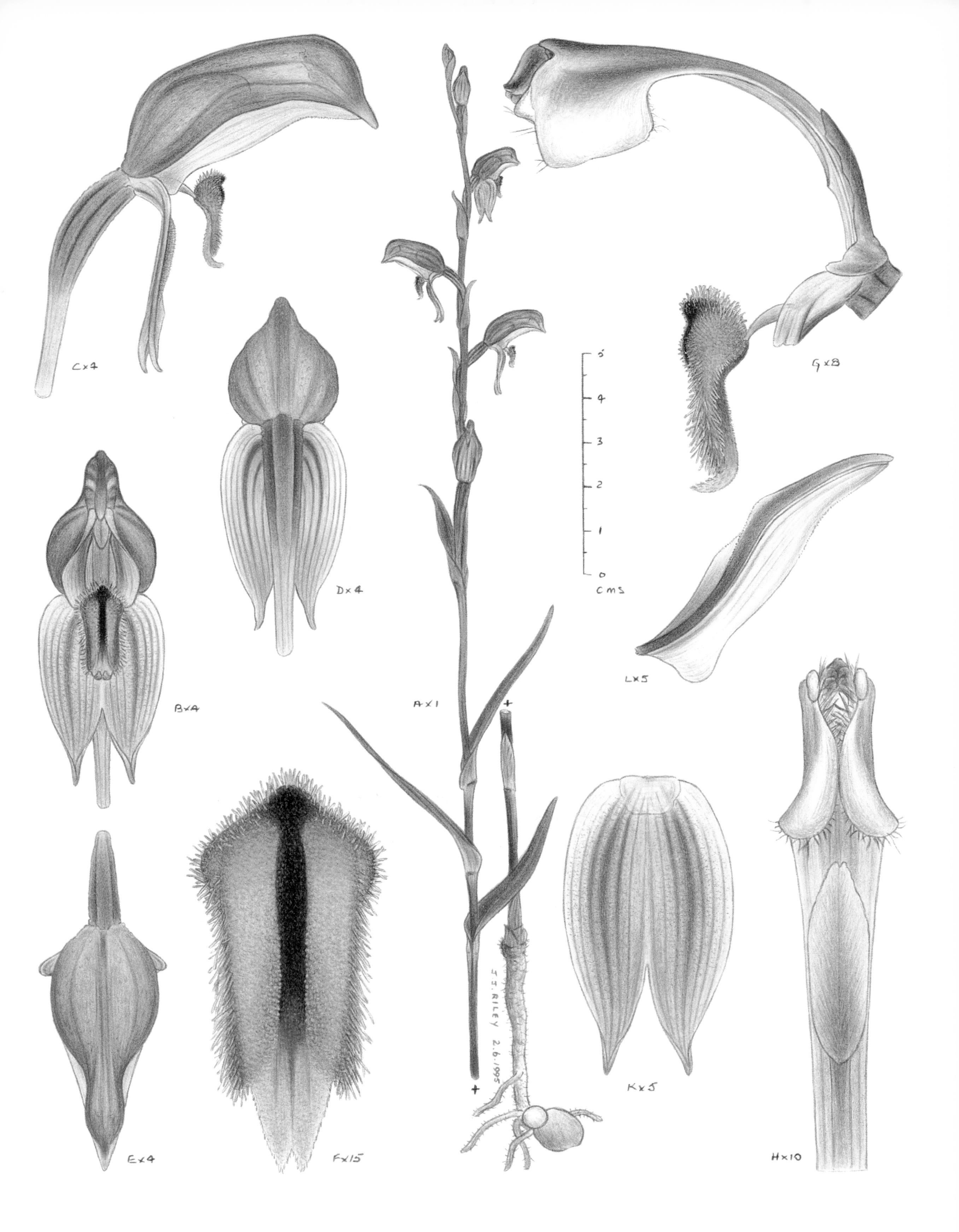
C×4
G×8
D×4
B×4
A×1
5
4
3
2
1
0
CMS
L×5
E×4
F×15
J.T. RILEY 2.6.1995
K×5
H×10

Pterostylis smaragdyna D.L. Jones & M.A. Clem.

1993 | *Muelleria* 8(1):82

TYPE LOCALITY Diamond Creek, Victoria

ETYMOLOGY *smaragdyna* — emerald green

FLOWERING TIME May to September

DISTRIBUTION Found in Victoria, from around the Melbourne area to the Grampians and Rushworth, and possibly in South Australia.

ALTITUDINAL RANGE 30 m to 200 m

DISTINGUISHING FEATURES *P. smaragdyna* is part of the *P. longifolia* complex. Flowering plants are tall-growing and can have up to a dozen flowers and several stem-clasping leaves without a rosette. Non-blooming plants have an untidy loose rosette of leaves that are held off the ground on a small stem. The petals have prominent flanges, giving the impression of tonsils. The labellum is broad, emerald green with a darker central stripe and a few white bristles on the base. The flowers are large for this complex and the seed capsule is distinctly ribbed.

HABITAT Found in sheltered sites in open woodland and forest growing in leaf litter and humus-rich soils among or close to small shrubs.

CONSERVATION STATUS Uncommon. Occurs in national parks and reserves. Secure. The spread of suburbia around Melbourne is putting pressure on this poorly known species.

DISCUSSION *P. smaragdyna* is a robust species with large, somewhat bulbous, green-striped flowers. It occurs as scattered individuals or in small groups. It reproduces from seed, with fungus gnats being the main pollinating agents. The labellum is highly sensitive and snaps back towards the reproductive parts of the bloom when touched. Populations in South Australia are under investigation.

Pterostylis smaragdyna
St Helena, Victoria

26 August 1993

A flowering plant
B non-flowering plant
C flower from front
D flower from side
E flower from rear
F flower from above
G labellum flattened
H labellum and column from side
J column from front

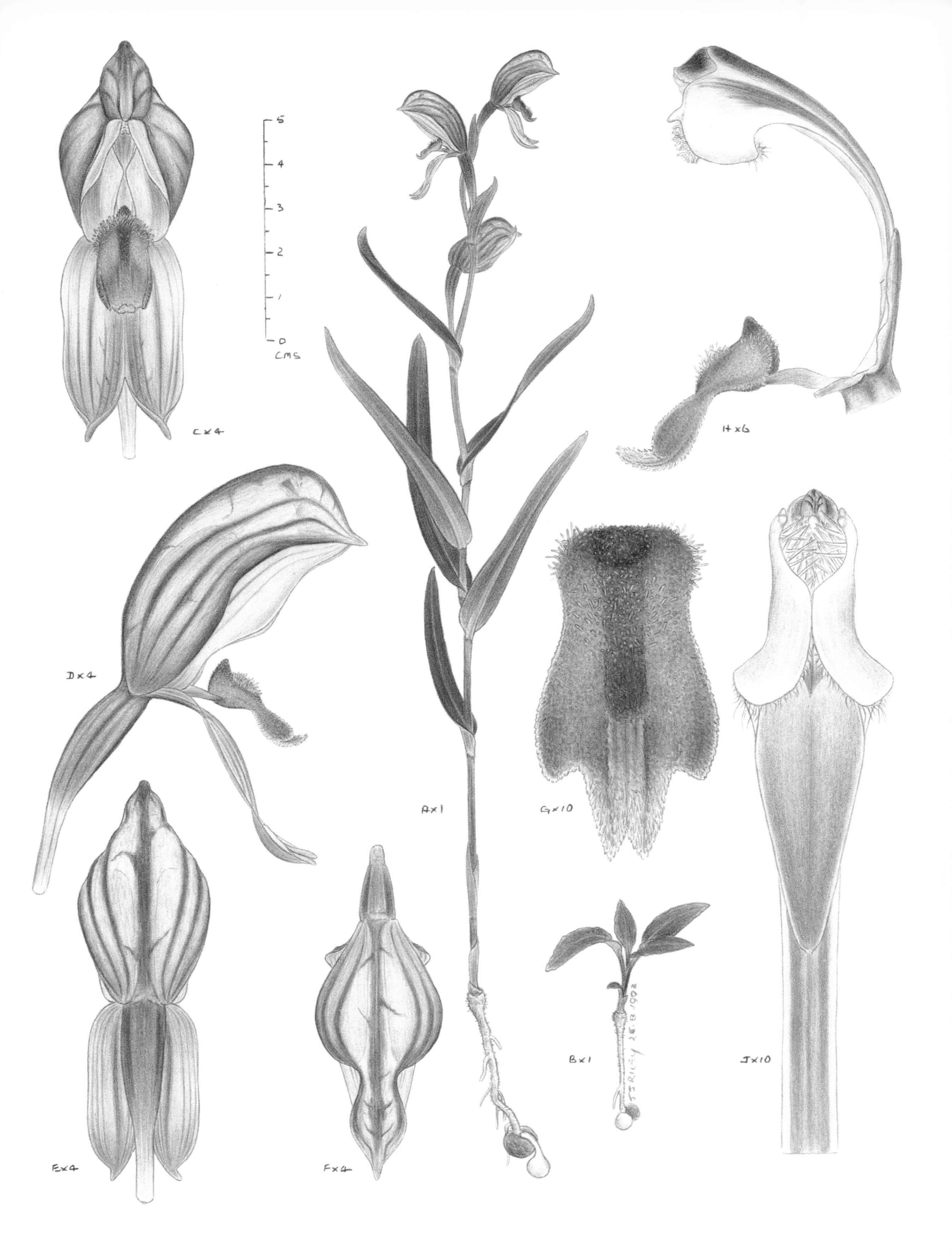

C×4
5
4
3
2
1
0
CMS
H×6
D×4
A×1
G×10
B×1
J×10
E×4
F×4

Pterostylis tunstallii D.L. Jones & M.A. Clem.

1989 | *Australian Orchid Research* 1:128

TYPE LOCALITY Tomerong, New South Wales

ETYMOLOGY After collector Ronald G. Tunstall

FLOWERING TIME June to August

DISTRIBUTION This species is found in New South Wales from the lower Blue Mountains to Narooma on the South Coast, to Genoa and Wilsons Promontory in Victoria, and across to the Furneaux Group, Tasmania.

ALTITUDINAL RANGE 50 m to 400 m

DISTINGUISHING FEATURES *P. tunstallii* is part of the *P. longifolia* complex. Flowering plants are tall-growing and have up to eight flowers and several stem-clasping leaves without a rosette. Non-blooming plants have an untidy loose rosette of leaves that are held off the ground on a small stem. This species has smaller flowers than related taxa and flowers are held almost horizontal. The lateral sepals are swept backwards. The labellum is oblong, tapering only slightly to the apex and is brown or brownish-green with some white bristles on the base.

HABITAT Usually found in woodland and forest with a fern and scrubby understorey. Often seen growing in leaf litter and humus-rich soils in sheltered sites.

CONSERVATION STATUS Uncommon. Conserved in national parks and reserves. Secure.

DISCUSSION *P. tunstallii* is a poorly known species with a somewhat sporadic distribution. It is most common in the Nowra area, often growing with other members of the *P. longifolia* complex, yet hybrids are unknown. It occurs as scattered individuals or in small groups. It reproduces from seed, with fungus gnats being the main pollinating agents. The labellum is highly sensitive and snaps back towards the reproductive parts of the bloom when touched.

Pterostylis tunstallii
Tomerong, New South Wales

8 July 1994

A flowering plant
B non-flowering plant
C flower from front
D flower from side
E flower from rear
F flower from above
G labellum flattened
H labellum and column from side
J column from front
K lateral sepals
L petal

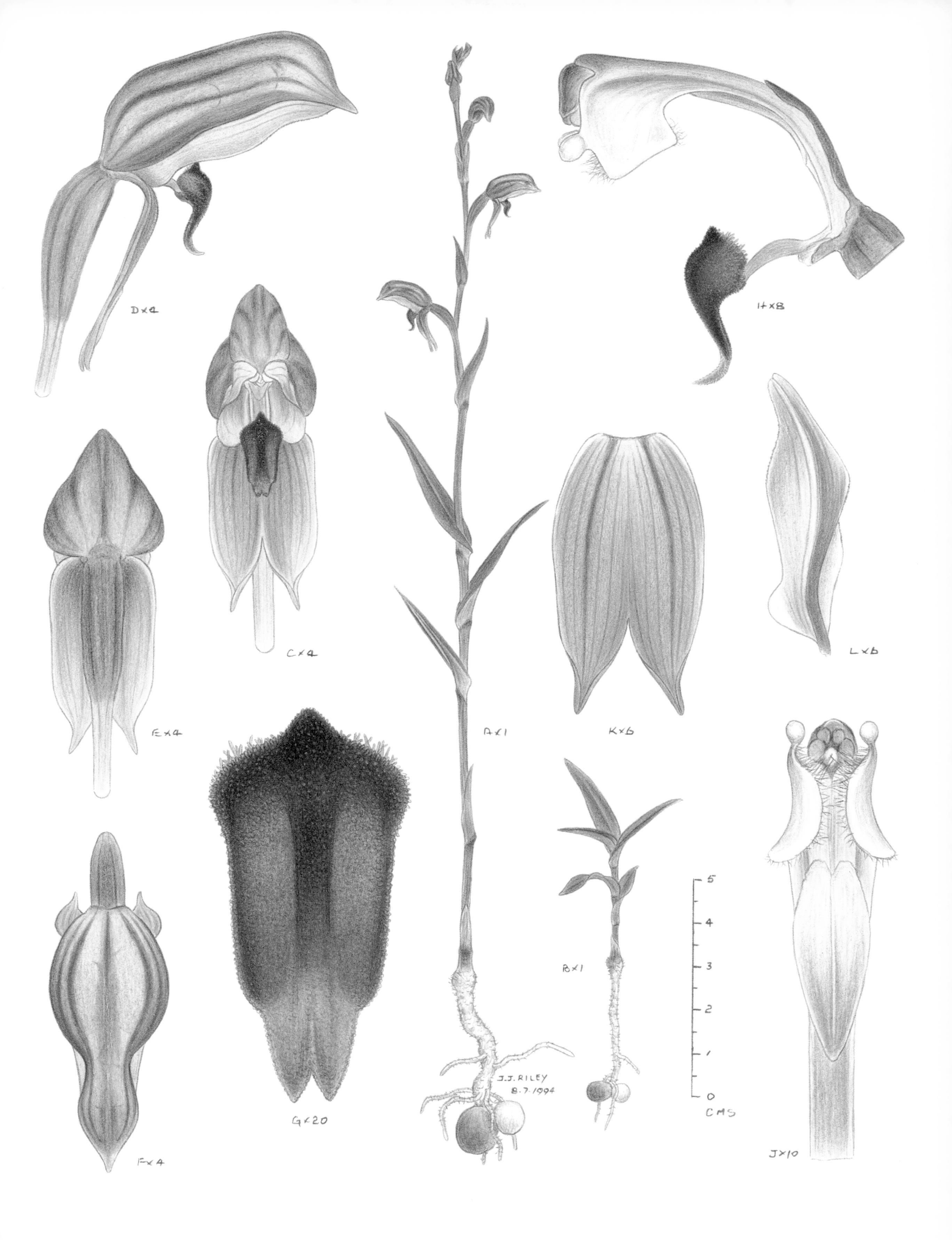
D×4
H×8
C×4
E×4
A×1
K×6
L×6
G×20
B×1
J.J. RILEY
8.7.1994
5
4
3
2
1
0
CMS
F×4
J×10

Pterostylis melagramma D.L. Jones

1998 | *Australian Orchid Research* 3:145

TYPE LOCALITY King Island, Tasmania

ETYMOLOGY *melagramma* — dark line

FLOWERING TIME August to November

DISTRIBUTION This species is found in Tasmania, including King Island and Flinders Island across to Victoria, where it occurs from around Melbourne and Sale to the Goldfields and Grampians.

ALTITUDINAL RANGE 10 m to 800 m

DISTINGUISHING FEATURES *P. melagramma* is part of the *P. longifolia* complex. Flowering plants are tall-growing and have multiple flowers and several stem-clasping leaves without a rosette. Non-blooming plants have an untidy loose rosette of leaves that are held off the ground on a small stem. It has small petal flanges that only partly block the throat of the flower. The labellum is oblong and constricted in the middle and is greenish-brown with a dark central stripe. Numerous white bristles fringe the basal third of the labellum.

HABITAT *P. melagramma* is found growing in a variety of vegetation communities, from moist forest and woodland to coastal scrubby heathland and scrubby woodland. Often growing in leaf litter and humus-rich soils in moist sheltered sites.

CONSERVATION STATUS Occurs in national parks and reserves. Secure.

DISCUSSION *P. melagramma* can be locally common, forming scattered groups of many individual plants. It reproduces from seed, with fungus gnats being the main pollinating agents. The labellum is highly sensitive and snaps back towards the reproductive parts of the bloom when touched. *P. melagramma* is also reported as occurring in southern New South Wales, far western Victoria and South Australia. Continuing research into the *P. longifolia* complex as a whole will aim to determine the taxonomic status of these populations.

Pterostylis melagramma
South Arm Peninsula, Tasmania

31 August 1995

A flowering plant
B flower from front
D flower from side
E flower from rear
F flower from above
G labellum flattened
H labellum and column from side
J column from front
K lateral sepals
L petal

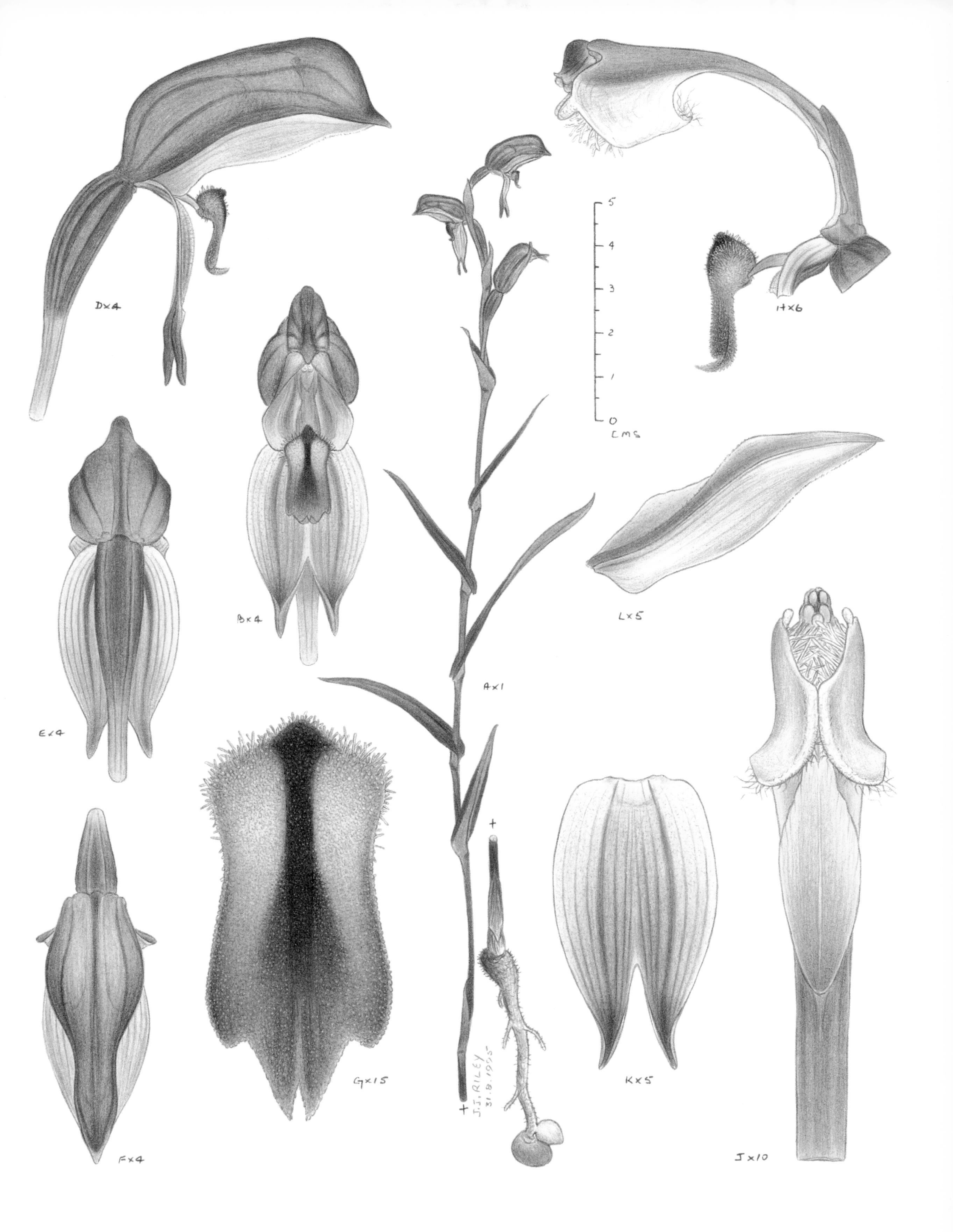

D×4
H×6
5
4
3
2
1
0
CMS
B×4
E×4
L×5
A×1
G×15
K×5
J.J. RILEY
31.8.1995
F×4
J×10

Pterostylis williamsonii D.L. Jones

1998 | *Australian Orchid Research* 3:157

TYPE LOCALITY Coles Bay, Tasmania

ETYMOLOGY After Ron Williamson

FLOWERING TIME July to September

DISTRIBUTION Endemic to Tasmania, found on the East Coast and extending north-west to the Tamar River near Launceston.

ALTITUDINAL RANGE Up to 250 m

DISTINGUISHING FEATURES *P. williamsonii* is part of the *P. longifolia* complex. Flowering plants are tall-growing and have multiple flowers and several stem-clasping leaves without a rosette. Non-blooming plants have an untidy, loose rosette of leaves that are held off the ground on a small stem. *P. williamsonii* is very similar to *P. tunstallii*. The lateral sepals of *P. williamsonii* are vertical and the labellum is very wide at the base, tapering noticeably towards the apex. It also lacks the white basal bristles found in *P. tunstallii*.

HABITAT Found in coastal *Casuarina* and tea tree scrub and woodland, *P. williamsonii* is also a common species in forest and moist woodland with a fern and scrub understorey. Soils are well drained, sandy or gravelly loams.

CONSERVATION STATUS Common. Occurs in nature reserves. Secure.

DISCUSSION *P. williamsonii* is locally common in the coastal parts of eastern Tasmania. It occurs as scattered individuals or in small groups. It reproduces from seed, with fungus gnats being the main pollinating agents. The labellum is highly sensitive and snaps back towards the reproductive parts of the bloom when touched. On South Arm Peninsula, normal green-flowered specimens of *P. williamsonii* can be seen growing with colour-recessive plants, with all the floral parts being a pale honey colour. The specimen illustrated is of the normal green form.

Pterostylis williamsonii
South Arm Peninsula, Tasmania

26 August 1995

A flowering plant
B flower from front
C flower from side
D flower from rear
E flower from above
F labellum flattened
G labellum and column from side
H column from front
J lateral sepals
K petal

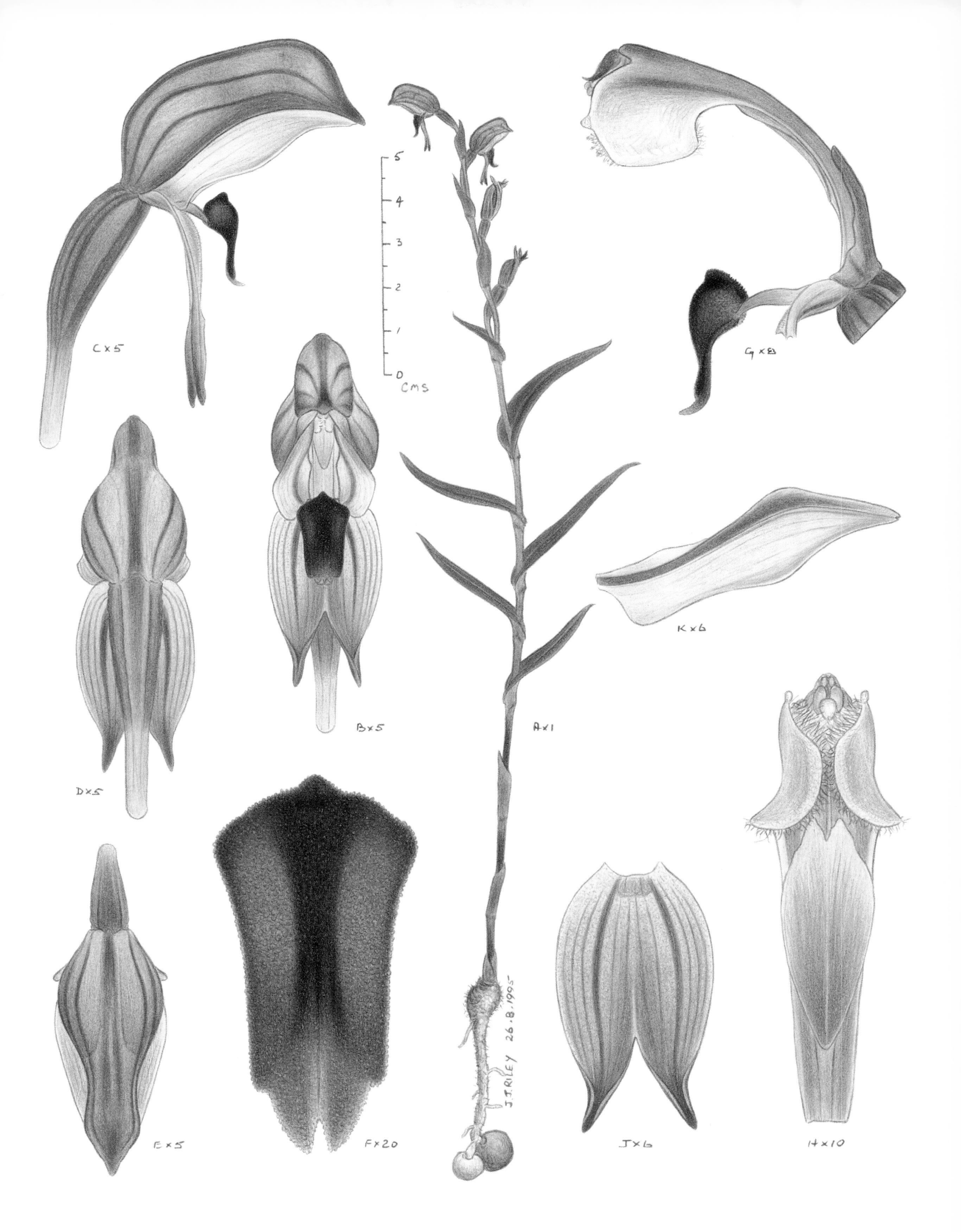

C×5
5
4
3
2
1
0
CMS
G×8
D×5
B×5
A×1
K×6
E×5
F×20
J.J.RILEY 26.8.1995
J×6
H×10

Pterostylis stenochila D.L. Jones

1998 | *Australian Orchid Research* 3:153

TYPE LOCALITY Brooks Bay, Tasmania

ETYMOLOGY *stenochila* — narrow lip

FLOWERING TIME July to September

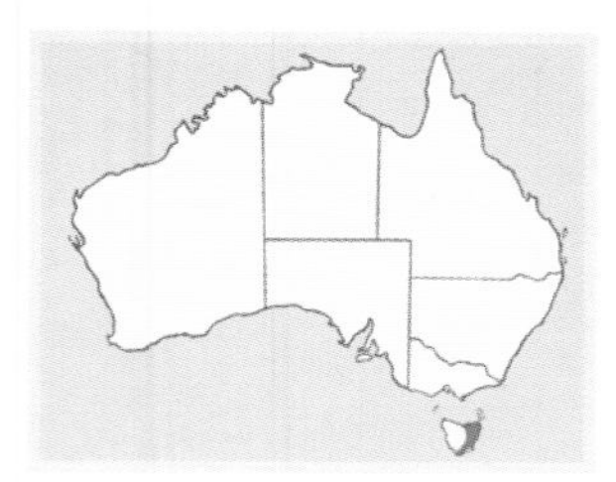

DISTRIBUTION This species is endemic to the eastern half of Tasmania, being most common around Hobart to the south-east.

ALTITUDINAL RANGE Up to 300 m

DISTINGUISHING FEATURES *P. stenochila* is part of the *P. longifolia* complex. Flowering plants are tall-growing and have multiple flowers and several stem-clasping leaves without a rosette. Non-blooming plants have an untidy, loose rosette of leaves that are held off the ground on a small stem. *P. stenochila* has dark green blooms, with the flanges on the petals giving the impression the flower has tonsils. The labellum is long, narrow and slightly constricted in the middle, and is lime green with a darker green, central stripe.

HABITAT Found in moderately dry, open forests and woodland with a shrub and fern understorey. It grows in leaf litter and humus. *P. stenochila* also occurs in sandy, humus enriched soils among shrubs in open situations.

CONSERVATION STATUS Common. Occurs in nature reserves. Secure.

DISCUSSION *P. stenochila* is a recently named species that is endemic to Tasmania, having the darkest green flowers of that state's recognised species. It occurs as scattered individuals or in small groups. It reproduces from seed, with fungus gnats being the main pollinating agents. The labellum is highly sensitive and snaps back towards the reproductive parts of the bloom when touched.

Pterostylis stenochila
Kingston, Tasmania

17 August 1995

A flowering plant
B flower from front
C flower from side
D flower from rear
E flower from above
F labellum flattened
G labellum and column from side
H column from front
J lateral sepals
K petal

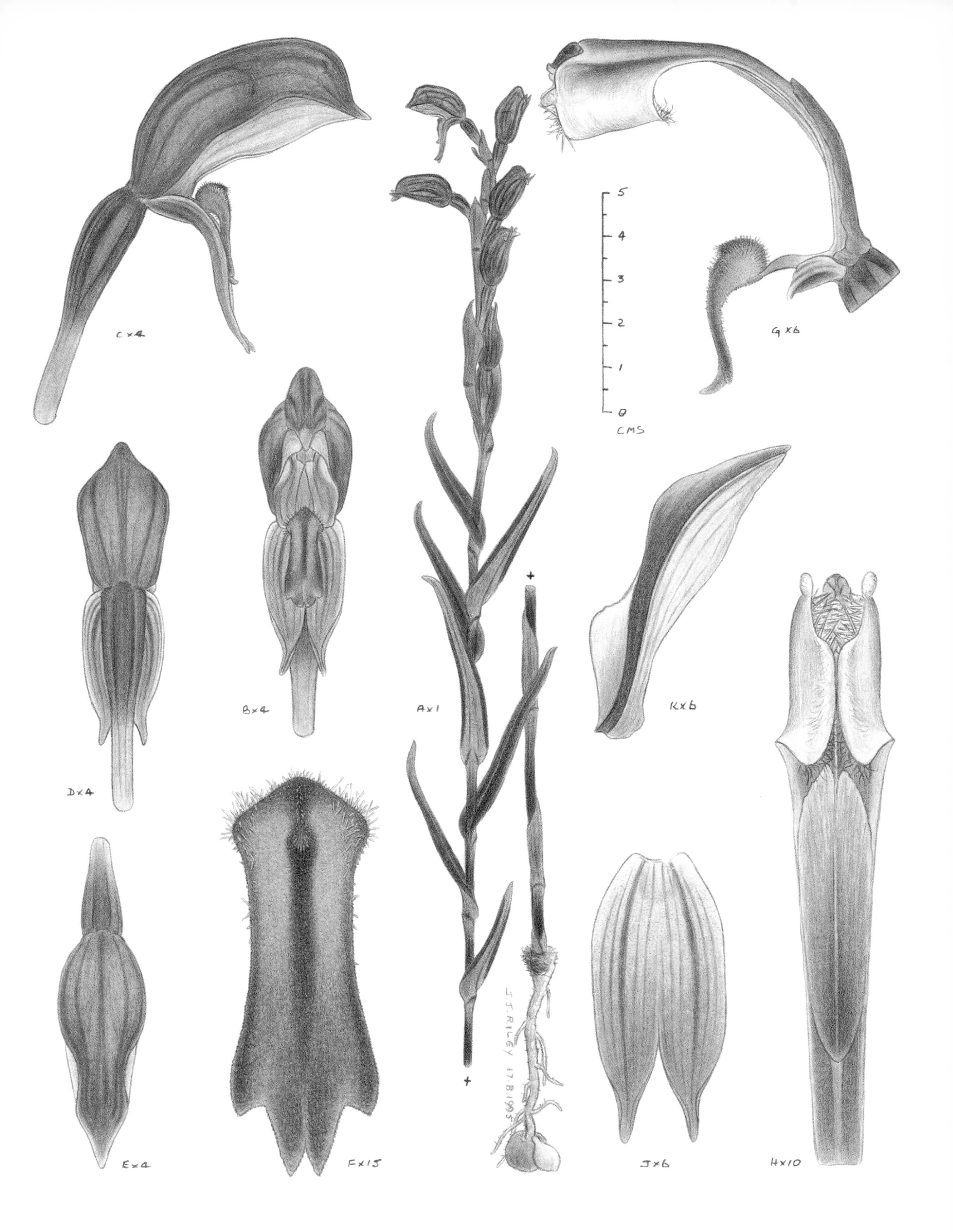
C×4
G×6
5
4
3
2
1
0
CMS
B×4
A×1
D×4
K×6
E×4
F×15
J.T. RILEY 17.8.1995
J×6
H×10

Pterostylis sargentii C.R.P. Andrews

1905 | *Journal of the Western Australian Natural History Society* 2(2):57

TYPE LOCALITY York, Western Australia

ETYMOLOGY After O.H. Sargent

FLOWERING TIME July to November

DISTRIBUTION This unique species is endemic to Western Australia, being found from Geraldton to east of Esperance.

ALTITUDINAL RANGE 50 m to 400 m

DISTINGUISHING FEATURES *P. sargentii* is a distinctive species without close relatives, that is easily recognised by its dark green flowers that are covered with sparkling white granules. It has a fleshy and intricate three-lobed labellum. The petals are unusual, being predominantly fringed on one side with long white cilia. Flowering plants have multiple blooms and several stem-clasping leaves without a rosette. Non-blooming plants have a loose rosette of leaves.

HABITAT Found in a variety of habitats in the wheat belt district of the south-west, from growing in sand on exposed rocky granite outcrops to woodland and scrub where it grows in thick leaf litter.

CONSERVATION STATUS Common. Conserved in national parks and reserves. Secure.

DISCUSSION *P. sargentii* is one of Western Australia's most widespread and easily recognised species, becoming colloquially known as the Frog Orchid. It occurs as scattered individuals or in small groups. It reproduces from seed, with fungus gnats being the main pollinating agents. The labellum is highly sensitive and snaps back towards the reproductive parts of the bloom when touched.

Pterostylis sargentii
Dragon Rocks Nature Reserve, Western Australia

31 July 1994

A flowering plant
B non-flowering plant
C flower from front
D flower from side
E labellum from above
F labellum viewed from its apex
G labellum and column from side
H column from front
J lateral sepals
K petal

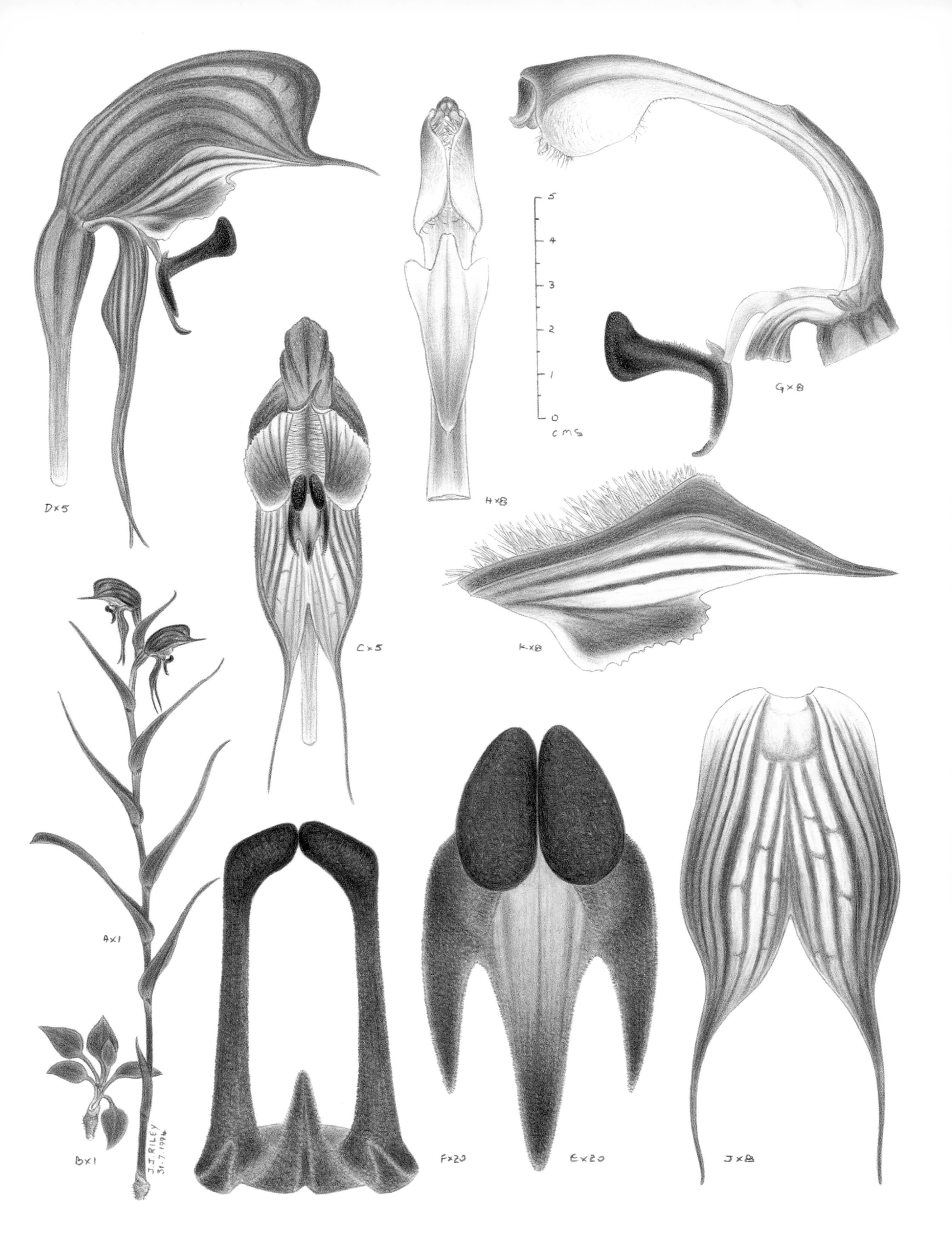
D×5
H×8
C×5
G×8
K×8
A×1
B×1
J.J. RILEY
31.7.1996
F×20
E×20
J×8
5
4
3
2
1
0
CMS

Pterostylis parviflora R. Br.

1810 | *Prodromus Florae Novae Hollandiae*: 327

TYPE LOCALITY Port Jackson, New South Wales

ETYMOLOGY *parviflora* — small flowers

FLOWERING TIME March to May

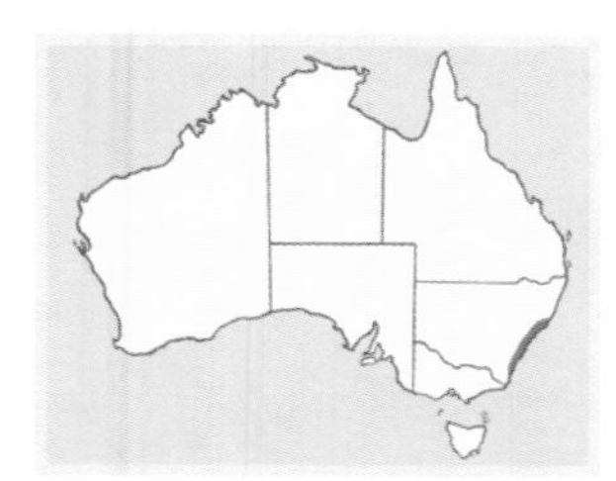

DISTRIBUTION *P. parviflora* is a species endemic to New South Wales, being found from Taree to Bega.

ALTITUDINAL RANGE 10 m to 900 m

DISTINGUISHING FEATURES *P. parviflora* is at the centre of a complex of diminutive multi-flowered species that have blooms that face towards the flowering stem. It has small, green and white-striped flowers that are somewhat triangular when viewed from the side. The petals have two greenish-brown stripes and the points on the sepals and galea apex are sometimes tinged honey brown. The labellum is not visible through the sinus. On blooming plants, up to three rosettes will appear on radical side growths, usually after flowering has finished. Sterile plants will produce their ground-hugging, bluish-green rosettes at the same time the flowering specimens emerge.

HABITAT *P. parviflora* can be found in stabilised sand dunes growing under *Banksia* spp. and *Leptospermum* spp. in coastal heath and open heathy woodland. It is most common around stunted shrubs in skeletal soils among outcropping rocks. It prefers the open areas with little ground cover between shrubs and trees, and within moss gardens and disturbed areas.

CONSERVATION STATUS Common and widespread. Occurs in national parks and reserves. Secure.

DISCUSSION Research is being conducted into the group of orchids that are presently known collectively as *P. parviflora*. It is expected this study will result in the identification and naming of taxa within the complex. The specimen illustrated is representative of plants growing around the type-site, and the distribution map reflects this. It occurs as scattered individuals or in small groups. It reproduces from seed, with fungus gnats being the main pollinating agents. The labellum is highly irritable and closes when the pollinator triggers the appendage at the base of the lip. Similar related taxa occur in coastal Victoria and Tasmania.

Pterostylis parviflora
Nerrigundah, New South Wales

16 April 1994

A flowering plant
B non-flowering plant
C flower from front
D flower from side
E flower from rear
F flower from above
G labellum flattened
H labellum and column from side
J column from front
K dorsal sepal
L lateral sepals
M petal

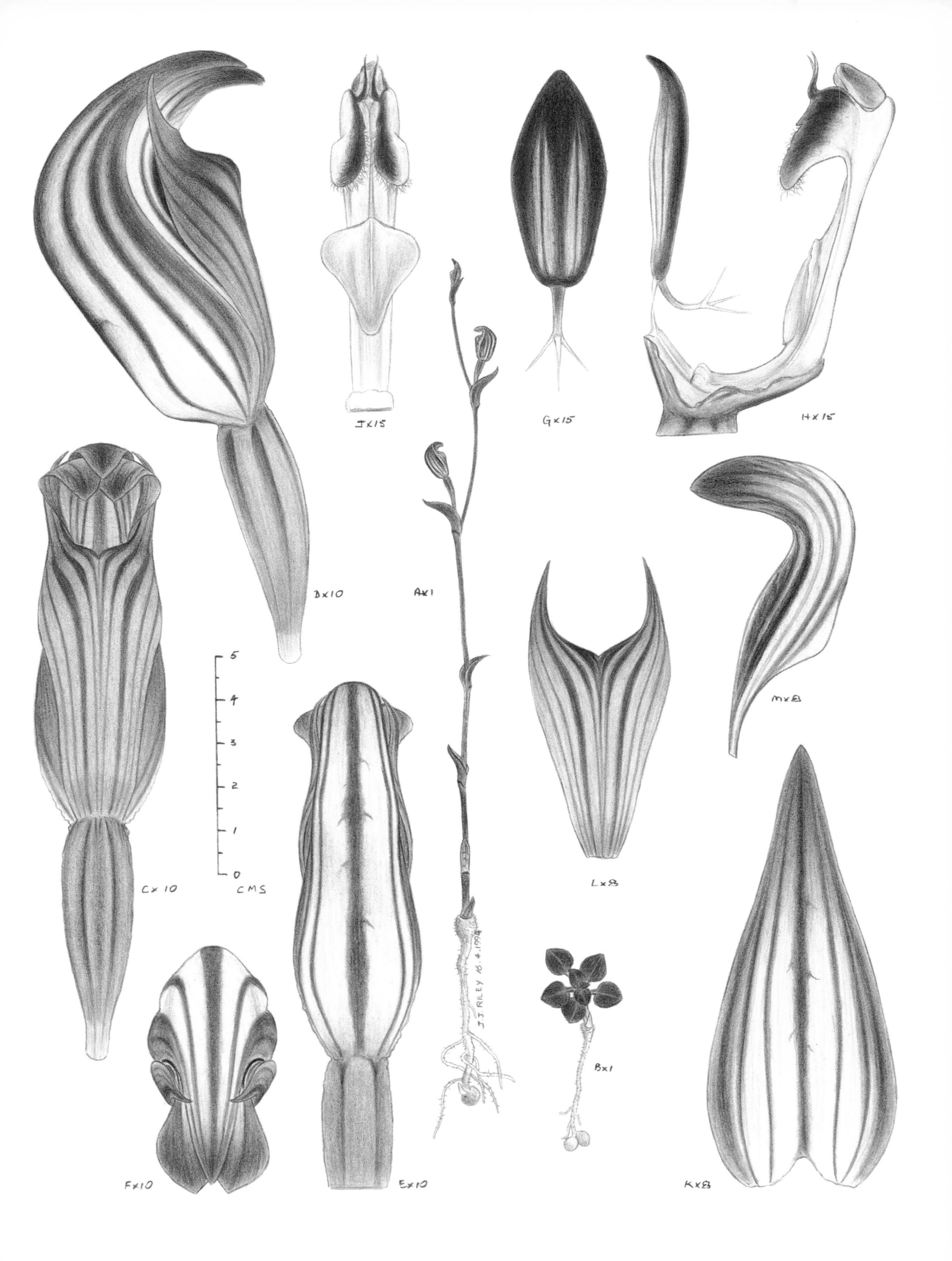
J×15
G×15
H×15
D×10
A×1
M×8
L×8
C×10
CMS
5
4
3
2
1
0
F×10
E×10
K×8
B×1
J.J. RILEY 16.4.1994

Pterostylis nigricans D.L. Jones & M.A. Clem.

1988 | *Austrobaileya* 2(5):550–51

TYPE LOCALITY Stradbroke Island, Queensland

ETYMOLOGY *nigricans* — black

FLOWERING TIME March to May

DISTRIBUTION This orchid is found north from Red Rock in northern New South Wales, to Fraser Island and the adjacent mainland in Queensland. It is a coastal species that also occurs on nearby islands.

ALTITUDINAL RANGE Up to 50 m

DISTINGUISHING FEATURES *P. nigricans* is part of the *P. parviflora* complex of multiflowered species that have blooms that face towards the flowering stem. The flowers have prominent, chocolate brown stripes and suffusions, and are more or less oval when viewed from the side. The united lateral sepals are widely flared. The labellum is not visible through the sinus. On blooming plants, up to three rosettes will appear on radical side growths usually after flowering has finished. Seedlings and sterile plants will produce their ground-hugging, dull green rosettes at the same time the flowering specimens emerge.

HABITAT *P. nigricans* is found on stabilised sand dunes dominated by *Banksia* spp. It also grows in sandy soils in the swampy heathland known as wallum.

CONSERVATION STATUS Uncommon. Occurs in national parks. Secure.

DISCUSSION *P. nigricans* is of sporadic occurrence and is found close to the coast and on the larger offshore islands. It occurs as scattered individuals or in small groups. It reproduces from seed, with fungus gnats being the main pollinating agents. The labellum is highly irritable and closes when the pollinator triggers the appendage at the base of the lip.

Pterostylis nigricans
Fraser Island, Queensland

1 May 1993

A flowering plant
B non-flowering plant
C flower from front
D flower from side
E flower from rear
F flower from above
G labellum flattened
H labellum and column from side
J column from front
K dorsal sepal
L lateral sepals
M petal

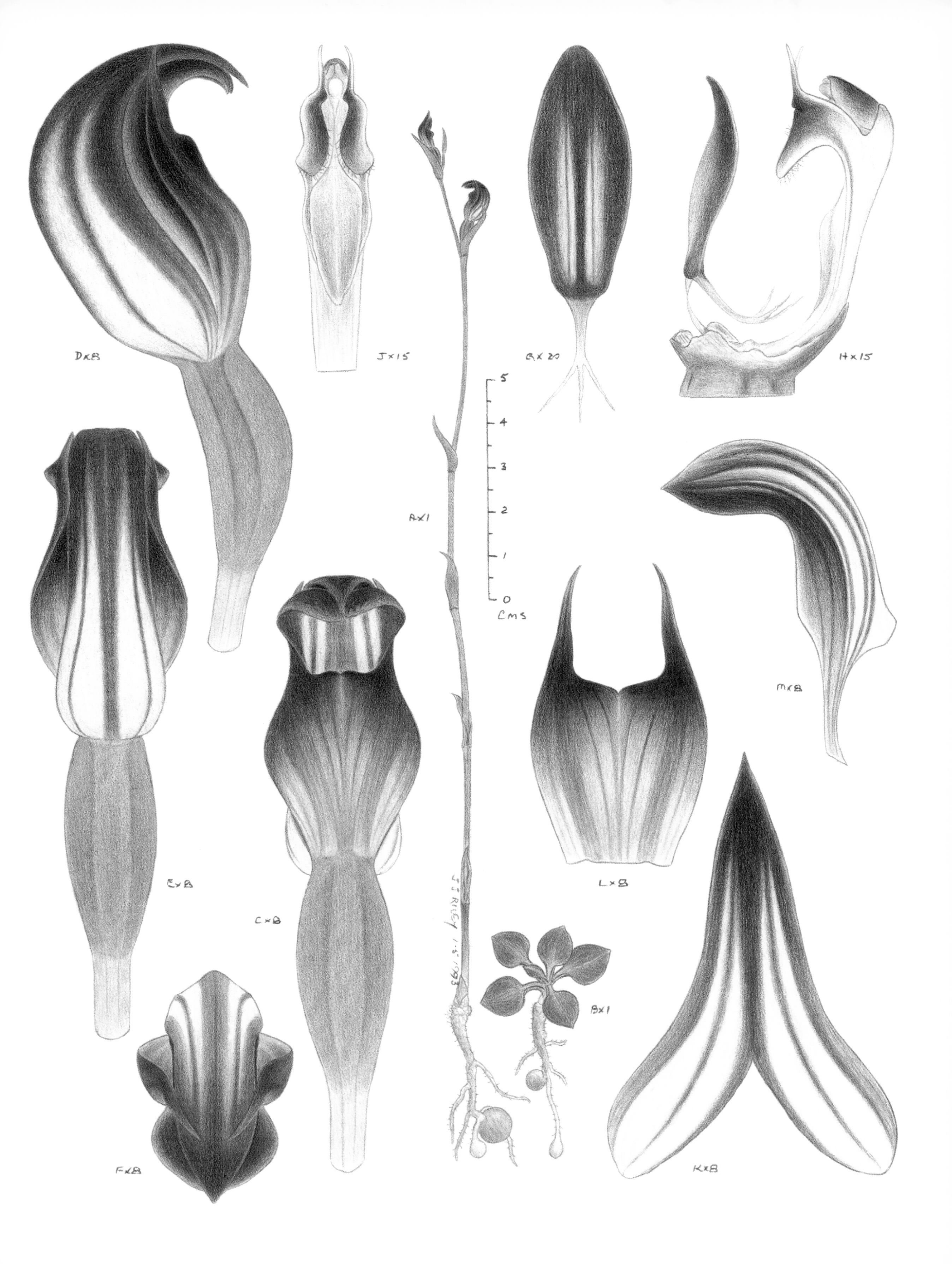

D×8
J×15
G×20
H×15
5
4
3
2
1
0
cms
A×1
M×8
E×8
C×8
L×8
B×1
F×8
K×8

Pterostylis uliginosa D.L. Jones

1998 | *Australian Orchid Research* 3:155

TYPE LOCALITY Appin, New South Wales

ETYMOLOGY *uliginosa* — growing in marshes

FLOWERING TIME December to February

DISTRIBUTION This species is mainly found on the coast and ranges of New South Wales, from the Werrikimbe National Park, south-west of Kempsey, with disjunct populations in Victoria, South Australia and Tasmania.

ALTITUDINAL RANGE 20 m to 1000 m

DISTINGUISHING FEATURES *P. uliginosa* is part of the *P. parviflora* complex of multiflowered species, and has three or four crowded blooms that face towards the flowering stem. On blooming plants, up to three rosettes will appear on radical side growths usually after flowering has finished. Seedlings and sterile plants will produce their ground-hugging, green rosettes at the same time the flowering specimens emerge. *P. uliginosa* is a summer-flowering species with small green and white flowers, that grows in or near swamps. The green labellum is not visible through the sinus.

HABITAT *P. uliginosa* is a moisture-loving species that grows in and on the margins of swamps and swampy heath. Plants also can be found among sedges in peaty seepage areas.

CONSERVATION STATUS Uncommon. Occurs in national parks and reserves. Secure.

DISCUSSION *P. uliginosa* is a self-pollinating species that reproduces from seed. This is the only species in the *P. parviflora* complex with a green labellum. It occurs as scattered individuals or in small groups. The dense, sedgy swamps with thickets of *Hakea*, *Melaleuca* and *Leptospermum* are very difficult to negotiate and, with *P. uliginosa* being small and blending into the background, this species is often missed. However, after fires it is surprising how many plants are growing in these swamps.

Pterostylis uliginosa
Penrose State Forest, New South Wales

21 January 1992

A flowering plant
B rosette of leaves
C flower from front
D flower from side
E flower from rear
F flower from above
G labellum flattened
H labellum and column from side
J column from front

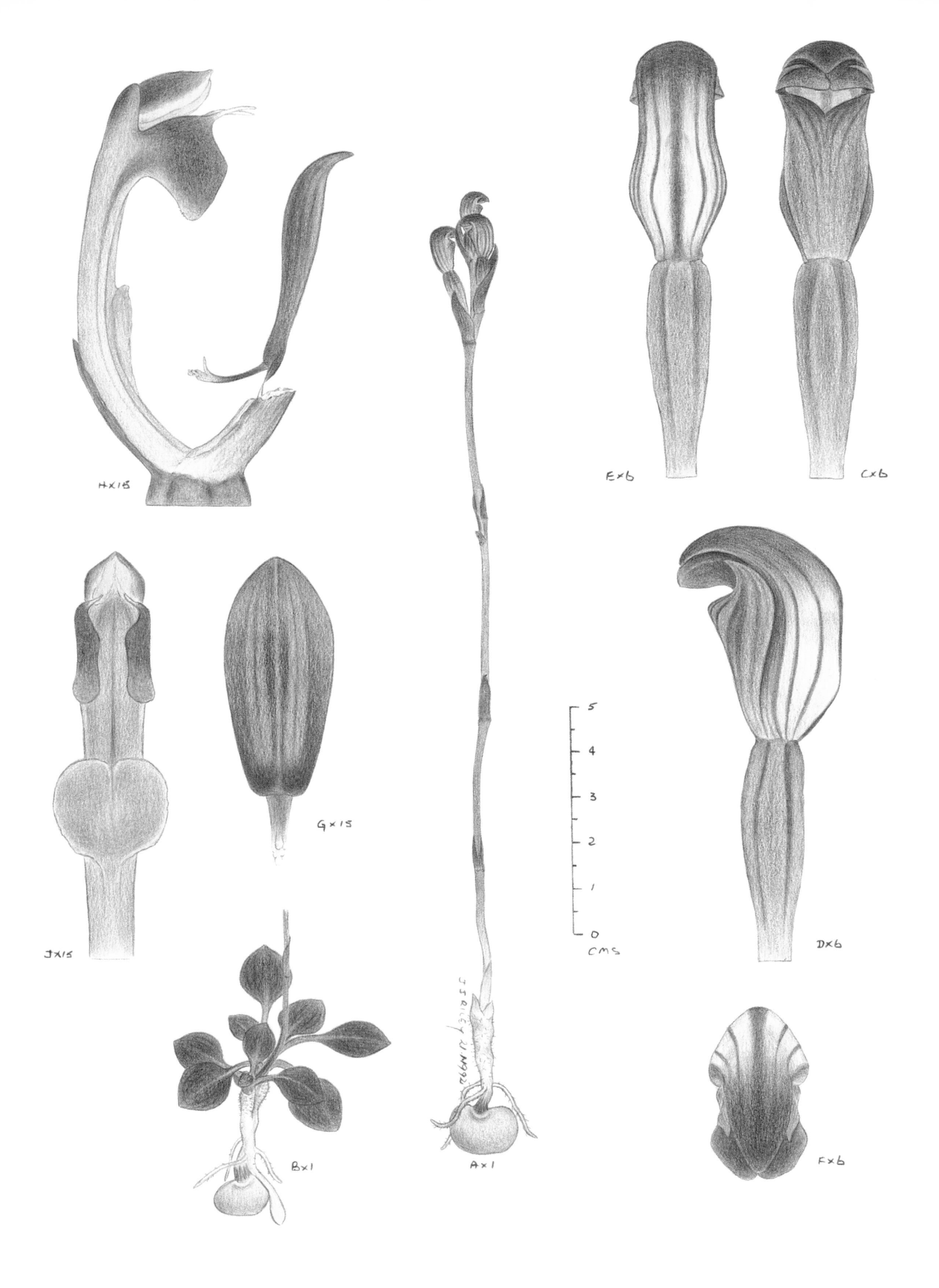

H×15
E×6
C×6
J×15
G×15
5
4
3
2
1
0
CMS
D×6
B×1
A×1
F×6

Pterostylis bicornis D.L. Jones & M.A. Clem.

1987 | *Proceedings of the Royal Society of Queensland* 98:124–26

TYPE LOCALITY Mount Maroon, Queensland

ETYMOLOGY *bicornis* — two-horned

FLOWERING TIME April to July

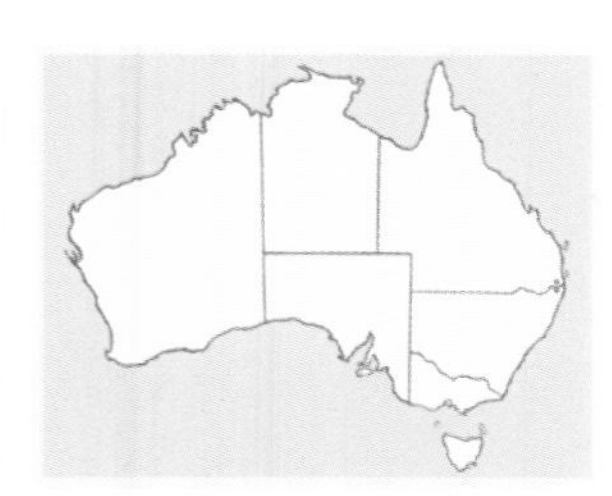

DISTRIBUTION This rare species is found in the McPherson Ranges in south-eastern Queensland and in far north-eastern New South Wales.

ALTITUDINAL RANGE 800 m to 900 m

DISTINGUISHING FEATURES *P. bicornis* is a distinctive member of the *P. parviflora* complex of multiflowered species that have blooms that face towards the flowering stem. It has small, pale yellowish-green flowers that have prominent, horn-like extensions to the petals. The labellum has an unusual folded apex and is visible when in the set position. On blooming plants, a single rosette will appear on a side growth after flowering has started. Seedlings and sterile plants will produce their ground-hugging, green rosettes at the same time the flowering specimens emerge.

HABITAT *P. bicornis* is found on the bare rocky outcrops of the high ranges on both sides of the Queensland–New South Wales border, where pockets of humus-rich soils support patches of stunted vegetation. It often grows in moss and coral lichens among shrubs and rocks.

CONSERVATION STATUS Rare. Conserved in national parks and reserves. Vulnerable.

DISCUSSION *P. bicornis* is a poorly known species with a limited distribution. There are many rocky outcrops in the high country around the border that would have suitable habitat and may harbour this species. Unfortunately the steep and rugged terrain, together with limited access, hinders extensive searches for this species. It occurs as scattered individuals or in small groups. It reproduces from seed, with fungus gnats being the main pollinating agents. Robust plants will reproduce vegetatively from daughter tubers formed at the end of stolonoid roots. The labellum is highly irritable and closes when the pollinator triggers the appendage at the base of the lip.

Pterostylis bicornis
Mount Maroon, Queensland

1 May 2000

A flowering plant
B non-flowering plant
C flower from side
D flower from front
E flower from above
F labellum and column from side
G labellum flattened
H column from front
J dorsal sepal
K lateral sepals
L petal

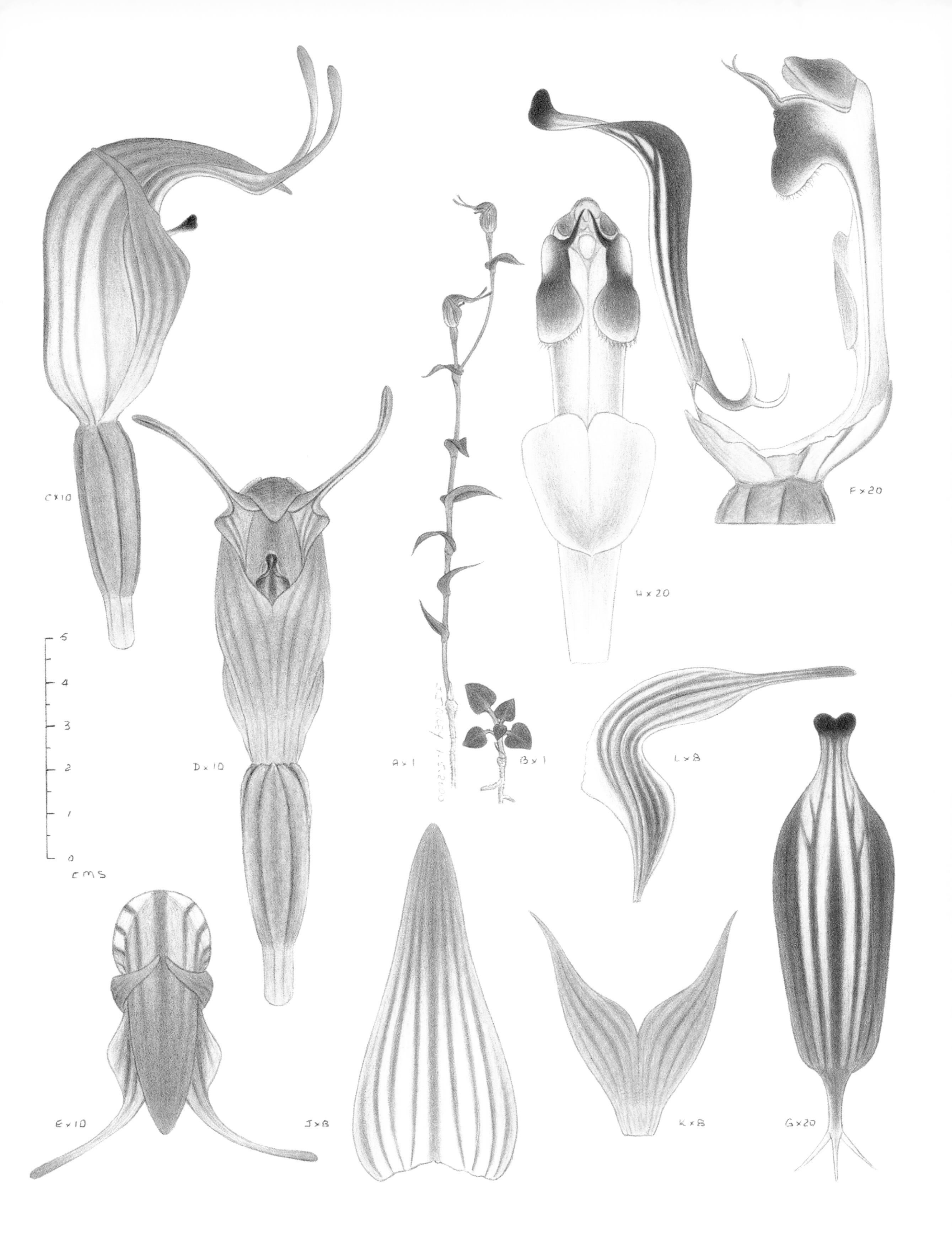

C×10
D×10
A×1
B×1
H×20
F×20
L×8
5
4
3
2
1
0
CMS
E×10
J×8
K×8
G×20

Pterostylis barbata Lindl.

1840 | *Edward's Botanical Register*, Appendix to Vols 1–23

TYPE LOCALITY Swan River, Western Australia

ETYMOLOGY *barbata* — bearded

FLOWERING TIME July to October

DISTRIBUTION This bizarre species is endemic to Western Australia, being found from the Darling Range near Perth to Albany.

ALTITUDINAL RANGE 10 m to 500 m

DISTINGUISHING FEATURES *P. barbata* has a single translucent flower with green stripes and hatchings. The petals have very dark maroonish stripes. The labellum is long, filiform and sparsely hairy. This is one of the species of bearded greenhoods that have a loose, stem-encircling rosette of variegated leaves and an intricate labellum bearing sparse, long hairs.

HABITAT *P. barbata* can be found in open *Eucalyptus* forest and *Casuarina* woodland. It grows in humus enriched, sandy soils among shrubs.

CONSERVATION STATUS Common. Occurs in national parks and reserves. Secure.

DISCUSSION Research has shown that *P. barbata* is a complex of related taxa. The type from the Swan River area is the most widespread and common of these and represents true *P. barbata*. The plant illustrated is this taxon. Both flowering and non-flowering plants have a pyramid-shaped rosette of white and green, distinctly veined leaves. It is a solitary species that reproduces from seed.

Pterostylis barbata
Serpentine, Western Australia

19 September 1992

A plant
B flower from front
C flower from side
D flower from rear
E labellum
F labellum and column from side
G column from front

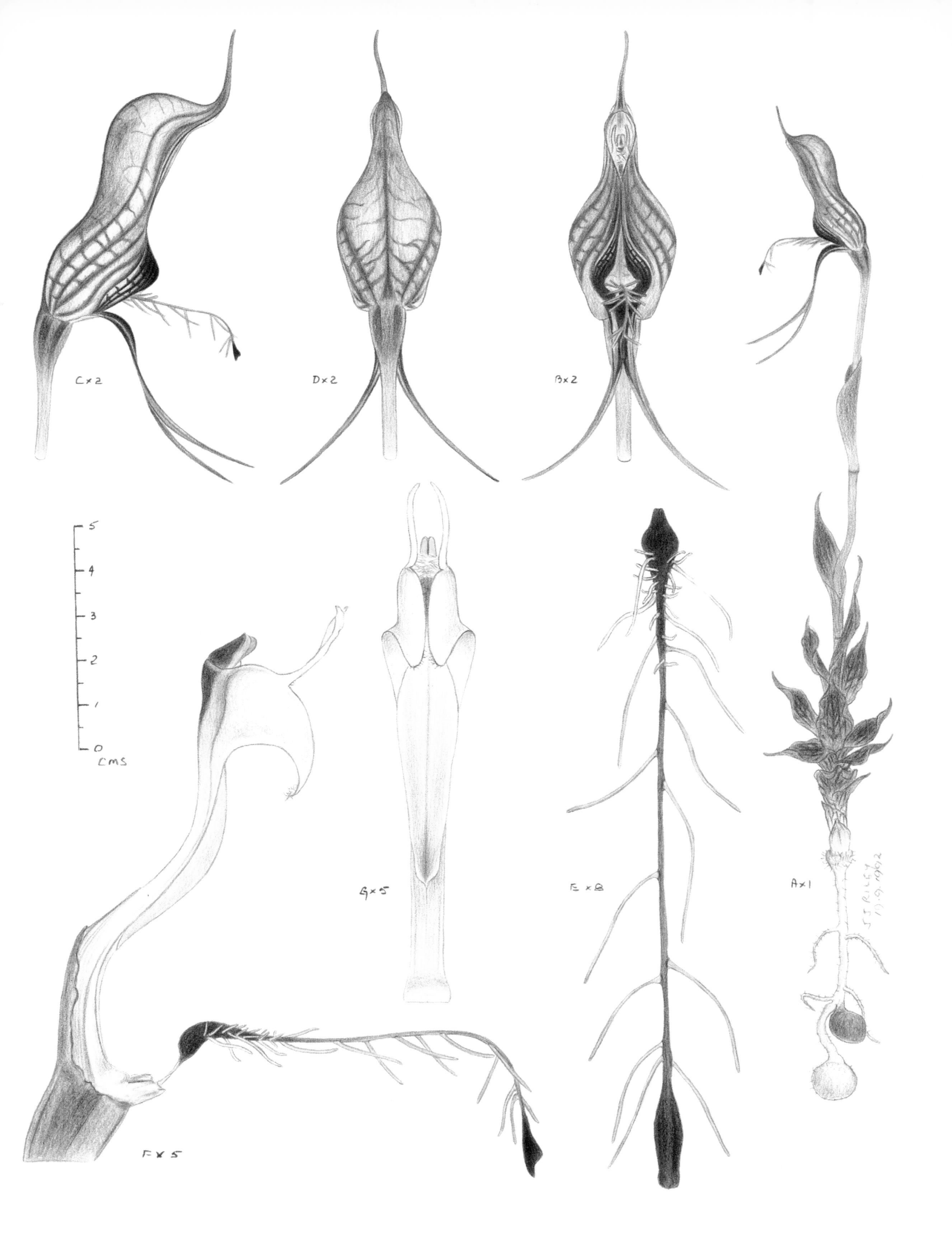
C×2
D×2
B×2
5
4
3
2
1
0
CMS
G×5
E×8
A×1
S J RILEY
10.9.1992
F×5

Pterostylis plumosa Cady

1969 | *Australian Plants* 5(39):138

TYPE LOCALITY Abercrombie Caves, New South Wales

ETYMOLOGY *plumosa* — feathery

FLOWERING TIME August to October

DISTRIBUTION True *P. plumosa* is endemic to New South Wales, being found in the Central Tablelands and Central Western Slopes.

ALTITUDINAL RANGE 300 m to 700 m

DISTINGUISHING FEATURES *P. plumosa* has a single translucent flower with green stripes and a reticulated pattern. The labellum is long and filiform with numerous, stiff, yellow hairs. This is one of the species of bearded greenhoods that have a loose, stem-encircling rosette of variegated leaves and an intricate labellum covered with long hairs.

HABITAT *P. plumosa* is found in open *Eucalyptus* woodland with a scrubby understorey on stony clay loams. It also grows among grasses and small shrubs on sheltered stony slopes in mixed *Eucalyptus* and *Callitris glaucophylla* woodland on clay loams. Occasionally, *P. plumosa* can also be found in more exposed situations in humus-rich soils tucked under low bushes.

CONSERVATION STATUS Occurs in national parks and reserves. Secure.

DISCUSSION *P. plumosa* occurs in small scattered to moderately large groups with a high proportion of flowering plants. At some sites it can be locally common. Both flowering and non-flowering plants have a pyramid-shaped rosette of green-veined leaves. It is a solitary species that reproduces from seed. Recent studies have shown that *P. plumosa* is a complex of related taxa, mostly undescribed. Recordings from other states, and outside the distribution area given here, are unlikely to be *P. plumosa*, but a related unnamed species.

Pterostylis plumosa
Kangarooby, New South Wales

14 September 1992

A plant
B flower from front
C flower from side
D flower from rear
E labellum
F labellum and column from side
G column from front

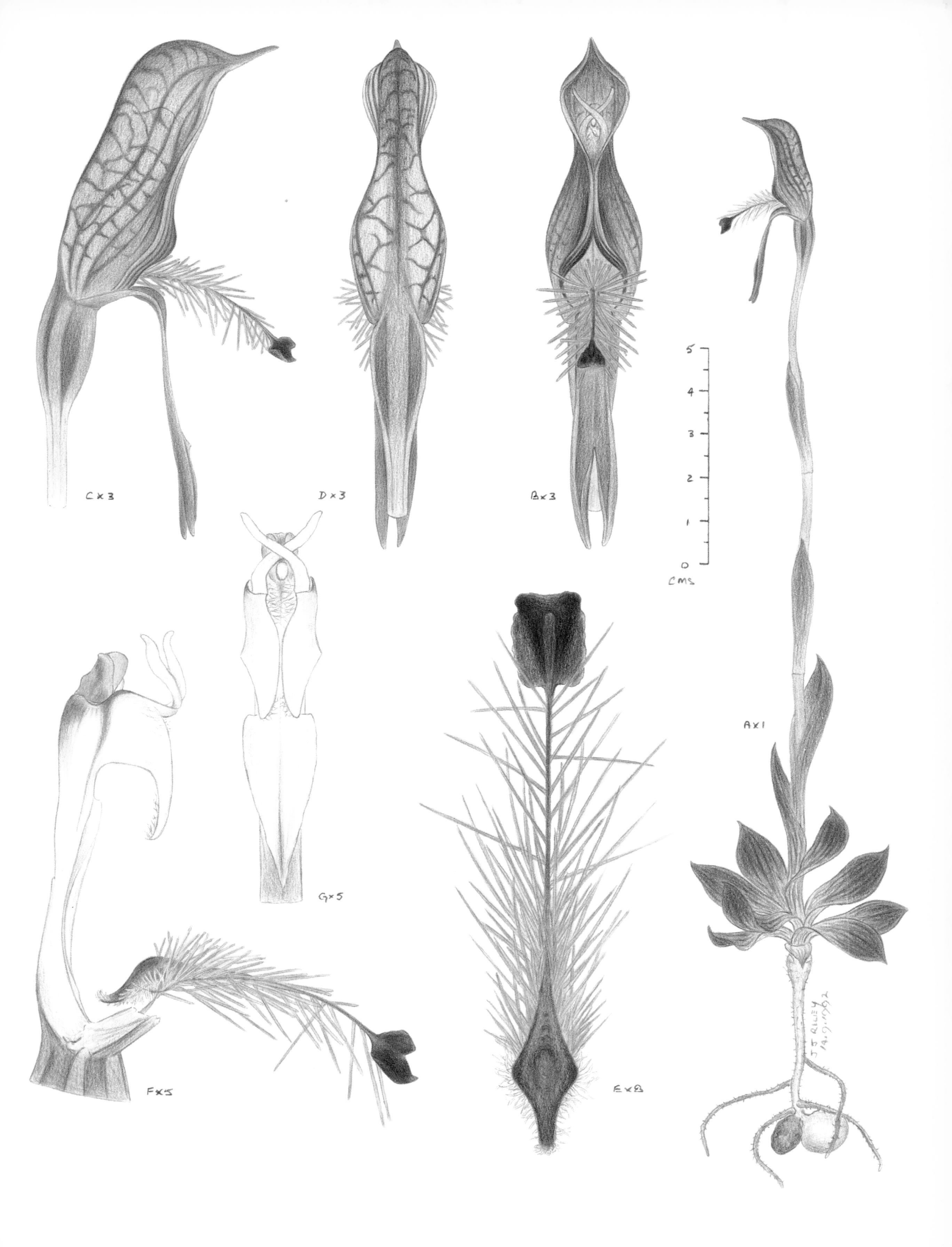

C×3
D×3
B×3
5
4
3
2
1
0
CMS
A×1
G×5
F×5
J J RILEY
14.9.1992

Pterostylis recurva Benth.

1873 | *Flora Australiensis* 6:360

TYPE LOCALITY Swan River, Western Australia

ETYMOLOGY *recurva* — curved backwards

FLOWERING TIME August to October

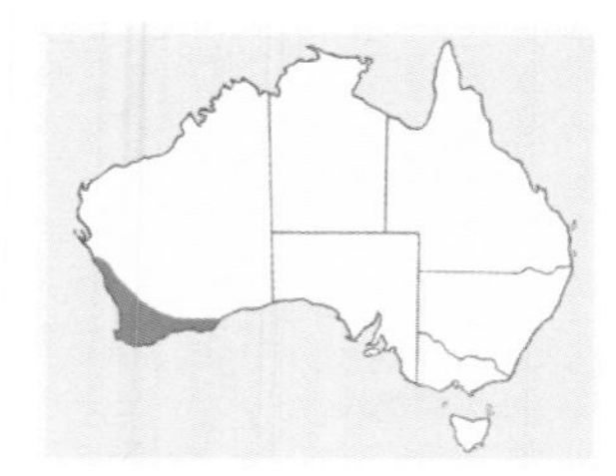

DISTRIBUTION This unique species is endemic to Western Australia, being found in the south-west corner from Geraldton to Israelite Bay.

ALTITUDINAL RANGE 10 m to 500 m

DISTINGUISHING FEATURES *P. recurva*, with its unusual arrangement of the petals and sepals, cannot be confused with any other species. The specific name refers to the lateral sepals that are reflexed downwards. It is a tall-growing species that can have up to three flowers.

HABITAT *P. recurva* is a common species with a wide distribution. It is found in many different plant communities and habitats with the exception of swamps and wet, seepage areas.

CONSERVATION STATUS Common and widespread. Occurs in national parks and reserves. Secure.

DISCUSSION *P. recurva* is an unusual species that is common and well known in Western Australia. The open, funnel-shaped flowers have given the plant its common name of Jug Orchid. It grows as scattered individuals or in small groups in relatively open situations or among scrub where the flowers can be seen above the low vegetation. It is a solitary species that reproduces from seed.

Pterostylis recurva
Canning Vale, Western Australia

26 August 1994

A flowering plant
B non-flowering plant
C flower from front
D flower from side
E flower from above
F labellum flattened
G labellum and column from side
H column from front
J petal

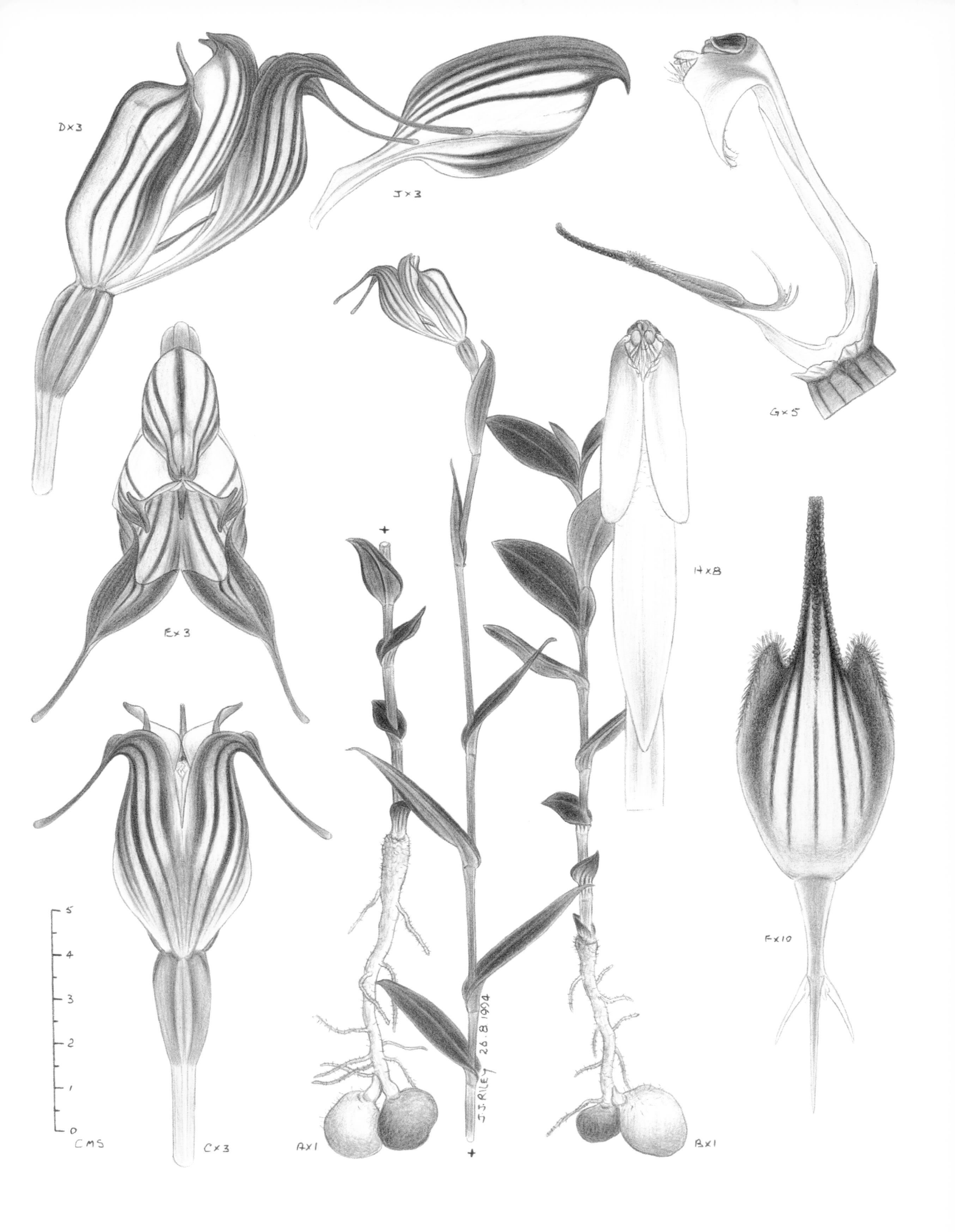

D×3
J×3
G×5
E×3
H×8
F×10
5
4
3
2
1
0
CMS
C×3
A×1
J.J. RILEY 26.8.1994
B×1

Pterostylis gibbosa R. Br.

1810 | *Prodromus Florae Novae Hollandiae*: 328

TYPE LOCALITY Toongabbie, New South Wales

RECENT SYNONYMS *Pterostylis ceriflora* Blackmore & Clemesha

ETYMOLOGY *gibbosa* — humped

FLOWERING TIME August to November

DISTRIBUTION This species is endemic to New South Wales, and is known to occur between Yallah and Nowra, with a disjunct population at Milbrodale in the Hunter Valley.

ALTITUDINAL RANGE 10 m to 250 m

DISTINGUISHING FEATURES *P. gibbosa* is part of the so-called *rufa* group of *Pterostylis*. *P. gibbosa* is a tall, elegant species with bright green flowers and a contrasting labellum that is brownish-purple to almost black. The wide lateral sepals are reflexed backwards, with the free points behind the ovary.

HABITAT Around Nowra, *P. gibbosa* grows in open forest dominated by Spotted Gums (*Eucalyptus maculata*), with a grass and shrub understorey on clay loams. In the Wollongong area it is found in open, grassy *Melaleuca* woodland on silty clay loams. At Milbrodale the habitat is open forest of *Eucalyptus* spp., and Black Cypress Pine (*Callitris endlicheri*), with a shrubby and grass understorey on sandy loams over sandstone.

CONSERVATION STATUS Rare. One population is known to occur in a national park. Vulnerable.

DISCUSSION The true identity of *P. gibbosa* has been subject to controversy and confusion in the past. Both *P. gibbosa* and *P. saxicola* are present on Robert Brown's type sheet of *P. gibbosa*. It would appear *P. gibbosa* disappeared from around Sydney very early, leaving the equally rare and then unnamed *P. saxicola*, which led to the assumption that this species was *P. gibbosa*. Based on this supposition, in the 1960s when plants were found near Wollongong, they were described as *P. ceriflora*. Research at the British Museum uncovered a colour illustration of *P. gibbosa* by Ferdinand Bauer, an artist who accompanied Brown. It was then apparent which taxon Brown had named as *P. gibbosa*. *P. ceriflora* became a synonym and it was realised that the species, now called *P. saxicola*, was as then unnamed. *P. gibbosa* occurs as scattered individuals and reproduces from seed.

Pterostylis gibbosa
Albion Park, New South Wales

2 October 1989

A plant
B flower from front
C flower from side
D flower from rear
E flower from above
F labellum
G labellum and column from side

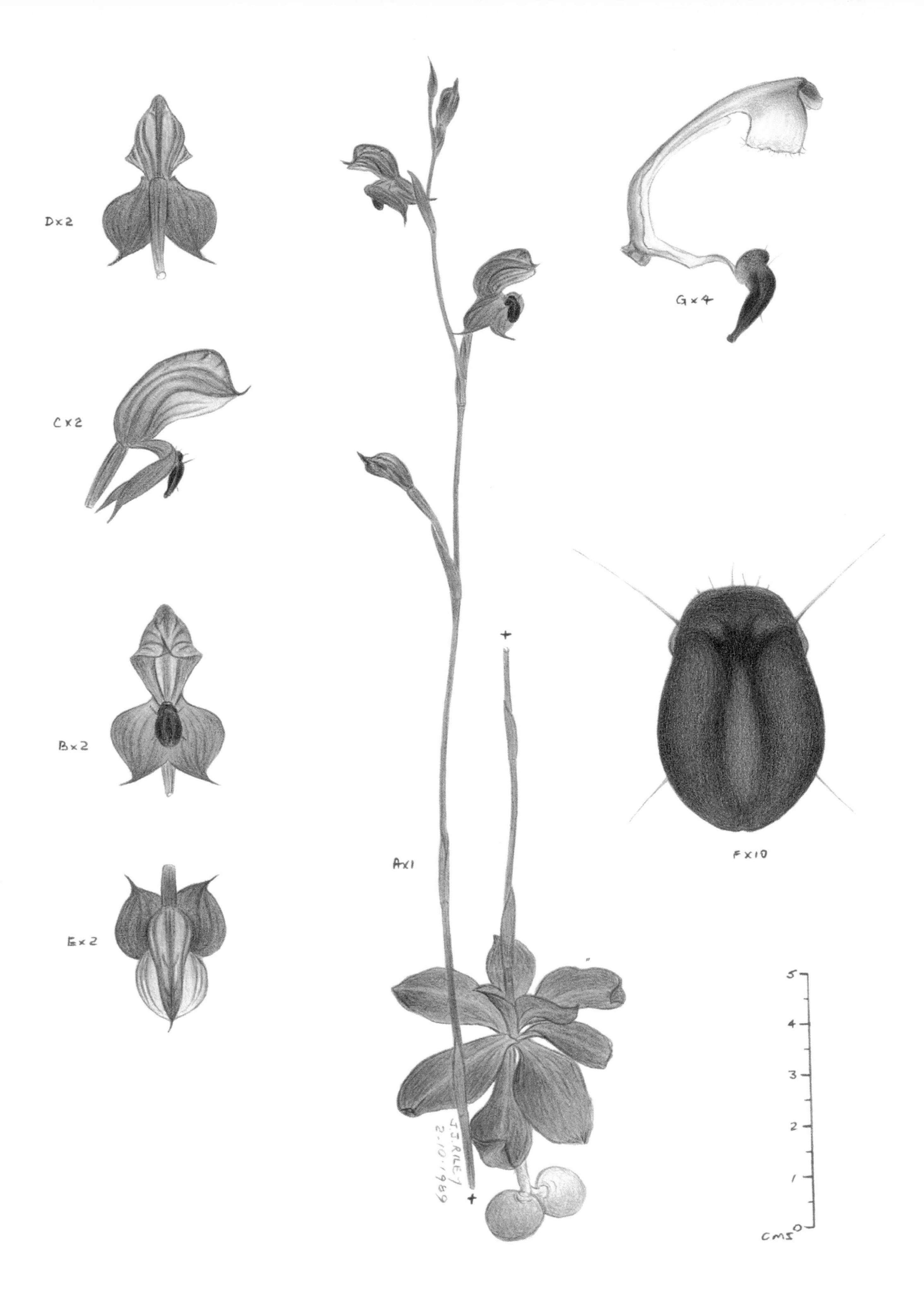
D×2
C×2
B×2
E×2
A×1
G×4
F×10
J.J. RILEY
2.10.1989
5
4
3
2
1
0
CMS

Pterostylis saxicola D.L. Jones & M.A. Clem.

1997 | *Orchadian* 12:132

TYPE LOCALITY Kentlyn, New South Wales

ETYMOLOGY *saxicola* — dwelling among rocks

FLOWERING TIME September to December

DISTRIBUTION This species is endemic to New South Wales, being found to the west and south-west of Sydney, from Wilberforce to Glenbrook in the lower Blue Mountains, and from Picnic Point to around Picton.

ALTITUDINAL RANGE 10 m to 200 m

DISTINGUISHING FEATURES *P. saxicola* is part of the *rufa* group of *Pterostylis*, having an upright inflorescence of multiple flowers emerging from a flat rosette of ground-hugging leaves, which is often withering by the blooming season. *P. saxicola* has shiny, translucent flowers with reddish-brown markings. The lateral sepals are entirely reddish-brown and hang straight down. The deeply channelled, ovate labellum has a swollen, basal lobe and is reddish-maroon.

HABITAT *P. saxicola* is most commonly found growing in open woodland and forest. Here it favours open positions among scattered tussocky grasses and shrubs. It grows in shallow soils with a covering of moss and coral lichens over sandstone shelves.

CONSERVATION STATUS Rare. Very few plants are known to occur in a national park. Vulnerable.

DISCUSSION While probably never common in the past, *P. saxicola* is now a rare orchid. Development, initially for agriculture, and later for housing to the west and south-west of Sydney, has destroyed most of its habitat. Today only small remnant populations remain. *P. saxicola* has been confused with *P. gibbosa* in the past. *P. saxicola* occurs as scattered individuals or small groups and reproduces from seed. Fungus gnats and mosquitoes that attempt to copulate with the labellum pollinate the flowers. The labellum is highly sensitive and snaps back towards the reproductive parts of the bloom when touched.

Pterostylis saxicola
Douglas Park, New South Wales

15 October 1989

A plant
B flower from side
C flower from front
D flower from back
E labellum and column from side
F labellum from above

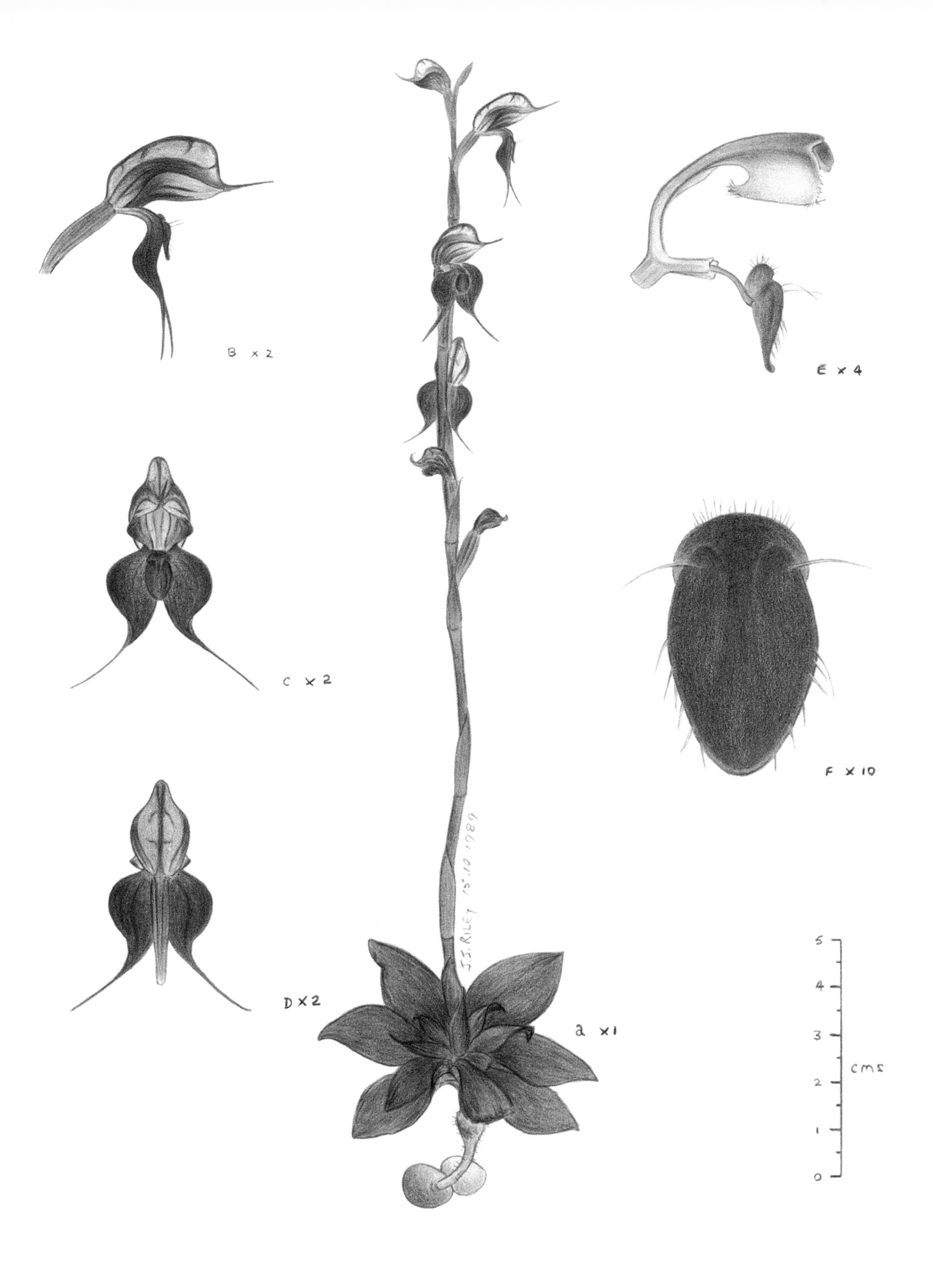

B x 2
C x 2
D x 2
a x 1
E x 4
F x 10
5
4
3
2
1
0
cms
J.I. RILEY 15.10.1989

Pterostylis woollsii Fitzg.

1876 | *Australian Orchids* 1(2)

TYPE LOCALITY Richmond, New South Wales

ETYMOLOGY After Rev. W. W. Woolls

FLOWERING TIME October to December

DISTRIBUTION This species occurs randomly on the tablelands and slopes of Queensland from the Carnarvon Gorge, along the Central and Northern Tablelands and slopes of New South Wales as far west as the Pilliga Scrub. Its range in southern New South Wales is unknown. It also occurs in the Rushworth area of Victoria.

ALTITUDINAL RANGE 10 m to 900 m

DISTINGUISHING FEATURES *P. woollsii* cannot be mistaken for any other species, with its extremely long filiform points to the lateral sepals. *P. woollsii* has the largest flowers of the *rufa* group of *Pterostylis*, having an upright inflorescence of multiple flowers emerging from a flat rosette of ground-hugging leaves, which is often withered by the blooming season.

HABITAT In New South Wales and Queensland, *P. woollsii* can mostly be found in open woodland dominated by *Callitris* spp., with some *Eucalyptus* spp. present and supporting a grassy understorey. Soils are red, sandy clay loams or gravelly loams derived from weathered granite. It can also be found growing among rocks and on rocky ridges in open *Eucalyptus* woodland. In Victoria, *P. woollsii* is a very rare species of limited distribution and grows in or close to mallee communities.

CONSERVATION STATUS Uncommon. Known to occur in national parks and nature reserves. Vulnerable.

DISCUSSION *P. woollsii* is a plant of the tablelands and western slopes. The type-site at Richmond, however, is on the coastal plain to the north-west of Sydney where it is now extinct. Peak flowering is in mid-summer, when temperatures can be high and, when accompanied by dry conditions, the flowers often abort. It occurs as scattered individuals and reproduces from seed. Fungus gnats and mosquitoes that attempt to copulate with the labellum pollinate the flowers. The labellum is highly sensitive and snaps back towards the reproductive parts of the bloom when touched.

Pterostylis woollsii
Moonbi Range, New South Wales

24 November 1990

A plant
B flower from front
C flower from rear
D flower from above
E labellum from above
F labellum and column from side

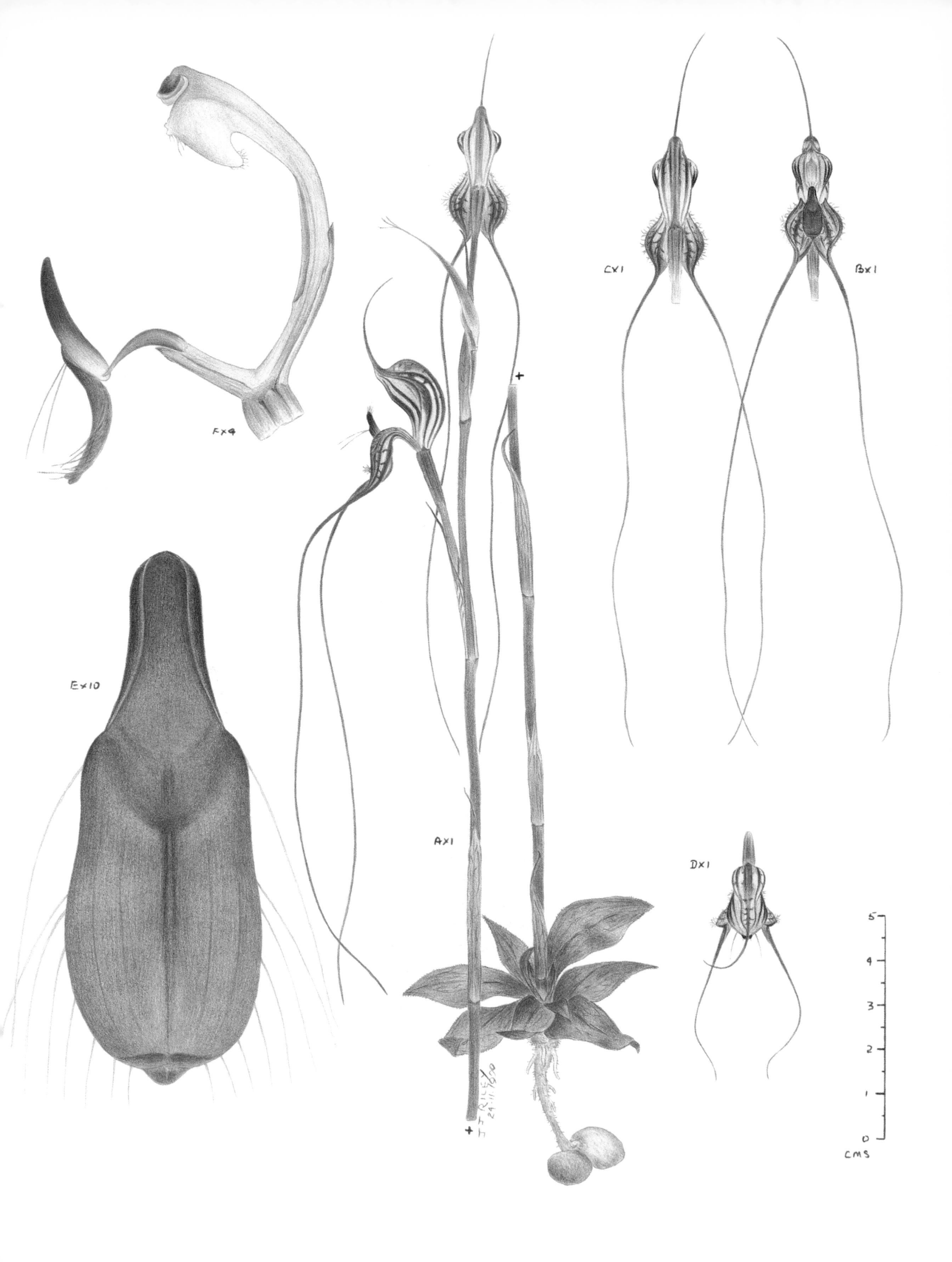

F×4
E×10
A×1
C×1
B×1
D×1
5
4
3
2
1
0
CMS

Pterostylis pusilla R.S. Rogers

1918 | *Transactions and Proceedings of the Royal Society of South Australia* 42:26

TYPE LOCALITY Geranium, South Australia

ETYMOLOGY *pusilla* — very small

FLOWERING TIME September to November

DISTRIBUTION Found in New South Wales on the western slopes of the Great Dividing Range, south from Reefton, the central and western parts of Victoria to the lower Flinders Ranges, Eyre Peninsula and the south-east corner of South Australia.

ALTITUDINAL RANGE 80 m to 400 m

DISTINGUISHING FEATURES *P. pusilla* is part of the *rufa* group of *Pterostylis*, having an upright inflorescence of multiple flowers emerging from a flat rosette of ground-hugging leaves, which is often withering by the blooming season. *P. pusilla* is a short-growing plant with small olive brown to reddish-green flowers. The lateral sepals are narrow with prominently folded margins.

HABITAT In western Victoria and South Australia, *P. pusilla* favours open mallee communities, where it can be seen growing in bare ground close to grass tussocks and shrubs. In central Victoria, it grows in dry open *Eucalyptus* woodland with a sparse understorey. In New South Wales, it occurs in both open mallee and dry mixed woodland of *Eucalyptus* and *Callitris* with little understorey competition. It favours open mossy lichen patches among the grass, shrubs and trees.

CONSERVATION STATUS Occurs in national parks and reserves. Secure.

DISCUSSION *P. pusilla* is the smallest of the *rufa* group of *Pterostylis*. It is easily overlooked in the wild on account of its small, somewhat dull blooms. Of sporadic distribution in the eastern half of its range, *P. pusilla* is widespread and common in the western half. It occurs as scattered individuals and reproduces from seed. Fungus gnats and mosquitoes that attempt to copulate with the labellum pollinate the flowers. The labellum is highly sensitive and snaps back towards the reproductive parts of the bloom when touched.

Pterostylis pusilla
Peake, South Australia

1 October 1998

A plant
B flower from front
C flower from side
D labellum from above
E labellum from below
F labellum and column from side
G column from front
H lateral sepals
J petal

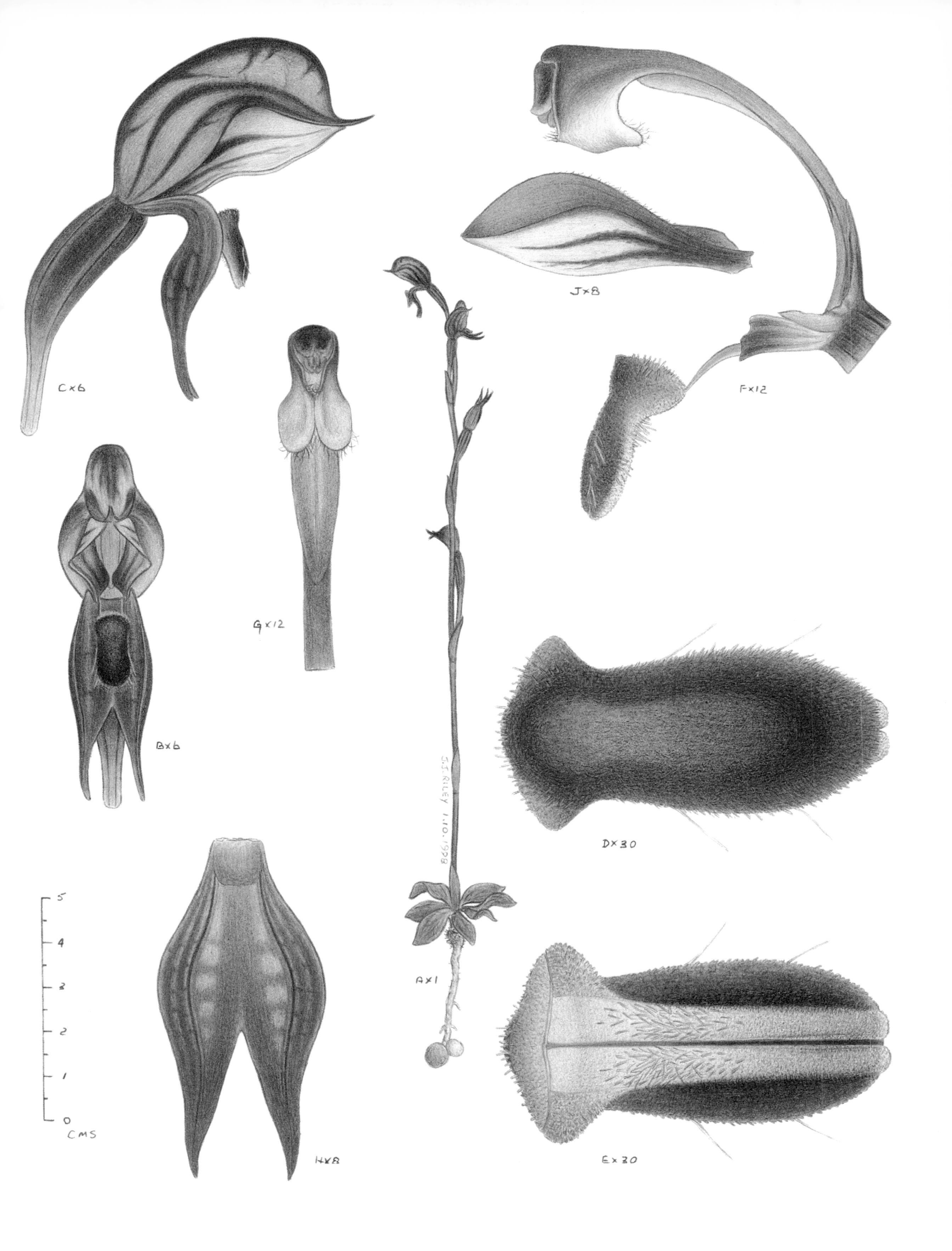
C×6
J×8
F×12
G×12
B×6
D×30
J.J. RILEY 1.10.1998
5
4
3
2
1
0
CMS
A×1
H×8
E×30

Pterostylis maxima M.A. Clem. & D.L. Jones

1989 | *Australian Orchid Research* 1:124

TYPE LOCALITY Bendigo, Victoria

ETYMOLOGY *maxima* — largest, greatest

FLOWERING TIME September to November

DISTRIBUTION This species is endemic to Victoria, being restricted to the Goldfields and Wimmera Districts.

ALTITUDINAL RANGE 100 m to 350 m

DISTINGUISHING FEATURES *P. maxima* has large, shiny flowers with distinct, reddish-brown stripes. The labellum is also reddish-brown and is broadly ovate with a prominent basal lobe. It has numerous short marginal setae with two, long, forward-pointing setae on the basal lobe. The long, filiform points of the lateral sepals project well forward. *P. maxima* is part of the *rufa* group of *Pterostylis*, having an upright inflorescence of multiple flowers emerging from a flat rosette of ground-hugging leaves, which is often withering by the blooming season.

HABITAT This species is found in dry, open woodland, forest and mallee communities. Here *P. maxima* prefers stony slopes and rocky outcrops where it can be seen growing in open ground among the tussock grasses, stunted bushes and rocks. It may also occur in open forest growing in sand.

CONSERVATION STATUS Uncommon. Poorly represented in national parks and reserves.

DISCUSSION *P. maxima* is a poorly known and uncommon species. In parts of the gold fields of central Victoria, *P. maxima* can be observed in moderate numbers. In some areas it has colonised revegetated mounds of old, alluvial gold workings. It occurs as scattered individuals and reproduces from seed. Fungus gnats and mosquitoes that attempt to copulate with the labellum pollinate the flowers. The labellum is highly sensitive and snaps back towards the reproductive parts of the bloom when touched. Similar populations from New South Wales and South Australia are not *P. maxima*, but appear to be undescribed taxa.

Pterostylis maxima
Bendigo, Victoria

16 November 1989

A plant
B flower from front
C flower from side
D flower from rear
E labellum from above
F labellum and column from side

C
F
5 mm
B
1 cm
E
2 mm
1 cm
J J RILEY
16 11 89
D
A

Pterostylis cobarensis M.A. Clem.

1989 | *Australian Orchid Research* 1:121

TYPE LOCALITY Cobar, New South Wales

ETYMOLOGY From the Cobar district

FLOWERING TIME September to November

DISTRIBUTION This species is endemic to New South Wales and is known with certainty from the Young, Cobar, Nyngan, Nymagee, Mount Hope and Broken Hill areas.

ALTITUDINAL RANGE 300 m to 550 m

DISTINGUISHING FEATURES *P. cobarensis* is a tall species with translucent, white flowers highlighted with greenish-brown markings. The labellum is narrow and fleshy with long, spreading marginal calli and short calli on the basal lobe. *P. cobarensis* is part of the *rufa* group of *Pterostylis*, having an upright inflorescence of multiple flowers emerging from a flat rosette of ground-hugging leaves, which is often withered by the blooming season.

HABITAT *P. cobarensis* is found on the *Eucalyptus* and *Callitris* covered ridges and slopes of low ranges and hills of the western plains. It grows in pockets of soil among rocks and in humus-rich soils around the base of *Callitris* trees, or on sheltered stony slopes among tussocky grasses and shrubs. It grows in a harsh climate, with cold winters, hot summers and unreliable rainfall.

CONSERVATION STATUS Uncommon. Not known to occur in a national park or flora reserve. Vulnerable.

DISCUSSION *P. cobarensis* is a poorly known, xerophytic species, only recently recognised as a distinct taxon. It will undoubtedly have a wider distribution than that given above, possibly extending into Queensland and South Australia. The area is remote, the distances great and the ranges and hills are isolated and difficult to access, all of which has limited searches to easily reached sites. It occurs as scattered individuals and reproduces from seed. Fungus gnats and mosquitoes that attempt to copulate with the labellum pollinate the flowers. The labellum is highly sensitive and snaps back towards the reproductive parts of the bloom when touched.

Pterostylis cobarensis
Nymagee, New South Wales

18 October 1994

A plant
B flower from front
C flower from side
D labellum from above
E labellum from below
F labellum and column from side
G column from front
H lateral sepals
J petal

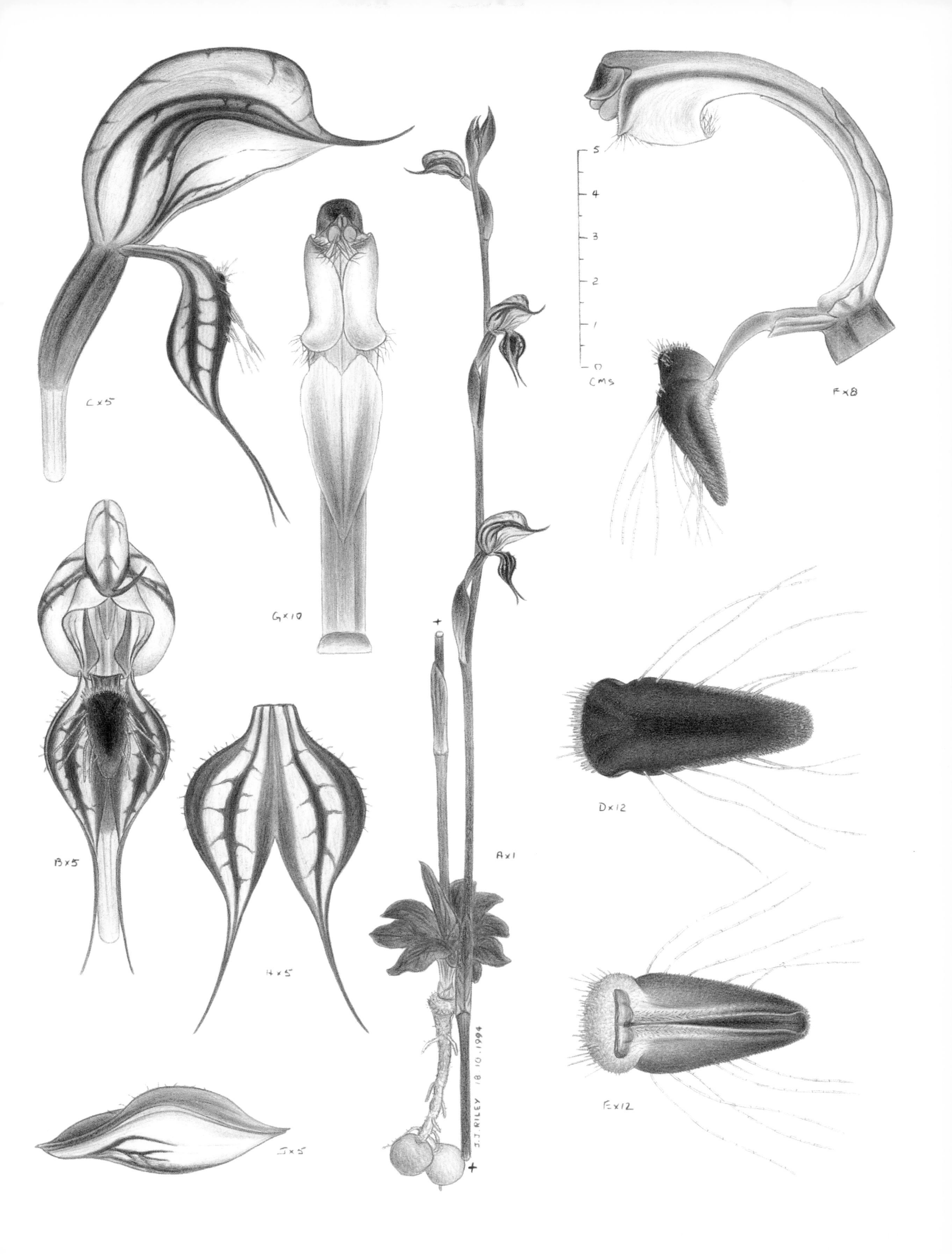
C×5
G×10
B×5
H×5
J×5
A×1
J.J. RILEY 18.10.1994
5
4
3
2
1
0
CMS
F×8
D×12
E×12

Pterostylis basaltica D.L. Jones & M.A. Clem.

1993 | *Muelleria* 8(1):75

TYPE LOCALITY Woorndoo, Victoria

ETYMOLOGY *basaltica* — growing on basalt soils

FLOWERING TIME November to January

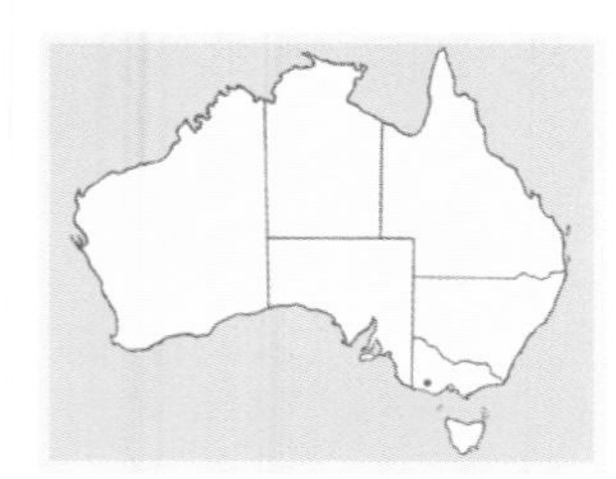

DISTRIBUTION This species is endemic to Victoria, being found between Lake Bolac and Mortlake.

ALTITUDINAL RANGE 150 m to 250 m

DISTINGUISHING FEATURES *P. basaltica* is part of the *rufa* group of *Pterostylis*, having an upright inflorescence of multiple flowers emerging from a flat rosette of ground-hugging leaves, which is often withering by the blooming season. Plants of *P. basaltica* are short and stout with translucent white flowers that have greenish-brown and reddish-brown markings. The galea and lateral sepals end in long, filiform points. The labellum is narrow and fleshy with about six pairs of spreading, white marginal setae and about four to six setae on the basal lobe.

HABITAT *P. basaltica* grows in small, shallow, mossy pockets of soil on low, rocky rises on the western basalt plains of Victoria. The soils of these plains are rich but shallow, supporting only grassland with some small, heathy shrubs and *Acacia* spp.

CONSERVATION STATUS Rare. Not known to occur in a national park or flora reserve. Endangered.

DISCUSSION The volcanic activity that formed the basalt plains occurred in relatively recent geological times. Weathering since then has produced very rich, shallow soils, ideal for agriculture. Very little of the original, native grassland and vegetation remains today. *P. basaltica* was undoubtedly more widespread but has been reduced to small remnant island populations. It occurs as scattered individuals and reproduces from seed. Fungus gnats and mosquitoes that attempt to copulate with the labellum pollinate the flowers. The labellum is highly sensitive and snaps back towards the reproductive parts of the bloom when touched.

Pterostylis basaltica
Woorndoo, Victoria

15 November 1997

A plant
B flower from front
C flower from side
D labellum from above
E labellum from below
F labellum and column from side
G column from front
H lateral sepals
J petal

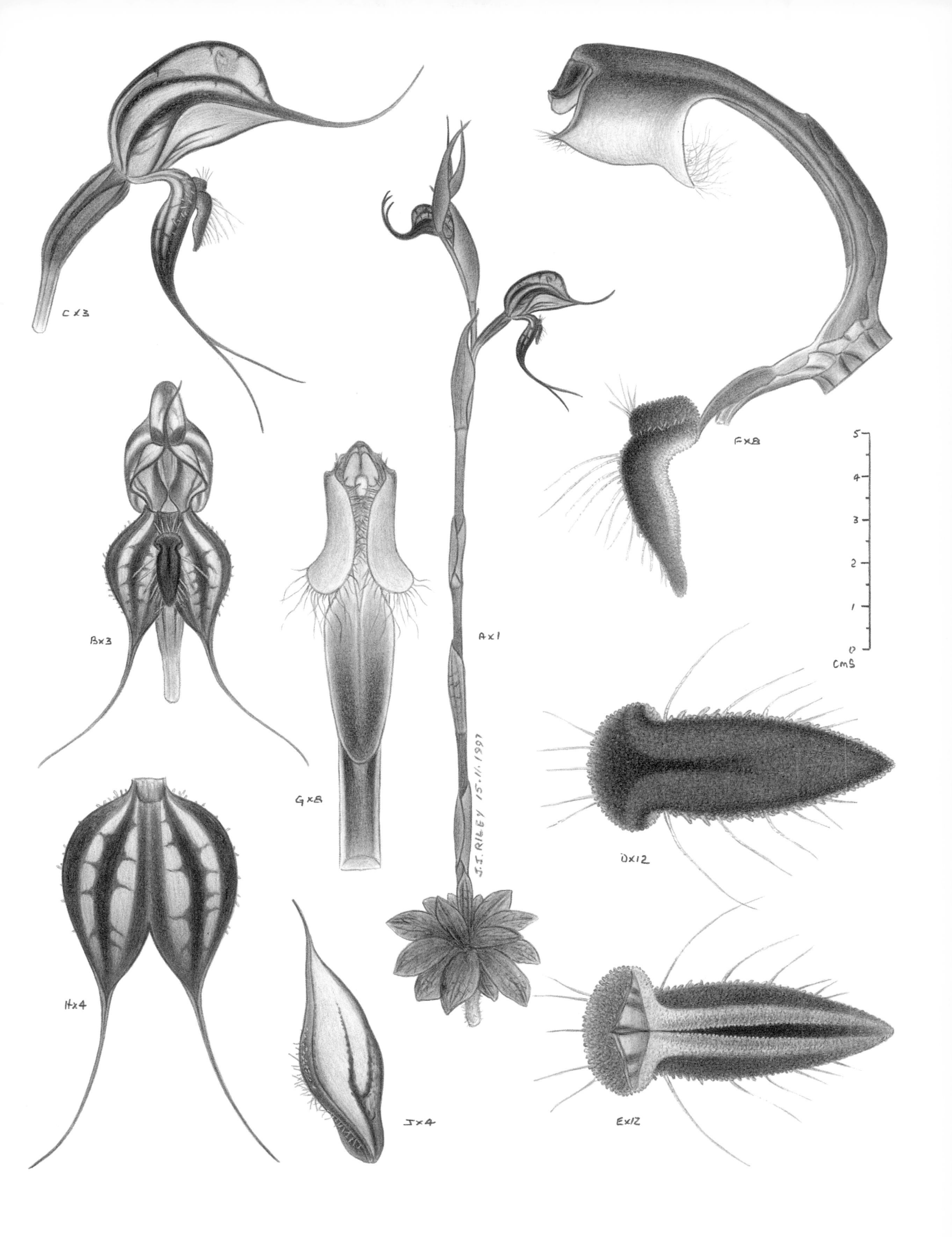
C×3
B×3
G×8
A×1
J.J. RILEY 15.11.1997
F×8
5
4
3
2
1
0
CMS
D×12
H×4
J×4
E×12

Pyrorchis nigricans (R. Br.) D.L. Jones & M.A. Clem.

1994 | *Phytologia* 77(6):449

TYPE LOCALITY Port Jackson, New South Wales

RECENT SYNONYMS *Lyperanthus nigricans* R. Br.

ETYMOLOGY *nigricans* — black

FLOWERING TIME August to November

DISTRIBUTION *P. nigricans* is a widespread species, being found in New South Wales, south from Forster, through southern and western Victoria, to southern South Australia including the Eyre Peninsula, across to the south-west of Western Australia. It also occurs on the Bass Strait islands and the north coast and parts of the East Coast of Tasmania.

ALTITUDINAL RANGE Up to 150 m

DISTINGUISHING FEATURES *P. nigricans* has a large, ground-hugging, fleshy leaf with reddish-black markings. The flowers are white with dark red stripes, with a hooded dorsal sepal and a prominently fringed labellum.

HABITAT This species is found in a range of habitats and vegetation communities, from coastal heath, heathy woodland and forest to mallee heathland and around granite outcrops. It can also be found in moderately well-drained sedgelands and close to swamp margins. Soils are often sand or sandy loams.

CONSERVATION STATUS Common and widespread. Conserved in national parks and reserves. Secure.

DISCUSSION *P. nigricans* forms extensive, often crowded colonies vegetatively from daughter tubers formed on the end of stolonoid roots. Without a bushfire the previous season, colonies numbering hundreds or even thousands will have few or no plants blooming. However, the year after a burn, *P. nigricans* produces memorable mass displays of hundreds of flowering plants. It also reproduces from seed, with small native bees pollinating the flowers.

Pyrorchis nigricans
Nelson Bay, New South Wales

13 September 1994

A plant
B flower from front
C flower from side
D labellum from side
E labellum flattened
F column from side
G column from front
H dorsal sepal
J lateral sepal
K petal

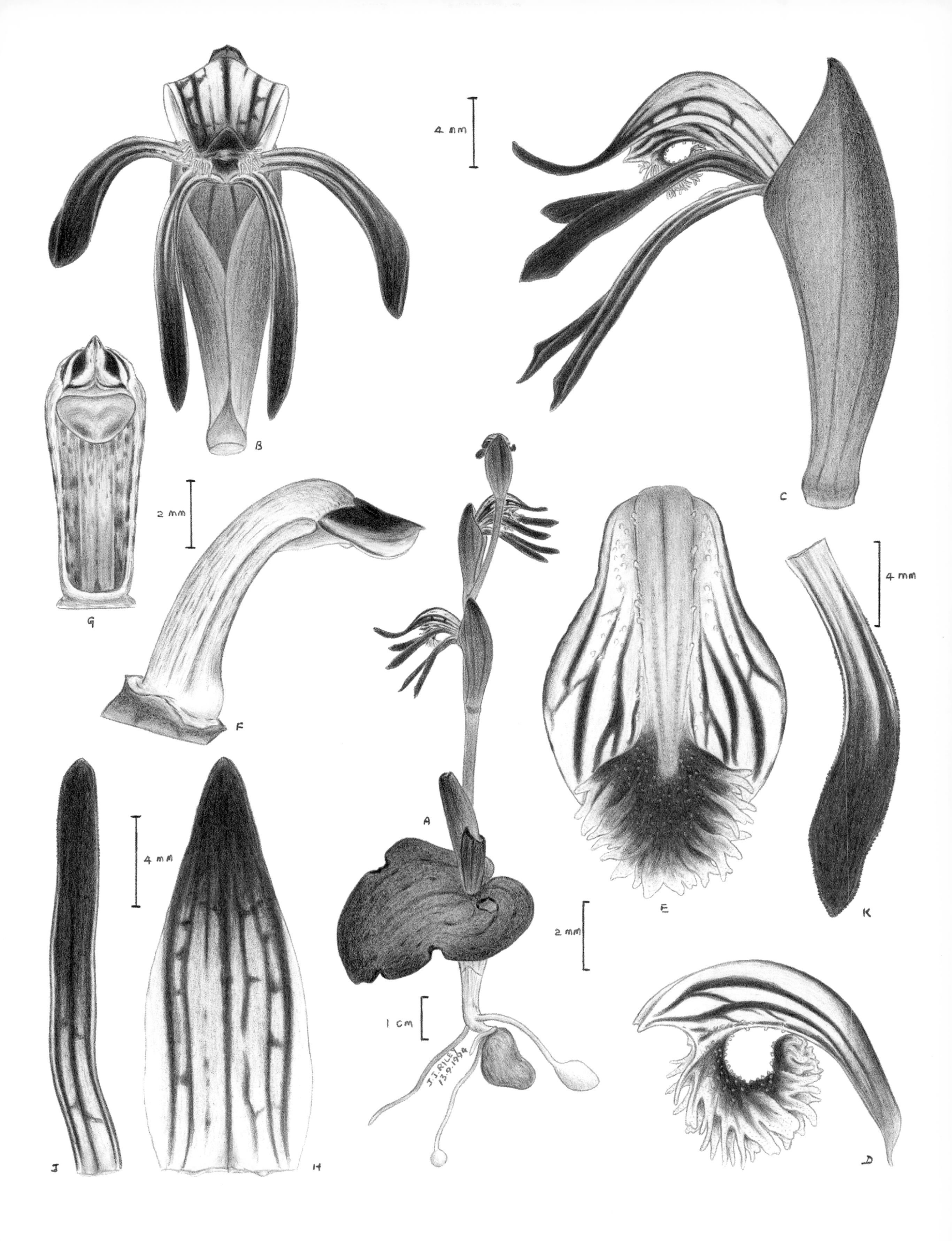

B
4 mm
C
G
2 mm
F
A
E
4 mm
K
4 mm
2 mm
1 cm
J.J. RILEY
13.9.1994
J
H
D

Rhizanthella slateri (Rupp) M.A.Clem & Cribb

1984 | *Kew Magazine* 1(2):84–91

TYPE LOCALITY Bulahdelah, New South Wales

RECENT SYNONYMS *Cryptanthemis slateri* Rupp

ETYMOLOGY After Edwin Slater

FLOWERING TIME August to December

DISTRIBUTION The Eastern Underground Orchid appears to be endemic to New South Wales, being found from Alum Mountain at Bulahdelah to at least as far south as the Nowra district, and inland to the Blue Mountains and the Braidwood district.

ALTITUDINAL RANGE Up to 600 m

DISTINGUISHING FEATURES *R. slateri* is a subterranean, saprophytic species that cannot be mistaken for any other orchid. The other named species, *R. gardneri*, occurs in Western Australia.

HABITAT This species has been found in a range of habitats. Ranging from shrubby, open *Eucalyptus* forest with a *Melaleuca* understorey on the slopes at the type-site of Alum Mountain, moist *Eucalyptus* forest in the Blue Mountains, wet sclerophyll forest at Wisemans Ferry to *Eucalyptus* woodland at Nowra where plants were recently discovered at the base of Scribbly Gums.

CONSERVATION STATUS Recorded from national parks and reserves. Status is uncertain, due to its subterranean habit.

DISCUSSION *R. slateri* is undoubtedly more plentiful than present accounts indicate, as almost all recordings have been accidental. Most common of these are excavations for housing or other development. One colony was uncovered after a storm had washed away the leaf litter and top soil. *R. slateri* grows from a subterranean, thick, fleshy, branching rhizome, which may wither in the centre forming small clonal groups of plants. The flowering heads rise from these rhizomes to the top of the soil, just below the leaf litter. Only the flowering heads are illustrated. Very little is known about its reproductive cycle. Recent research indicates that recordings from the Lamington Plateau in southern Queensland represent an undescribed taxon.

Rhizanthella slateri
Bulahdelah, New South Wales

28 August 1994

A flowering portion of plant
B flowering portion of plant
C flowering head from above
D flowering head from side with bracts removed
E bract
F flower from side
G flower from front
H lateral sepal
J dorsal sepal
K petal
L labellum from side
M labellum from above
N column from side
O column from front

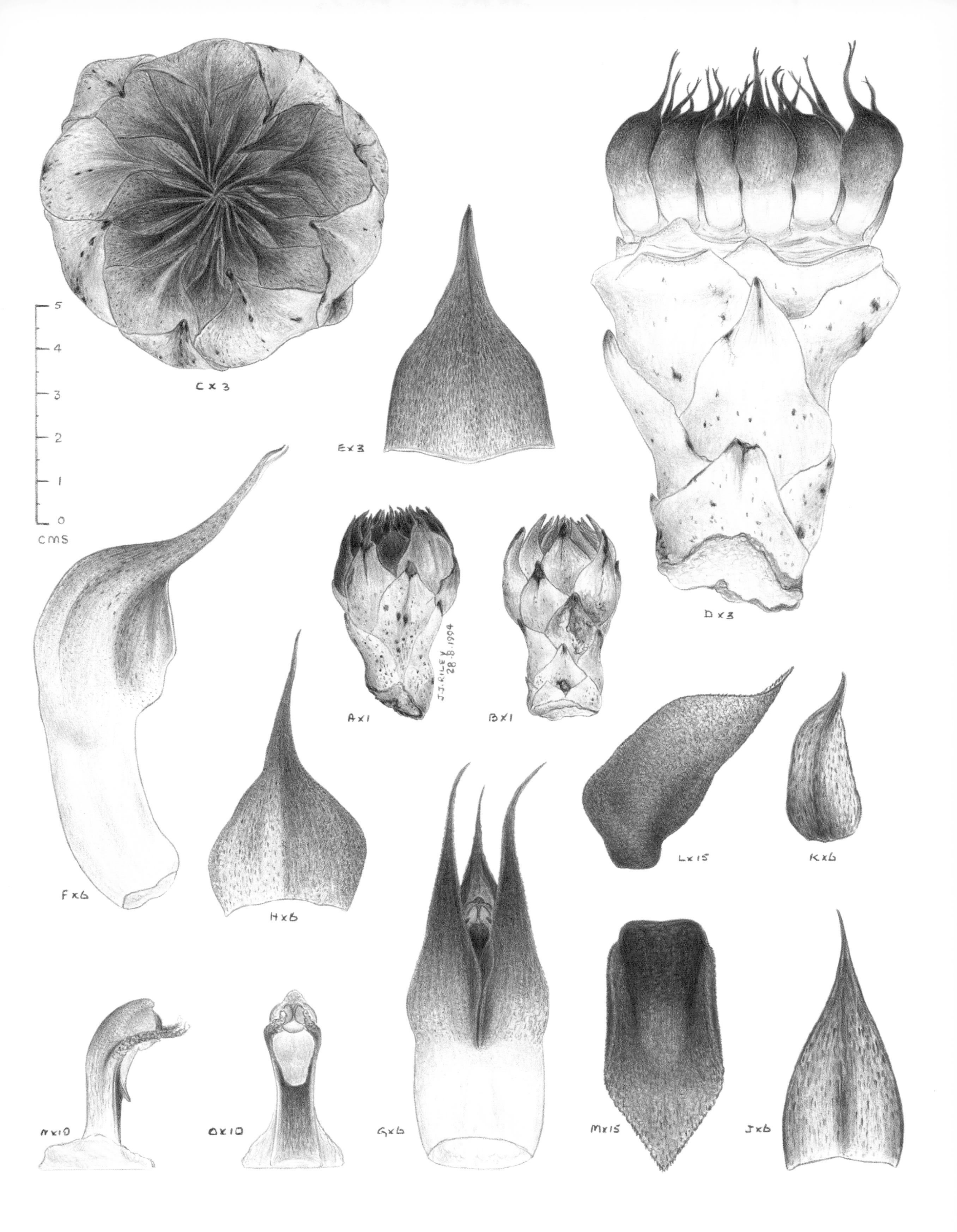

C x 3
E x 3
D x 3
5
4
3
2
1
0
cms
A x 1
B x 1
J.J. RILEY
28.8.1994
F x 6
H x 6
L x 15
K x 6
N x 10
O x 10
G x 6
M x 15
J x 6

Rimacola elliptica (R. Br.) Rupp

1942 | *Victorian Naturalist* 58:188

TYPE LOCALITY Port Jackson, New South Wales

ETYMOLOGY *elliptica* — ellipse-shaped

FLOWERING TIME November to February

DISTRIBUTION This species is endemic to New South Wales, being found from Mount White to the Fitzroy Falls and into the Blue Mountains. It occurs close to the coast with random populations at Terrey Hills and Galston Gorge but is more plentiful in the higher ranges.

ALTITUDINAL RANGE 50 m to 900 m

DISTINGUISHING FEATURES *R. elliptica* is a monotypic species that cannot be confused with any other orchid when in bloom. The pendent leaves and roots are superficially similar to those of the unrelated *Cryptostylis subulata*. It always grows in sandstone rock crevices.

HABITAT Found growing in horizontal, clay-filled fissures and cracks that are always moist. Occasionally growing in isolation, but is most often seen among species of *Blechnum* and *Gleichenia* ferns and the Forked Sundew, *Drosera binata*.

CONSERVATION STATUS Occurs in national parks and reserves. Secure.

DISCUSSION The flowers of *R. elliptica* are long-lasting, even retaining their shape and colour well after they have been pollinated and are about to disperse their seed. The rugged, precipitous gorges and cliffs of the Sydney Sandstone with their horizontal bedding have many, moist, clay-filled seams that are ideal for *R. elliptica*. Due to the difficult accessibility to such sites, this unusual and beautiful orchid can be studied at close quarters in only a few locations that are easily reached. In the upper Blue Mountains, *R. elliptica* is commonly found growing in rock crevices above *Adenochilus nortonii,* that is in the wet dark soils below. *R. elliptica* reproduces from seed and is pollinated by a small black wasp.

Rimacola elliptica
Terrey Hills, New South Wales

4 December 1989

A plant
B flower from side
C flower from above
D labellum and column from side
E labellum and column from front
F labellum flattened
G column from side
H column from front

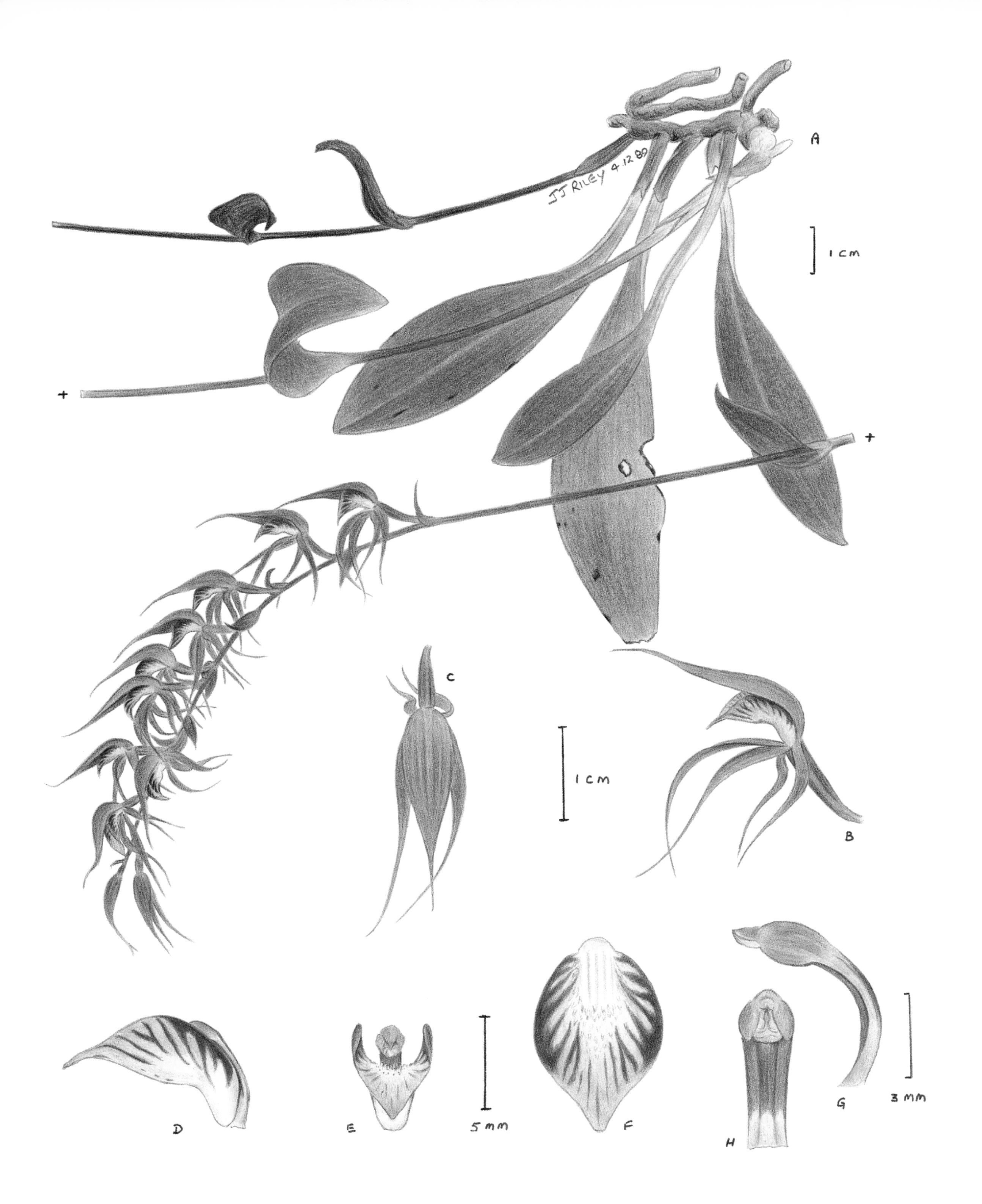

A
J J RILEY 4.12.89
1 CM
C
1 CM
B
5 mm
3 MM
D
E
F
G
H

Sarcochilus hillii (F. Muell.) F. Muell.

1860 | *Fragmenta Phytographiae Australiae* 2:94

TYPE LOCALITY Moreton Bay, Queensland

ETYMOLOGY After Walter Hill

FLOWERING TIME October to February

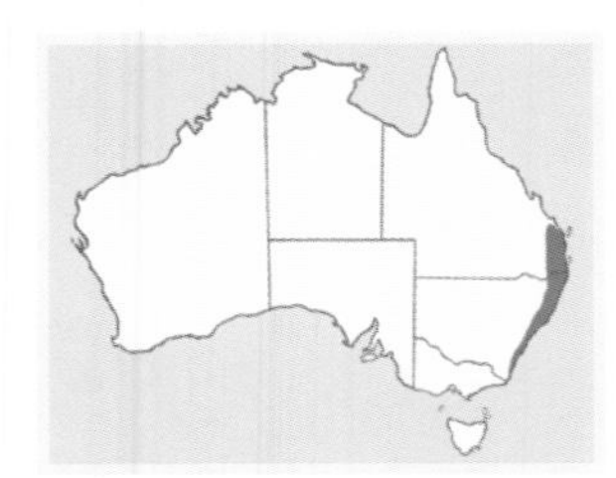

DISTRIBUTION Found along the coast and extending into the ranges from Bega on the South Coast of New South Wales, north to Gympie in Queensland.

ALTITUDINAL RANGE 10 m to 800 m

DISTINGUISHING FEATURES *S. hillii* is similar to the closely related *S. minutiflos*. The labellum of *S. hillii* has small sidelobes and the midlobe is hairy on the outside rim as well as the inside. It also has three, large, conspicuous, orange-coloured calli, with the two outside ones higher than the central calli.

HABITAT *S. hillii* is one of the twig epiphytes and grows on the outer branches of trees and larger shrubs. It is often present in dry coastal rainforest and can be seen in large numbers near or overhanging streams in open gullies. Further inland it is common in dry scrub and vine forest.

CONSERVATION STATUS Common and widespread. Occurs in national parks and reserves. Secure.

DISCUSSION *S. hillii* is a small plant that can be locally common. It favours the twigs and small branches and is rarely seen on the trunks of trees. It rarely grows on rocks. The roots travel a considerable distance along the host, and are often sighted before the actual plant is. It is almost always found in humid, open situations with dappled light and plenty of air movement. *S. hillii* is a sequentially flowering species that may be in bloom for many weeks. Its northern distribution limit is uncertain, due to confusion with *S. minutiflos*, as the plants look similar when not seen in flower.

Sarcochilus hillii
Paterson, New South Wales

5 January 1999

A plant
B flower from front
C flower from side
D labellum from front
E labellum from rear
F labellum from above
G longitudinal section of labellum
H column from front
J column from side
K dorsal sepal
L lateral sepal
M petal

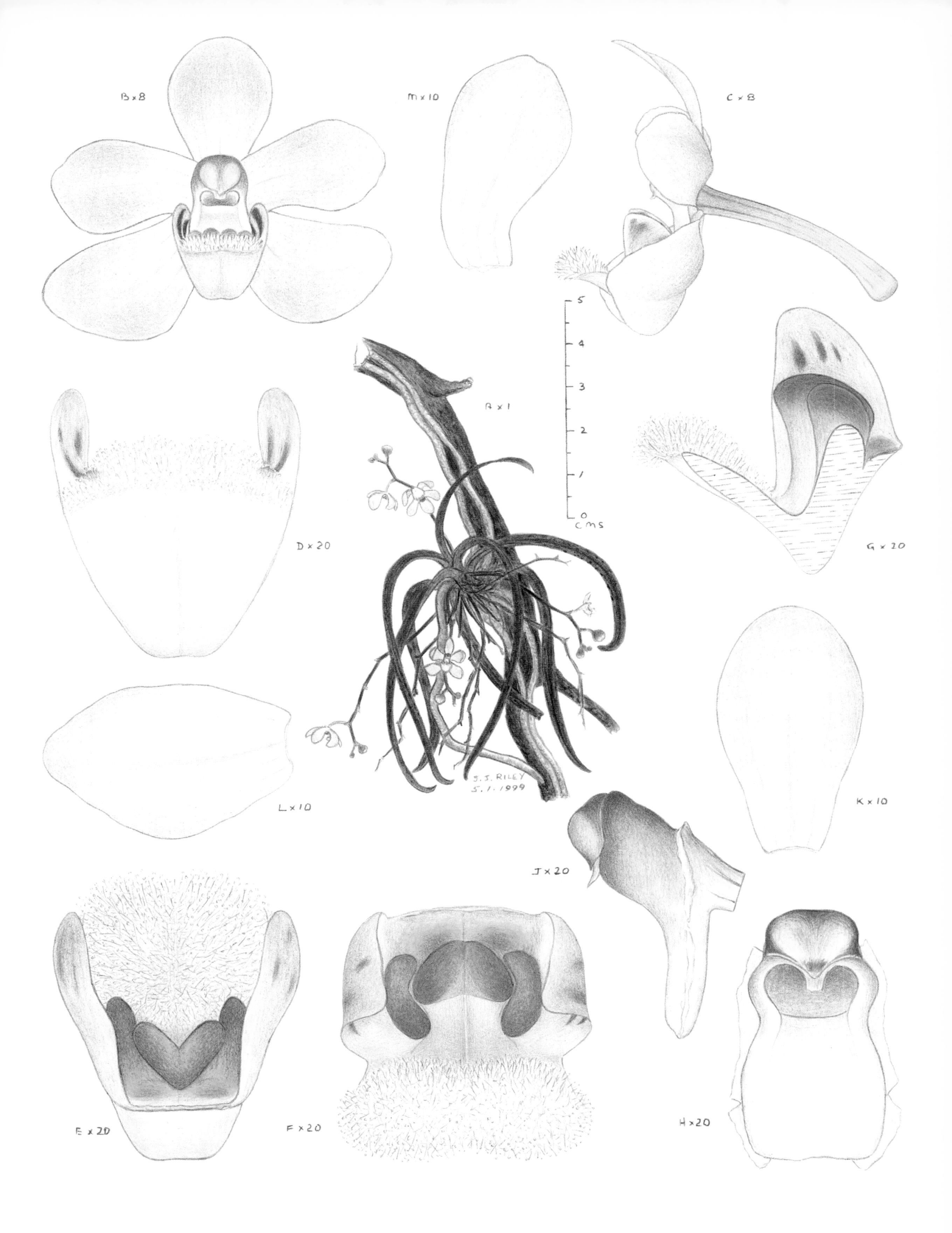
B x 8
M x 10
C x 8
A x 1
5
4
3
2
1
0
CMS
D x 20
G x 20
J. J. RILEY
5. 1. 1999
L x 10
K x 10
J x 20
E x 20
F x 20
H x 20

Sarcochilus australis (Lindl.) Rchb. f.

1861 | *Annalen Botanices Systematicae* 6:501

TYPE LOCALITY Circular Head, Tasmania

ETYMOLOGY *australis* — southern

FLOWERING TIME October to December

DISTRIBUTION This widespread species is found in northern and eastern Tasmania and the islands of the Furneaux Group, the Otways in Victoria, and north to at least as far as Nowendoc and Dorrigo in New South Wales. There are other sporadic recordings from northern New South Wales, including Glen Innes and Kyogle.

ALTITUDINAL RANGE Up to 900 m

DISTINGUISHING FEATURES *S. australis* is a twig epiphyte that has pendulous inflorescences with up to sixteen, pleasantly fragrant flowers. The petals and sepals are apple to olive green in colour. The lateral lobes of the labellum are as broad as long and have a rounded apex. It is possible that some of the plants allegedly from northern New South Wales may have been the related *S. spathulatus*.

HABITAT In the southern parts of its range, *S. australis* occurs in temperate rainforest, moist ferny gullies and wet sclerophyll forest. In this habitat many plants are seen on the main branches and trunks of rainforest trees and shrubs, as well as the outer twigs. In New South Wales, *S. australis* can be found growing on trees close to or overhanging creeks in moist, sheltered gullies in tall open forest and close to mangroves in estuarine areas. It is locally common in the numerous small gorges that dissect the Sydney Sandstone. These gullies are mostly cool, with moderately high light intensity and excellent air movement.

CONSERVATION STATUS Common. Grows in national parks and reserves. Secure.

DISCUSSION *S. australis* is a very attractive species when seen in bloom. Plants growing on slender twigs rarely grow into large specimens, as storms and strong winds invariably dislodge the epiphytes, which often fall to the ground and perish. The largest plants are, therefore, seen on major branches or the trunks of trees. It is a very rare species in northern New South Wales, with its distribution and taxonomic status in need of investigation.

Sarcochilus australis
Marrawah, Tasmania

13 November 1994

A plant
B flower from front
C flower from side
D flower from rear
E labellum from front
F labellum from rear
G labellum from above
H column from side
J column from front
K column detail
L dorsal sepal
M lateral sepal
N petal

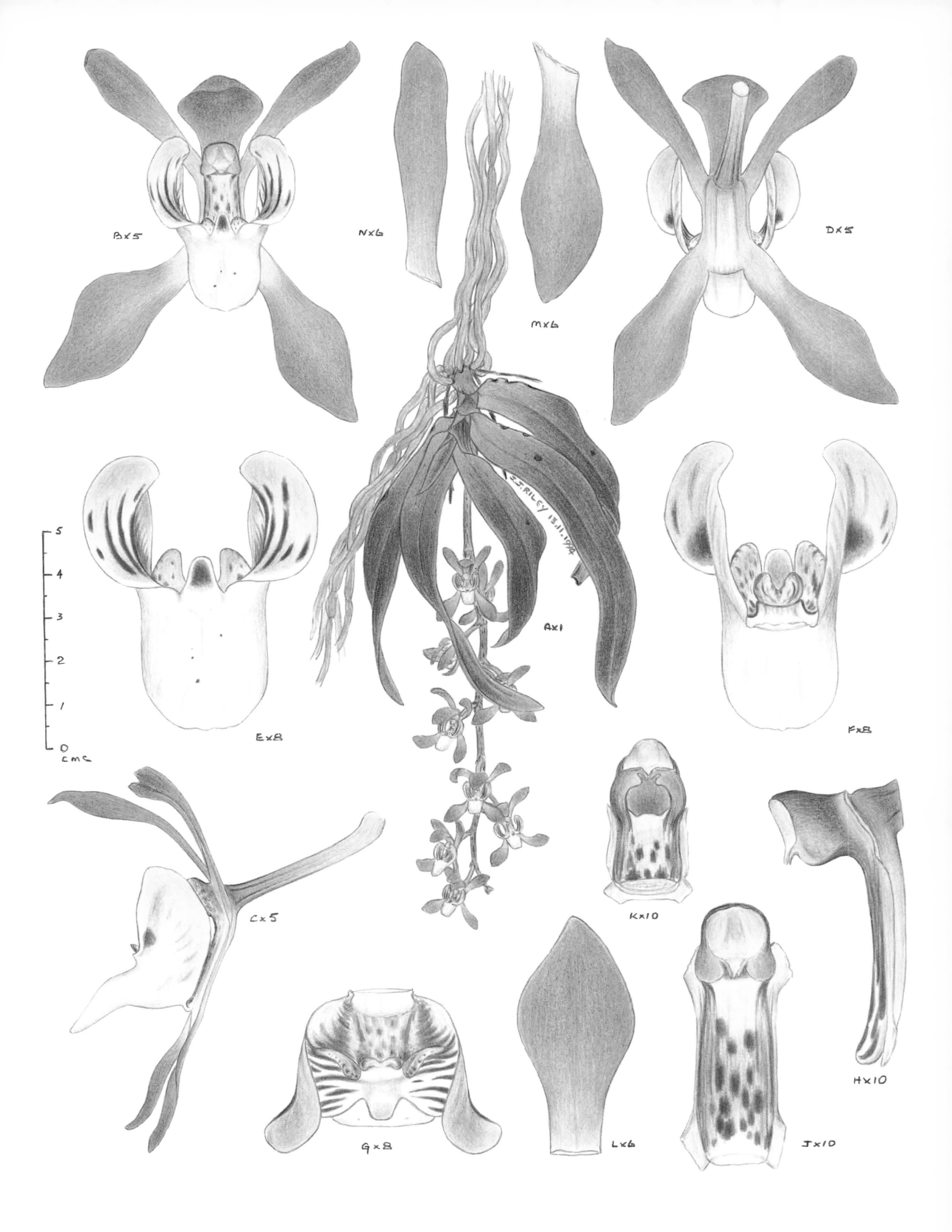

B×5
N×6
D×5
M×6
A×1
J.J. RILEY
E×8
F×8
5
4
3
2
1
0
cms
C×5
K×10
H×10
G×8
L×6
J×10

Sarcochilus olivaceus Lindl.

1839 | *Edward's Botanical Register* 25:32

TYPE LOCALITY Sydney, New South Wales

ETYMOLOGY *olivaceus* — olive green

FLOWERING TIME September to December

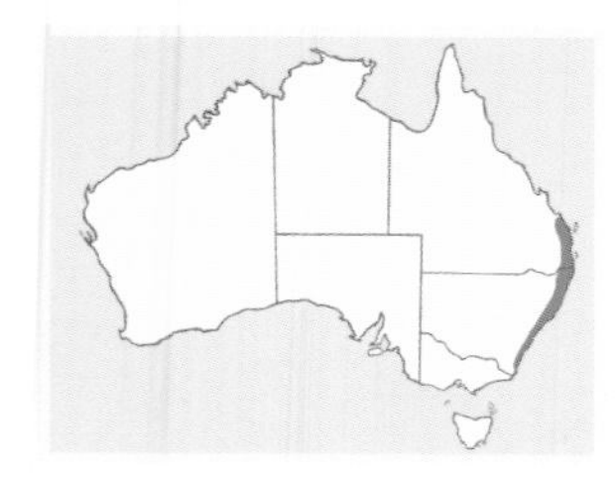

DISTRIBUTION Found along the coast and adjacent ranges and tablelands from near Eden on the South Coast of New South Wales, north to around Gympie in Queensland.

ALTITUDINAL RANGE 10 m to 900 m

DISTINGUISHING FEATURES *S. olivaceus* is a pendulous epiphyte that has somewhat untidy, greenish-yellow to olive green, fragrant flowers with narrow petals and sepals. The labellum is always yellowish-green with some pale brown blotches near the apex. *S. olivaceus* has thin-textured, dark green leaves that last on the plant for many years.

HABITAT *S. olivaceus* generally prefers moist, shady areas close to creeks in rainforest or wet sclerophyll forest. It can also occur in sheltered gullies near open woodland where there is heavy shade and high humidity. It grows on the major branches and trunks of rainforest trees and on woody lianas. Occasionally it will occur on rocks and in some locations can be quite common growing as a lithophyte. This species can tolerate very low light levels.

CONSERVATION STATUS Common. Occurs in national parks and reserves. Secure.

DISCUSSION True *S. olivaceus* always has a yellowish-green labellum and the original plants used for the type came from the Sydney area. There is some confusion over the northern distributional limits of *S. olivaceus*, as some populations may have been mistaken for the closely related *S. parviflorus*. There is need for more research into the *S. olivaceus* complex, to clarify the nomenclatural status of the Queensland plants and the numerous disjunct populations that occur both along the coast and in the ranges.

Sarcochilus olivaceus
Ourimbah, New South Wales

12 October 1999

A plant
B flower from front
C flower from side
D labellum from side
E labellum from front
F labellum from rear
G labellum from above
H column from front
J column from side
K dorsal sepal
L lateral sepal
M petal

B × 3
C × 3
A × 1
5
4
3
2
1
0
cms
J × 10
M × 4
K × 4
L × 4
D × 8
H × 10
E × 8
F × 8
G × 8

Sarcochilus hirticalcar (Dockr.) M.A. Clem. & B.J. Wallace

1989 | *Australian Orchid Research* 1:133

TYPE LOCALITY McIlwraith Range, Queensland

RECENT SYNONYMS *Parasarcochilus hirticalcar* Dockr. *Pteroceras hirticalcar* (Dockr.) Garay

ETYMOLOGY *hirticalcar* — with a hairy spur

FLOWERING TIME October to January

DISTRIBUTION This species is confined to the McIlwraith Range in North Queensland.

ALTITUDINAL RANGE 450 m to 650 m

DISTINGUISHING FEATURES *S. hirticalcar* is an epiphytic species that has very attractive flowers with distinctive, red-brown markings. It cannot be confused with any other monopodial, epiphytic orchid. The blooms are produced sequentially with two or three generally out at the same time. The flowers are long-lasting and a heathy plant may produce up to a dozen flowers and be in bloom for many weeks.

HABITAT *S. hirticalcar* grows in rainforest on the trunks and major branches of trees along water courses in shaded situations. Occasionally it will grow as a twig epiphyte where the plants are exposed to higher light levels, humidity is always high, and there is constant circulation of fresh air.

CONSERVATION STATUS Rare. Known to occur in a national park. Vulnerable.

DISCUSSION *S. hirticalcar* is a rare epiphytic species with a restricted distribution that was first discovered in 1966. Plants occur as scattered individuals, or rarely in small colonies. Its thick textured flowers are waxy and usually creamy yellow to yellowish-green. All the segments are marked or outlined with a rich red-brown colour. It is because of this striking contrast that this species has become known colloquially as the Harlequin Orchid.

Sarcochilus hirticalcar
McIlwraith Range, Queensland

22 December 1997

A plant
B flower from front
C flower from side
D labellum from side
E labellum from above
F longitudinal section of labellum
G column from side
H column from front
J dorsal sepal
K lateral sepal
L petal

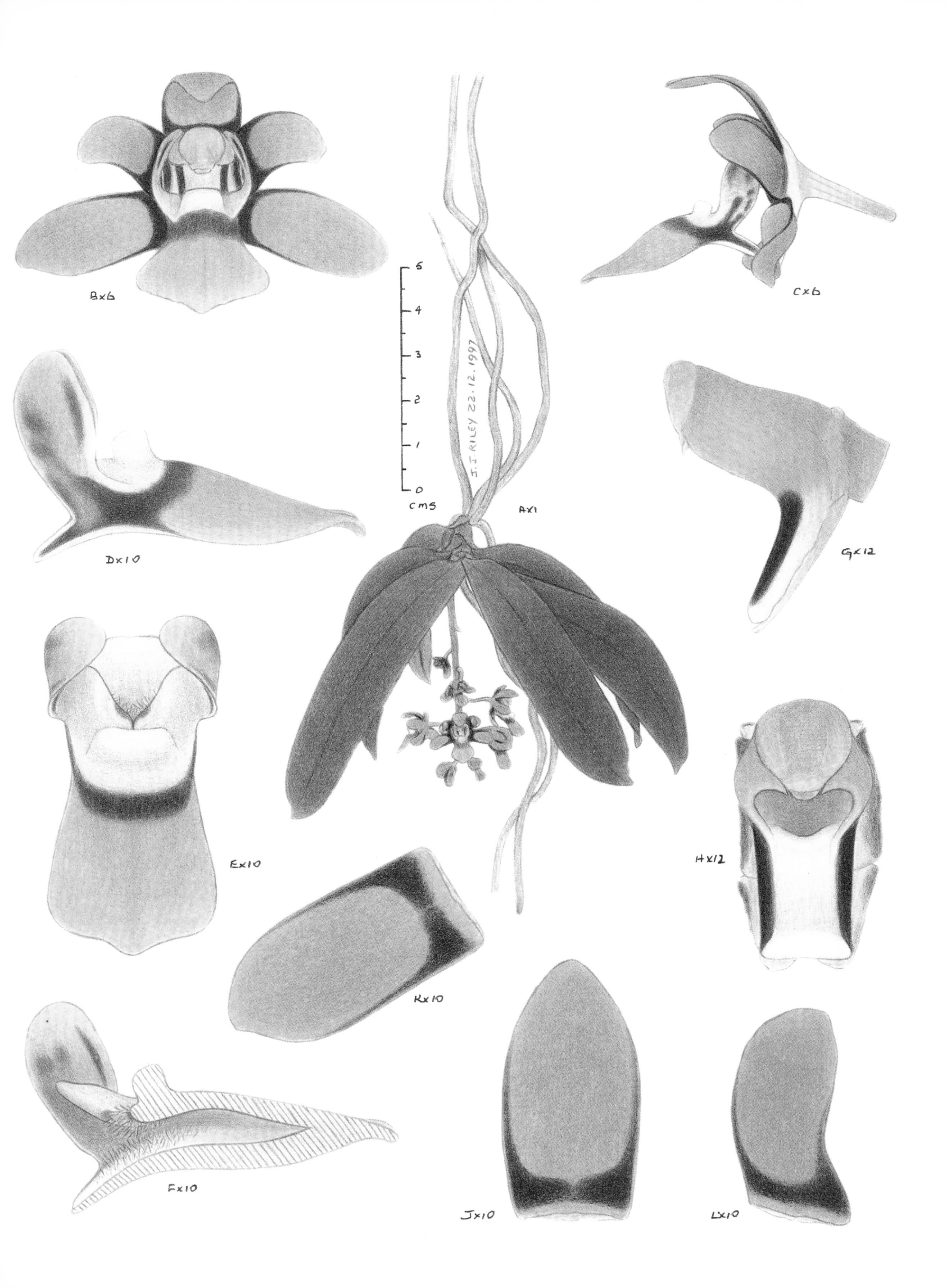
B×6
C×6
D×10
E×10
F×10
G×12
H×12
K×10
J×10
L×10
A×1
cms
0
1
2
3
4
5
J.J. RILEY 22.12.1997

Thelymitra cyanea (Lindl.) Benth.

1873 | *Flora Australiensis* 6:323

TYPE LOCALITY Circular Head, Tasmania

ETYMOLOGY *cyanea* — dark blue

FLOWERING TIME November to February

DISTRIBUTION This widespread species is found in Tasmania and King Island, the Fleurieu Peninsula of South Australia, the Eastern Highlands of Victoria with a disjunct population in the west, north from the Victorian border to around Point Lookout, New South Wales.

ALTITUDINAL RANGE 30 m to 1500 m

DISTINGUISHING FEATURES *T. cyanea* is one of the blue sun orchids with smaller flowers than the related *T. venosa*. The column of the bloom is nearly as wide as it is long. The flowers close in dull, cold conditions.

HABITAT *T. cyanea* is a moisture-loving, terrestrial species. It is more plentiful in montane to alpine meadows and frost hollows, where it grows with *Sphagnum* and amongst sedges, in and around swamps and areas that are permanently wet. Soils are sandy to peaty loams.

CONSERVATION STATUS Widespread. Occurs in national parks and reserves. Secure.

DISCUSSION In the north of its range, *T. cyanea* is a plant of the ranges and tablelands, with only isolated occurrences at low altitudes. In Tasmania, it is widespread and common in both the lowlands and montane areas. It is a most attractive species with freely opening flowers and is one of few sun orchids that has a labellum that is wider than the perianth segments. Under ideal conditions, *T. cyanea* can be found in large, scattered groups that reproduce vegetatively from daughter tubers formed on the ends of stolonoid roots. This species also reproduces from seed and is pollinated by small native bees.

Thelymitra cyanea
Boyd Plateau, New South Wales

24 December 1989

A plant
B flower from front
C flower from side
D flower from rear
E column from side
F column from front
G column from rear

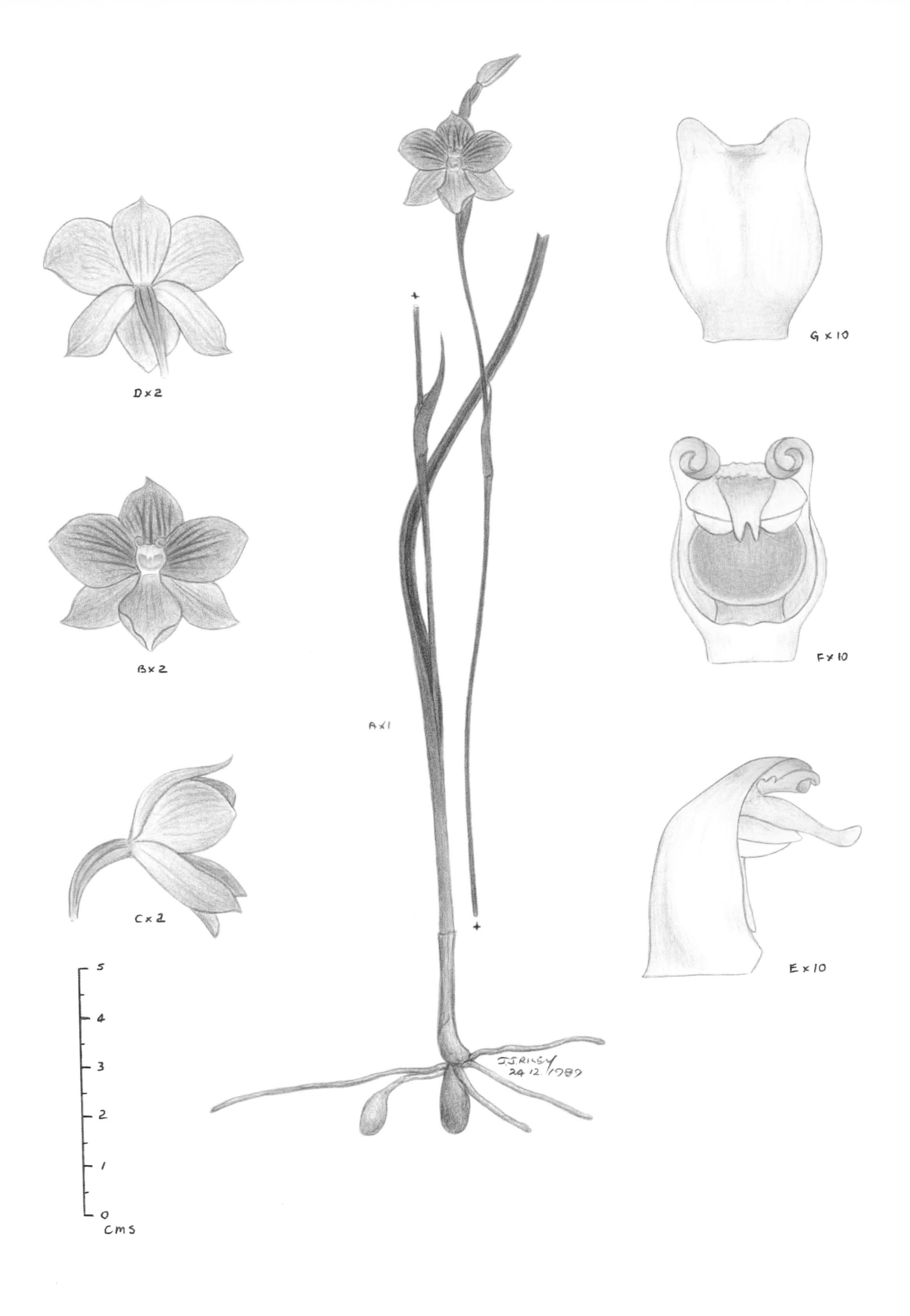
D×2
B×2
C×2
A×1
G×10
F×10
E×10
J.J. RILEY
24.12.1989
5
4
3
2
1
0
cms

Thelymitra venosa R. Br.

1810 | *Prodromus Florae Novae Hollandiae*: 314

TYPE LOCALITY Botany Bay, New South Wales

ETYMOLOGY *venosa* — veined

FLOWERING TIME October to December

DISTRIBUTION This species is endemic to New South Wales, being found along the coast and adjacent ranges from south of Sydney to Robertson and the upper Blue Mountains.

ALTITUDINAL RANGE 30 m to 950 m

DISTINGUISHING FEATURES *T. venosa* has larger flowers than the sister species, *T. cyanea*. The column of the bloom is twice as long as it is wide. The flowers stay open in dull, cold conditions, often into the evening, an unusual feature for a sun orchid.

HABITAT At Wentworth Falls in the Blue Mountains, *T. venosa* grows on moist sandstone ledges and in clay filled crevices, often with *Rimacola elliptica*. In this situation, the leaves of the *Thelymitra* grow downwards, with the inflorescence doing a full bend to become upright. It also occurs in other parts of the upper Blue Mountains in the numerous hanging swamps of the area. Away from the mountains, *T. venosa* grows in permanently moist areas in open forest, woodland and on the margins of swamps.

CONSERVATION STATUS Uncommon. Occurs in national parks. Secure.

DISCUSSION *T. venosa* has large, blue flowers that are distinctly veined. Unlike most *Thelymitra* species, the blooms of *T. venosa* remain open in dull conditions and often well into the night. It is another species where the labellum is noticeably different to the other segments. The colour of the flowers can vary from pale blue to an intense blue with very dark blue stripes. All are very eye-catching and attractive with the darker clones spectacular. The plant illustrated is a lighter coloured example. *T. venosa* slowly increases vegetatively from daughter tubers formed on the ends of stolonoid roots. It also reproduces from seed, being pollinated by small native bees. It is one of very few, true blue, orchid species.

Thelymitra venosa
Wentworth Falls, New South Wales

10 November 1992

A plant
B flower from front
C flower from side
D flower from rear
E labellum
F column from side
G column from front
H column from rear
J dorsal sepal
K lateral sepal
L petal

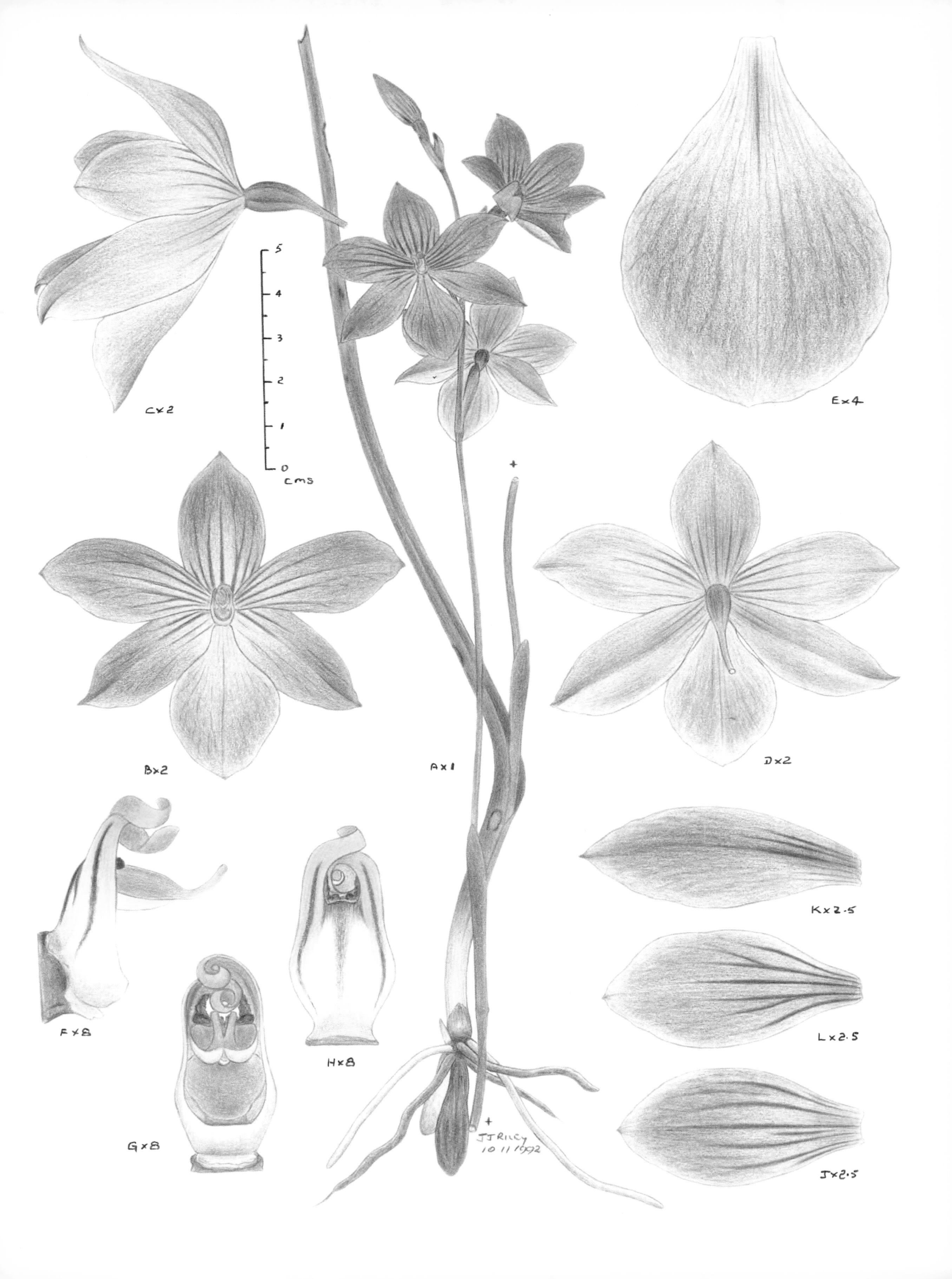

C×2
5
4
3
2
1
0
cms
E×4
B×2
A×1
D×2
F×8
H×8
G×8
K×2.5
L×2.5
J×2.5
JJ Riley
10 11 1992

glossary

aerial growth	a young plant that develops on older pseudobulbs, also known as a keiki
aerial roots	adventitious roots that are not supported
anther	the pollen-bearing part of the stamen
apex	tip of flower or leaf
apical	of the apex
apomictic	producing seeds without fertilisation
appendage	an attachment
basal	attached or grouped at the base
bilobed	with two lobes
bract	a leaf-like structure
bud	an unopened flower or a new shoot in its early stages
calli	plural of callus
callus	a thickening in the labellum, being a bump, club or ridge
capsule	a fruit containing many seeds
cauline	producing leaves along the stem
channelled	grooved
cilia	fine hairs
ciliate	fine hairs forming a fringe
clavate	club-like
cleistogamous	flowers that self-pollinate without opening
clone	a group of genetically identical plants, propagated vegetatively from one individual
column	the central, fleshy part of the orchid flower comprising the sexual organs
complex	related group
concave	sunken or hollow like a basin
conspicuous	obvious, striking
corolla	the petals of a flower collectively
daughter tuber	a newly formed tuber, produced vegetatively off the parent plant
deciduous	the seasonal shedding of leaves, which live for less than a year
decurved	curved down
disjunct	discontinuous, relating to distribution
distal	towards the tip of a structure
dorsal sepal	the solitary top sepal, behind the column
elliptic	like a stretched out circle, with two axes of symmetry
endemic	restricted to a particular country, region or area
ephemeral	short-lived
epiphyte	a plant that grows on trees, but is not parasitic
evergreen	retaining leaves throughout the year
filamentous	very fine and thread-like
filiform	very fine and thread-like
fimbriate	a fringe of long teeth
flange	overhang, extension
fused	joined or growing together
galea	hood-like structure in *Pterostylis*, being a combination of the petals and dorsal sepal
genera	plural of genus
genus	a group of related species, a taxonomic category above species
glabrous	lacking hairs, smooth
gland	an embedded or projecting structure that may secrete nectar or fragrance
globose	ball-shaped
greenhood	common name for many *Pterostylis* species
habit	the general appearance of a plant
habitat	the environment in which a plant grows
hatching	criss-cross markings

herbarium	a botanical collection of pressed plant specimens
incurved	bent inwards
indigenous	native to a country, region or area
inflorescence	the flowering stem of a plant
irritable	highly sensitive
labellum	a modified petal, also known as the lip
lamina	an expanded portion of a leaf or petal
lanceolate	long and narrow, being broadest at the base
lateral sepal	the two sepals on either side of the labellum, sometimes fused together
linear	long and thin
lip	the labellum
lithophyte	a plant that grows on rocks
lobe	a segment
margin	edge or side
mentum	a chin-like extension at the base of a flower
midlobe	the central of three lobes of the labellum
midrib	a primary vein, often prominently raised or depressed
monopodial	having one main or primary stem, producing leaves and flowers along that stem
monotypic	a genus with a single species
montane	growing in the mountains
natural hybrid	a rare naturally occurring hybrid between two different species
nomenclature	the rules governing valid names
open forest	forest with an almost continuous canopy, dominated by trees
osmophore	a scent-producing gland
ovary	the structure directly behind tthe flower that develops into the fruit or capsule
ovate	egg-shaped
ovoid	egg-shaped in three dimensions
papillate	with tiny pimple-like bumps
papillose	with tiny pimple-like bumps
pendent	hanging
pendulous	hanging
perianth	the petals and sepals, not including the labellum
petal	one segment of the corolla or inner whorl of the flower
petaline	of the petals
pollinia	a group of pollen grains massed together
pseudobulb	the thickened or bulb-like stem of a sympodial orchid
pseudocopulation	the ploy used when a flower mimics a female insect in order to be pollinated by a male insect attempting to mate with the bloom
recurved	curved backwards and often downwards
reflexed	bent suddenly backwards
reticulate	forming a network of veins
rhomboidal	having a three-dimensional diamond shape
rosette	a group of leaves which radiate from a central point
rufa group	referring to species related to *Pterostylis rufa*
saprophyte	a leafless, non-green plant that lives off decaying plant matter
sclerophyll	a plant with hard, stiff leaves
sclerophyll forest	forest dominated by *Eucalyptus*, with an understorey of hard-leafed plants
seedling	a young plant raised from seed, that has yet to flower

sepal	one segment of the outer leaf-like structures surrounding the flower
serrated	finely and regularly saw-toothed
sessile	without a stalk
setae	the fine bristle-like hairs on the labellum margins of some *Pterostylis*
shoot	a term used for a new growth
sidelobe	lobes on the side of the labellum
sinus	a cavity or recess between two segments
sister species	a very closely related species
slender	tall and thin
solitary	occurring singularly
species	a taxonomic group uniting individuals that have certain similarities and breed freely among themselves to produce fertile offspring
sporadic	random, irregular
spur	a slender sac-like appendage, often secreting nectar
stamen	the male part of a flower producing pollen
stigma	the sticky receptive part of the column, which accepts pollen
stolon	an underground stem or shoot
stolonoid root	a long horizontal root capable of forming a new apical tuber
subspecies	taxonomic rank below species, often used to describe distinct geographical populations
succulent	fleshy or juicy
suffusions	markings
sympodial	a growth habit whereby each stem has limited growth and new shoots develop from the base of previous ones
synonym	a previous invalid name for a species
taxa	plural of taxon
taxon	a term used to describe a taxonomic group such as species, subspecies or variety
taxonomy	the classification of living organisms
tepals	the petals and sepals, not including the labellum
terete	slender, cylindrical and round in cross-section
terminal	at the tip
terrestrial	a plant that grows in or on the ground
translucent	see-through, semi-transparent
tuber	a thickened, underground storage body
tussock	a large clump or tuft
twig epiphyte	a species that grows on the very outermost branches of shrubs and trees
type	the original representative of a species, genus or other taxon
type-site	the location from which the type was named
type specimen	the original specimen used for the formal description, often preserved
variegated	green with white markings or stripes
variety	taxonomic rank below species, often used for minor differences or colour forms
vegetative	parts of a plant not directly involved with sexual reproduction
vegetatively	multiplying without sexual reproduction
xerophytic	drought-tolerant

bibliography

Backhouse, G & Jeanes, J (1995) *The Orchids of Victoria.* Miegunyah Press, Melbourne.

Banks, DP (1996a) Notes on the distribution of *Adenochilus nortonii. Orchadian* 12(1): 26–27.

—— (1996b) Notes on the 'Prawn Greenhood', *Pterostylis pedoglossa. Orchadian* 12(2): 68.

—— (1997) Southern range extension for *Liparis swenssonii. Orchadian* 12(5): 236.

—— (2001) The genus *Dockrillia*, and its use in hybrids. *Orchadian* 13(8): 347–56.

Bates, RJ & Weber, JZ (1990) *Orchids of South Australia.* Government Printer, South Australia.

Bates, RJ (2000) The dipodiums of South Australia and Victoria. *Orchadian* 13(6): 262–66.

Bishop, T (2000) *Field Guide to the Orchids of New South Wales and Victoria* (second edition). UNSW Press, Sydney.

Carr, G (1991) New taxa in *Caladenia*, *Chiloglottis* and *Gastrodia* (Orchidaceae) from south-eastern Australia. *Indigenous Flora and Fauna Association Miscellaneous Paper*, no.1.

Clements, MA (1989) Catalogue of Australian Orchidaceae. *Australian Orchid Research.* Volume 1. Australian Orchid Foundation, Melbourne.

Dockrill, AW (1992) *Australian Indigenous Orchids.* Volumes 1 & 2 (revised edition). Surrey Beatty & Sons, Sydney.

FitzGerald, RD (1977) *Australian Orchids* Volumes 1 & 2 (facsimile edition). Lansdowne Editions, Melbourne.

Harden, G (ed.) (1993) *Flora of New South Wales.* Volume 4. UNSW Press, Sydney.

Harrison, M (2001) *Bulbophyllum* species in Australia. *Australian Orchid Review* 66(6): 4–19.

Hoffman, N & Brown, AP (1998) *Orchids of South-west Australia* (revised second edition). UWA Press, Perth.

Hopper, SD & Brown, AP (2001) Contributions to Western Australian orchidology: 2. New taxa and circumscriptions in *Caladenia. Nuytsia* 14(1–2): 27–307.9.

Jones, DL (1988) *Native Orchids of Australia.* Reed, Sydney.

—— (1991) New taxa of Australian Orchidaceae. *Australian Orchid Research.* Volume 2. Australian Orchid Foundation, Melbourne.

—— (1997a) Towards a revision of the *Caladenia dilatata* R. Br. (Orchidaceae) complex — 1: The *Caladenia dilatata* alliance. *Orchadian* 12(4): 157–71.

—— (1997b) Two rare new species of *Pterostylis*, allied to *Pterostylis alveata. Orchadian* 12(4): 180–87.

—— (1997c) Six new species of *Pterostylis* (Orchidaceae) from Australia. *Orchadian* 12(6): 245–58.

—— (1998) Contributions to Tasmanian Orchidaceae. 1–9. *Australian Orchid Research.* Volume 3. Australian Orchid Foundation, Melbourne.

—— (1999) Eight new species of *Caladenia* (Orchidaceae) from eastern Australia. *Orchadian* 13(1): 4–24.

—— (2000a) Ten new species of *Prasophyllum* (Orchidaceae) from south-eastern Australia. *Orchadian* 13(4): 149–173.

—— (2000b) Four new names in *Caladenia* (Orchidaceae), and a note on *Caladenia carnea* var. *subulata. Orchadian* 13(6): 255–57.

Jones, DL & Clements, MA (1997) Characterisation of *Pterostylis gibbosa* and description of *P. saxicola*, a rare new species from New South Wales. *Orchadian* 12(3):128–35.

Jones, DL, Clements, MA, Sharma, IK & Mackenzie, AM (2001) A new classification of *Caladenia* (Orchidaceae). *Orchadian* 13(9): 389–419.

Jones, DL, Wapstra, H, Tonelli, P & Harris, S (1999) *The Orchids of Tasmania*. Melbourne University Press, Melbourne.

Lavarack, PS & Gray, B (1985) *Tropical Orchids of Australia*. Thomas Nelson, Melbourne.

Lavarack, PS, Harris, WF & Stocker, G (2000) *Dendrobium and its Relatives*. Kangaroo Press, Sydney.

Mayr, H & Schmucker, M (1998) *Orchid Names and their Meanings*. Gantner Verlag, Vaduz.

Nicholls, WH (1969) *Orchids of Australia* (complete edition). Thomas Nelson, Melbourne.

Rupp, HMR (1969) *The Orchids of New South Wales* (facsimile edition). Government Printer of New South Wales, Sydney.

Smedley, DI (1998) Notes on the cultivation of *Sarcochilus hirticalcar*. *Orchadian* 12(9): 418–19.

Stephenson, AW (1997) A new recording of the Eastern Underground Orchid, *Rhizanthella slateri*, at sea level. *Orchadian* 12(4): 188.

Szlachetko, DL (2001) Genera et species Orchidalium. 1. *Polish Botanical Journal* 46 (1): 11–26.

Upton, WT (1992) *Sarcochilus Orchids of Australia*. 'Double U' Orchids, West Gosford.

Woolcock, C & Woolcock, D (1984) *Australian Terrestrial Orchids*. Thomas Nelson, Melbourne.

index of species names

Bolded page numbers denote that the species description appears on the page.